Cosmic Ray Physics

This book introduces you to the physics of cosmic rays, charged particles which reach us from known – and maybe unknown – sources in the cosmos. Starting from a brief history of this fascinating field, it reviews what we know about the creation of elements in the Big Bang and inside stars. It explains cosmic accelerators reaching fabulous energies. It follows the life cycle of cosmic rays all the way from their sources to detection near, on or below Earth. The central three chapters cover what we know about them at the level of the solar system, the Milky Way and the Universe at large. Up-to-date experimental results are presented in detail, showing how they are obtained and interpreted.

The book provides an accessible overview of this lively and diversified research field. It will be of interest to undergraduate physics students beginning their studies on astronomy, cosmology, and particle physics. It is also accessible to the general public by concentrating mathematical and technical detail into Focus Boxes.

Cosmic Ray Physics
An Introduction to the Cosmic Laboratory

Veronica Bindi
Mercedes Paniccia
Martin Pohl

CRC Press
Taylor & Francis Group
Boca Raton London New York

CRC Press is an imprint of the
Taylor & Francis Group, an **informa** business

Image Credit: NASA

First edition published 2023
by CRC Press
6000 Broken Sound Parkway NW, Suite 300, Boca Raton, FL 33487-2742

and by CRC Press
4 Park Square, Milton Park, Abingdon, Oxon, OX14 4RN

CRC Press is an imprint of Taylor & Francis Group, LLC

ISBN: 978-1-032-00391-7 (hbk)
ISBN: 978-1-032-02001-3 (pbk)
ISBN: 978-1-003-18138-5 (ebk)

DOI: 10.1201/9781003181385

Typeset in CMR10
by KnowledgeWorks Global Ltd.

Publisher's note: This book has been prepared from camera-ready copy provided by the authors.

En algún lugar debe haber un basural donde están amontonadas las explicaciones. Una sola cosa inquieta en este justo panorama: lo que pueda ocurrir el día en que alguien consiga explicar también el basural.[1]

Julio Cortázar, *Un Tal Lucas* [260]

---

[1] Somewhere there must be a garbage dump where the explanations are piled up. There is only one worry in this fair scenario: what might happen the day someone manages to explain also the dump.

Contents

Preface

The stars began to crumble and a cloud of fine stardust fell through space.
James Joyce, *A Portait of the Artist as a Young Man* [306, p. 87][2]

My first encounter with cosmic ray physics was Bruno Rossi's book "Cosmic Rays" [169] which I read when I was an undergraduate student in the 1970s. I consulted it because – in contrast to other authors – in this and his previous more general book [141], Rossi did not shy away from discussing experimental techniques and their influence on the quality of physics results. No wonder, since Rossi at MIT was one of the inventors of the coincidence technique, which helped converting a mostly anecdotal way of studying high-energy particles into a systematic science. During this period, cosmic rays also ceased to be the prime source of high-energy particles. Accelerators and later colliders provided controllable conditions to study particle production and interactions. Instead, cosmic rays became themselves a subject of research, as messengers from the cosmic laboratory.

In later years as a post-doc, I occasionally (mis-)used collider detectors at PETRA (DESY, Germany) and LEP (CERN, Switzerland), suitably modified, to make on-ground and underground measurements of cosmic muons, the penetrating heavy component of air showers caused by cosmic rays in the atmosphere. But it was only by the year 2000 that I got seriously involved in a dedicated cosmic ray experiment, when I learned about the successful flight of the prototype Alpha Magnetic Spectrometer (AMS-01) on Space Shuttle Discovery in 1998. I joined the collaboration constructing the final version of the detector, installed on the International Space Station in 2011. At the time, I thought that this would be a temporary involvement, while waiting for the Large Hadron Collider at CERN to start up. But it turned out to keep me busy for the rest of my professional life.

Constructing particle physics instruments for the use in space is a challenging task. Not only are there tough mechanical conditions during the rough ride on a rocket. The environment in orbit is even more demanding, as far as temperature excursions and radiation levels are concerned. Running and analysing space experiments over extended periods is equally challenging. Occasions for repair are rare or inexistent, in-orbit calibration and the tracking of varying conditions are complex tasks. However, as recent experiments on free-flying satellites and on the ISS show, these difficulties can be overcome such that particle physics in space reaches statistical and systematic accuracy similar to standards set by accelerator experiments.

Do these data make sense in terms of particle physics? Can a complete picture of cosmic rays from their production to their detection near Earth be constructed? And most importantly, do we know all their sources? The answer is probably "not quite yet", there is no standard model of cosmic rays. But the current status of interpreting high quality cosmic rays data is still worth describing. This is the purpose of this book. It wants to get you on the hook, just as Rossi's book got me on the hook 50 years ago. Thus one of its baits is to go into details about the experimental methods that lead to high quality data. And it will also not gloss over things we do not understand, so that you can drive (or at least follow) further progress in the field.

I want you, the reader, to decide what level of detail you wish to explore. I have thus separated the text into a continuous narrative, with only as much mathematical and technical detail as necessary, and Focus Boxes, which provide just these details basically at textbook level. A common joke about this approach is that all arguments which require math beyond

---

[2]Quoted by permission from Oxford University Press.

the rule of three are relegated to boxes. To fully appreciate their contents, it will probably be useful to have some background in modern particle physics. However, these notions are accessibly described in standard introductory textbooks on nuclear and particle physics, or in online courses like the one by Mercedes Paniccia and me.[3]

The three Chapters 8 to 10 are the heart of this book. Chapter 8 has been written by Veronica Bindi, describing radiation coming from our Sun and what happens to cosmic rays inside the heliosphere. Help with this chapter by Cristina Consolandi from University of Hawaii is gratefully acknowledged. Chapter 9 covers latest results on galactic cosmic rays and what has been learned from them so far. It has been contributed by Mercedes Paniccia. Chapter 10 has been written by me, even though its subject is a little outside my field of experience. I am very grateful to Bill Hanlon from University of Utah for spotting the flaws in the first draft of this chapter. Thanks are also due to Friedrich-Karl Thielemann for precious advice on the complex nuclear physics in Chapter 4. Carlos Maña Barrera from CIEMAT Madrid has been kind enough to review the first complete draft. He found many mistakes I should have spotted myself, thus sparing me a fair amount of embarrassment.

I wish to thank my co-authors for their patience and the many colleagues I consulted for the invaluable help and encouragement I received. Without them, there would be even more errors and misconceptions in this book than you will doubtlessly identify.

Martin Pohl
October, 2022

Authors

Veronica Bindi is Full Professor and Chair of the Physics and Astronomy Department at the University of Hawaii in Manoa (USA) where she is working since 2012. She obtained her PhD in 2006 at the University of Bologna (Italy) and worked for several years at INFN (Italy). For more than 10 years, she has been part of the team at CERN (Switzerland) that led the construction, integration and operation of the Alpha Magnetic Spectrometer (AMS) installed on the International Space Station in May 2011 to search for dark matter and study galactic cosmic rays. She is working on AMS data analysis with a particular focus on solar modulation, heliophysics and space radiation. She has received a National Science Foundation career award and a grant from NASA supporting future manned missions to Mars.

Mercedes Paniccia is Senior Research Associate at University of Geneva. She has done her studies in Physics at University of Rome "La Sapienza" in Italy, where she obtained a Master's degree in Particle Physics, and at University of Geneva, where she obtained a PhD in Physics in 2008 with a study on solar activity effects on cosmic-ray fluxes detected in near-Earth space by the prototype AMS-01 detector. She has worked at INFN (Italy) on the detection of neutrino oscillations at the OPERA experiment, and at CNRS (France) on the AMS experiment. She is member of the AMS Collaboration since 2003. She has contributed to the construction and integration of the semiconductor detectors for the silicon tracker, to the calibration and the commissioning of the electromagnetic calorimeter, to the operation of the AMS detector, and to data reconstruction and analysis. Since 2017, she leads the AMS data analysis group at University of Geneva whose research focuses on the measurement of cosmic-ray nuclei fluxes and isotopic composition of light nuclei, with the aim of studying the propagation mechanism of cosmic rays in the galaxy.

Martin Pohl is professor emeritus at University of Geneva. He obtained his PhD from RWTH Aachen (Germany) in the 1970s. He has been working on experimental particle physics at the PETRA (DESY, Hamburg Germany), LEP and LHC (CERN, Geneva Switzerland) colliders before turning to astroparticle physics in space. His projects are: AMS, a magnetic spectrometer measuring galactic cosmic rays in the GeV to TeV regime and the POLAR X-ray polarimeter for the Chinese space laboratory Tiangong 2. He has been the director of the department for particle and nuclear physics (DPNC) at University of Geneva and head of the Physics Department. He was the head of the University of Geneva group for AMS until his retirement in 2017. He is the author of a text book on particle physics, as well as the main author of two introductory online courses on the same subject (Coursera). He has recently published "Particles, Fields, Space-Time: From Thomson's Electron to Higgs' Boson" (CRC Press, 2020).

1 Cosmic Rays and Us

"Cosmic rays" is a broad term covering charged particles generated in cosmological and astrophysical phenomena. They travel through space from their origin, undergoing acceleration and transformation by interaction with the interstellar medium, including matter, magnetic fields and photons. They impinge onto the Earth's atmosphere at a rate of about ten thousand per m^2 and second at the peak of their flux, which occurs at a kinetic energy of a few hundred MeV. After attenuation by the atmosphere, there is still about one charged particle per cm^2 and minute left. Cosmic rays are thus an important part of our environment and influence human activities and technology.

Their flux and composition are not constant neither in time nor geographically. The activity of the Sun and the dynamics of its magnetic field influence cosmic rays both coming from the Sun and from outside the solar system. The Earth's magnetic field shields us from low-energy cosmic rays, more so around the equator than at the poles. The magnetic field directs low-energy particles towards the polar regions where they ionise the atmosphere and cause the beautiful aurorae borealis and australis.

Of course, the physicist's main interest lies with the physical information that cosmic rays carry: information about high-energy astrophysical and cosmological phenomena of which they are witnesses. In particular, cosmic rays may provide clues to important mysteries like the composition of dark matter. All this will be the subject of this book and covered in detail in the following chapters. In this chapter, we will rather give a glimpse of other areas where cosmic rays have an influence, ranging from weather and climate on Earth to life on ground, in space and on other planets.

The biosphere is well shielded by the Earth's magnetic field and atmosphere, and this in fact has allowed life to develop successfully where we live. However, high-energy cosmic rays interact with atmospheric molecules, about 78% nitrogen, 21% oxygen and about 1% other gases. The primary interaction causes air showers of rich particle content, the tails of which reach down to low altitudes.

As far as charged particles are concerned, the most important contributions to incoming cosmic radiation near Earth are the following, in order of increasing energy:

- The so-called solar wind consists of a plasma of electrons and ions with a velocity of 250 to 750 km/s, which is constantly emitted by the Sun. It deforms the Earth's magnetic field, squashing it in the direction facing the Sun and blowing a tail on the opposite side. The solar wind is not usually included in cosmic rays, since its particles have low energy and normally do not reach Earth. It is, however, important for the low-energy end of the cosmic ray flux. Its properties and its influence on cosmic rays are covered in Chapter 8.

- Transient violent events in the Sun's energy production, called solar flares, cause large increases in the flux of solar wind and allow it to invade the Earth's magnetosphere. They may extend the energy range to the so-called solar energetic particles. Their origin and properties are covered in Chapter 8.

- High-energy cosmic rays, mostly hydrogen (89%) and helium nuclei (9%), with a small admixture of electrons and heavy nuclei come from sources inside the Milky Way. The properties of these galactic cosmic rays are covered in Chapter 9.

- The much lower flux of extragalactic cosmic ray is too low in intensity to influence the biosphere but reaches energies many orders of magnitude higher than will ever

DOI: 10.1201/9781003181385-1

be reached by man-made accelerators. These ultra-high-energy cosmic rays are introduced in Chapter 10.

In addition, particles can be trapped and remain stored in the Earth's radiation belts, since the geomagnetic field works like a magnetic bottle (see Focus Box 5.1). The day-to-day shape and storage capacity of the bottle depends on the dynamic geomagnetic configuration. Stable examples are the Van Allen belts, regions populated with stored protons and electrons (see Section 8.9.1). An interesting feature of trapped particles is the so-called South Atlantic Anomaly (SAA). Due to the eccentricity and misalignment of the magnetic axis with respect to the rotational axis of the Earth, the inner Van Allen belt, mainly populated by protons, comes closer to the Earth's surface in the south Atlantic region and causes a large local increase in the flux of low-energy cosmic rays. This effect is also appreciable in Low Earth Orbit[1] (see Chapter 8).

The flux of solar and galactic cosmic rays near Earth depends on the Sun's activity level and varies generally with an 11-year periodicity. In addition, the solar magnetic field changes direction in a 22-year cycle. Solar events normally last minutes, hours or days, but the ensuing fluxes exceed the normal level by orders of magnitude. The most violent events are caused by coronal mass ejections (CME) on the Sun's surface. They can project vast amounts of plasma towards Earth, carrying strong magnetic fields and influencing the Earth's magnetosphere. This generally increases the flux of solar wind and reduces the low-energy part of the galactic component, as described in Chapter 8.

Violent events like coronal mass ejection can impact human technology more or less directly. The most important one on record was the so-called Carrington event lasting from September 1 to 2, 1859 [372]. A powerful coronal mass ejection hit the atmosphere causing a geomagnetic storm. This means that on a short time scale the magnetic field on the Earth's surface varies notably, inducing voltages in power and telegraphic lines. In the Carrington event, these were high enough to disrupt telegraphic services, give electric shocks to operators and cause sparks on transmission lines. Aurorae borealis and australis were observed close to the equator. They were reported to be as bright as daylight. Another geomagnetic storm, caused by a coronal mass ejection on March 9, 1989, hit the Hydro-Quebec power network on March 13, causing a wide-spread power failure. On a regular basis, less important magnetic disturbances interrupt the high frequency radio connections between aircrafts and air traffic control. This is particularly important for the about 7500 annual flights over polar routes, which are not covered by satellite navigation. In these cases, air traffic must be diverted to lower latitudes. Geomagnetic disturbances have also detonated US naval mines off the coast during the Vietnam war [652].

When interacting with the atmosphere, cosmic rays transfer energy and alter chemical components, aerosols and their growth in the troposphere. They may thus influence cloud formation, weather and climate [386, 667]. Correlations of cloud coverage and intensity of galactic cosmic rays [322] have been observed for low level clouds on the time scale of the 11-year solar cycle. On the time scale of days [309], correlations have been observed for high altitude clouds, accompanying the so-called Forbush decrease in galactic cosmic rays flux (see Section 8.7.3) following coronal mass ejections from the Sun. The geographic distribution of these two effects appears to be different, the former concerning mainly low latitudes, the latter preferentially the polar regions. The physical mechanism at work is not entirely clear, but the results of the CLOUD experiment at CERN [573] point out the role of ions in the nucleation of aerosols in the form of droplets and microscopic ice crystals – in addition to the influence of sulphuric acid and organic molecules of biological origin.

---

[1]Low Earth Orbit (LEO) is defined by orbital periods less than 128 minutes and eccentricities less than 0.25. This corresponds to space craft whose altitude never exceeds about 2000 km.

The **radiation dose** D is defined as the energy ΔE deposited by radiation per unit mass ΔM of the target substance:

$$D = \frac{\Delta E}{\Delta M}$$

Its SI unit is the Gray, with $1\,\text{Gy} = 1\,\text{J/kg} \simeq 6.242 \times 10^9\,\text{MeV/g}$. Charged projectiles transfer energy to a target material via excitation and ionisation of its atoms. The energy dE deposited by a particle in a target layer of thickness dx is called the **stopping power** dE/dx. Its mean is given by the Bethe-Bloch formula, discussed in Focus Box 1.2. In the context of radiation damage, dE/dx is also often called linear energy transfer. For a flux of Φ projectiles per unit area and time, the relation between dose and stopping power is then:

$$D = \int \Phi \frac{dE}{dx} \frac{1}{\rho}\, dt$$

with the target mass density ρ and an integration over the exposure time t.

Biological samples are not equally sensitive to different types of ionising radiation. Therefore, a useful though less stringently defined quantity to judge the harmfulness of irradiation is the **equivalent dose** $H = W_R D$, including a relative weight factor W_R. It varies with the projectile type, and is usually taken as $W_R = 1$ for electrons and photons, $W_R = 2$ for protons and charged pions, and $W_R = 20$ for alpha particles and heavy nuclei [389]. The SI unit of the equivalent dose is the Sievert, with $1\,\text{Sv} = 1\,\text{J/kg}$.

The harmfulness of radiation for different human tissue types can also differ substantially. Another weighting factor, W_T, tries to take this into account. The ICRP recommended factors [389] are largest for sensitive organs like lung, stomach and breasts, with $W_T = 0.12$, smaller for other tissues.

Focus Box 1.1: Ionisation, radiation dose and dose rate

Due to attenuation by the atmosphere, the radiation dose rate for organisms on Earth is mostly benign. In Europe, it amounts on average to about $320\,\mu\text{Sv/a}$ [608]. Only the 0.02% of the population living at high altitude are exposed to a dose rate in excess of mSv/a, a small fraction of the annual dose rate allowed for professional exposure in most countries. Radiation doses and dose rates are explained in Focus Box 1.1.

Radiation levels do go up with increasing altitude and latitude. Low-energy particle fluxes are particularly strong near the Earth's magnetic poles. Galactic cosmic rays have a lower flux but extend to high energies and contain a highly ionising component of heavy ions. When flying over the polar route, where low-energy cosmic rays are most abundant, an airplane passenger is exposed to a few $\mu\text{Sv/h}$ [305]. The annual doses accumulated by aircraft personnel are non-negligible, but remain at a level of a few mSv/a [311], in any case much below the limit of $20\,\text{mSv/a}$ recommended by the Federal Aviation Administration (FAA).

When leaving the Earth's protective atmosphere completely, radiation exposure again increases. This is a concern for space craft electronics and impacts manned space travel in a major way. Crew members on the International Space Station collect a dose of close to $80\,\text{mSv}$ during a six month stay [396], exceeding the allowed limit for workers in contact with radioactivity by a factor of four. About 80% of the organ dose comes from the highly penetrating galactic cosmic rays, which thus represent a major threat to astronaut health. Much higher dose rates would be incurred by an extended stay on the Moon surface, transit to Mars and an extended stay on the Mars surface [425]. This is of course due to the absence of a protective atmosphere. A summary of NASA's assessment of expected doses in shown in Figure 1.1. Comparison of simulated and measured dose rates shows that these can be reliably estimated, with the obvious exception of random solar events.

No solar energetic particle events have so far happened during manned space missions. However, a large event happened between two lunar missions, Apollo 16 and Apollo 17, in August 1972. Had this event happened during a manned mission, it would have put the astronauts' life in danger. The importance of these and other so-called space weather effects are pointed out in Section 8.10.

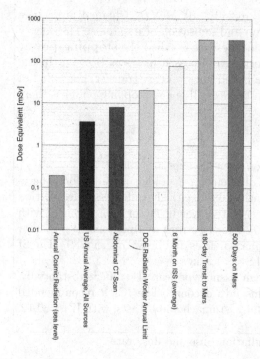

Figure 1.1 Estimate of the radiation astronauts would be exposed to on an expedition to Mars without major solar events, based on dosimetry on the Mars Rover. The total dose of a mission to Mars with an extended stay on its surface behind the usual space craft shielding would amount to about 1 Sv. This is compared here to other doses from human activity. (Credit: NASA/JPL-Caltech/SwRI)

The influence of high radiation exposures on electronics performance is readily measured at accelerators [537]. One distinguishes the slow degradation of performance by an extended exposure on one hand, from the impact of single large energy deposits, the so-called single event upsets, on the other hand. The former is usually counteracted by appropriate shielding. The latter would for example change bits in a memory and cause malfunction, mostly recoverable by appropriate countermeasures in circuit and software design.

The harmfulness of ionising radiation to humans is less easily assessed. Reliable data are only available for survivors of the nuclear bomb attacks on Hiroshima and Nagasaki, as well as workers in nuclear power plants and patients undergoing high dose therapeutical treatment. These have mostly been exposed to irradiation by neutrons and photons, or by radioactive fallout. It is thus not obvious how to apply the analysis of their pathologies to irradiation with cosmic rays. In the context of future manned missions to the Moon and Mars, the risks have been analysed in detail by NASA [425]. The largest risk appears to be the development of cancer by radiative cell damage. The risk of acute radiation syndrome, e.g. during solar energetic particle events, appears to be manageable, if the mission takes place in solar minimum and extravehicular activities are prohibited during solar storms. A long term risk is present by degenerative damage to human tissue through long term exposure at small dose rates, about which not much is known.

Of the organ dose incurred during missions on the International Space Station, 80% has been attributed to galactic cosmic rays, with only a small contribution by trapped protons [425]. It typically amounts to around 100 mSv. However, on a mission to Mars, every cell of an astronaut's body would be hit by an ionising radiation every few days, and by a high-energy, high-charge nucleus every month [374], unless the spacecraft is properly

Table 1.1

Length units in astronomy and astrophysics

	m	AU	ly	pc
m	1	6.668×10^{-12}	1.057×10^{-16}	3.240×10^{-17}
AU	1.496×10^{11}	1	1.581×10^{-5}	4.848×10^{-5}
ly	9.461×10^{15}	6.324×10^{4}	1	0.3066
pc	3.086×10^{16}	2.063×10^{5}	3.262	1

shielded. NASA has estimated the risk of developing fatal cancer on a 180 day lunar mission to be less than 1%, but several percent for an extended mission to Mars. Both are calculated during solar minimum, when the intensity of galactic cosmic rays is highest and behind an aluminium shield of $5\,\mathrm{g/cm^2}$. Shielding by a material with lower nuclear charge like polyethylene may be slightly more effective [346]. What looks more promising is an active electromagnetic shielding, sweeping away the lower energy part of galactic cosmic rays [369]. It is clear, however, from Figure 1.1 that any long term stay on a planet without atmosphere is severely endangered by galactic cosmic rays. Any settlement on Mars would have to be underground, not a very attractive prospect.

The question if and how cosmic radiation might have influenced the origin and evolution of life on Earth has been raised soon after its discovery [132]. In his Nobel lecture [86], Victor Hess said: "The investigation into the possible effects of cosmic rays on living organisms will also offer great interest". While the current flux in the biosphere stands little chance of being a major player in creating genetic diversity [264], there may have been epochs in the early phases of life on Earth where the flux was high enough to trigger an enhanced lightning rate, favouring the formation of simple organic molecules [518]. It is known since the famous work of Stanley Miller and Harold Urey [154] at the University of Chicago that electrical shocks can trigger the synthesis of complex molecules. Since the cosmic ray composition is dominated by muons with a charge ratio $N(\mu^+)/N(\mu^-) \simeq 1.25$ on the ground and $\simeq 1.4$ at high altitudes, it has even been argued that their handedness induces the chirality observed in DNA and its building blocks [762]. More experimental work is definitely needed to establish a causality chain and substantiate such claims. In any case, the biochemistry of living organisms is complex enough even for very simple cells to allow for a multitude of influences on the development of life on Earth [759].

1.1 DISTANCE, TIME AND ENERGY

In cosmic ray physics, very different length scales matter. For the particle physics arguments we will use, length scales of well below the size of a hydrogen nucleus matter, a bit less than a femtometer, $10^{-15}\,\mathrm{m}$. For astrophysical arguments, length scales range between an astronomical unit, $1\,\mathrm{AU} \simeq 1.5 \times 10^{11}\,\mathrm{m}$ or roughly the distance of Earth from the Sun, and a parsec, $1\,\mathrm{pc} \simeq 3 \times 10^{16}\,\mathrm{m}$. In-between the two scales, a lightyear , $1\,\mathrm{ly} \simeq 9.5 \times 10^{15}\,\mathrm{m}$ measures the distance travelled by light in one standard year. We provide a conversion table of length units customary in astrophysics in Table 1.1.

To give you a feeling about orders of magnitude in astronomy, the following rough numbers may be useful:

- The solar system has a diameter of about $10^2\,\mathrm{AU}$ or about $10^{-4}\,\mathrm{ly}$.

- The nearest star to the Sun is Proxima Centauri, at about $4.25\,\mathrm{ly}$ or $1.3\,\mathrm{pc}$. Typical distances between stars are about $10^2\,\mathrm{ly}$.

Energy loss by ionisation and excitation of atoms dominates the passage through matter for particles heavier than electrons. The energy dE deposited by a particle with electric charge ze and velocity $\beta = v/c$ in a target layer of thickness dx is called the **stopping power** dE/dx. At sufficiently high energies ($> 1\,\mathrm{MeV}$) its mean is given by the Bethe-Bloch formula:

$$\left\langle -\frac{dE}{dx} \right\rangle_{ion} = K\rho \frac{Z}{A}\frac{z^2}{\beta^2}\left[\log \frac{2m_e c^2 \beta^2}{I(1-\beta^2)} - \beta^2 - \frac{\delta}{2} \right]$$

K is a universal constant and m_e the electron mass. The target material is characterised by its nuclear charge Z, mass number A, mass density ρ and ionisation potential I. The latter depends on the nuclear charge and is empirically found to be:

$$\frac{I}{Z} = \begin{cases} 12 + 7/Z & \mathrm{eV}; \quad Z < 13 \\ 9.76 + 58.8 Z^{-1.19} & \mathrm{eV}; \quad Z \geq 13 \end{cases}$$

The correction δ takes into account the polarisation of the target material by the incoming charge. This polarisation effectively reduces the charge of distant atoms and thus the energy loss, i.e. $\delta > 0$.

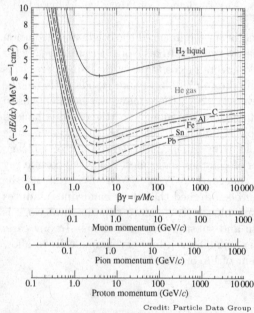

Credit: Particle Data Group

The stopping power is usually given per target mass surface density, $(1/\rho)dE/dx$, in units of $\mathrm{MeVcm^2/g}$. It is shown on the left as a function of velocity and momentum for common projectiles and materials [765]. It has a broad minimum at projectile momenta $p/m = \beta\gamma \simeq 3$ to 3.5, with a value between 1 and $2\,\mathrm{MeVcm^2/g}$. Above, it increases slowly according to the logarithmic factor. The actual energy loss is distributed around its most probable value by the non-elementary Landau distribution, roughly approximated by the analytic Moyal function:

$$L(\lambda) \simeq \frac{1}{\sqrt{2\pi}} e^{-\frac{1}{2}(\lambda + e^{-\lambda})}$$

The argument of the function, $\lambda = (\Delta E - \Delta E_W)/\xi$ with a width parameter ξ, quantifies the deviation of the actual energy loss ΔE_W in material of thickness Δx from the most probable one $\Delta E = \langle -dE/dx \rangle \Delta x$.

The validity of the Bethe-Bloch formula is limited due to the fact that it only takes into account energy loss by electromagnetic ionisation and excitation of target atoms. For electrons and positrons, additional interactions are important, as shown in Focus Box 2.7. Occasionally the struck atomic electron can retain sufficient energy to ionise the material further. Such electrons are called knock-on electrons or δ-rays. More details are found in [765, Sec. 34] and in the Nuclear Data Services of the IAEA at https://www-nds.iaea.org/stopping/.

Focus Box 1.2: Ionisation energy loss

Relativistic quantum systems are characterised by two natural constants, Planck's constant of action $\hbar$ and the speed of light c:

$$\hbar \equiv \frac{h}{2\pi} \simeq 1.055 \times 10^{-34} \, \mathrm{Js} \;\; ; \;\; c \simeq 2.998 \times 10^{8} \, \mathrm{m/s}$$

Their dimensions are $[\hbar] = ML^2/T$ and $[c] = L/T$, where M denotes mass, L length and T time. The so-called natural system of units makes both constants dimensionless and equal to unity, $\hbar = c = 1$; it defines $\hbar \equiv 1$ as the unit of action and $c \equiv 1$ as the unit of velocity. Both constants thus disappear from all equations. This permits e.g. to measure all masses (M), momenta (Mc) and energies (Mc^2) with the same units:

$$[E, M, p] = \frac{ML^2}{T^2} = \mathrm{GeV} = 10^9 \, \mathrm{eV} \simeq M_{\mathrm{p}}$$

The base unit electronvolt (eV) is defined as the energy gained by a unit charge particle traversing a potential difference of 1 V. A billion of these units, 1 GeV, corresponds roughly to the proton mass M_{p} and thus forms the natural scale of high-energy physics processes. When calculating ordinary quantities, one obviously needs to convert results back into SI units. Useful conversion constants are given in the table below:

Quantity	Conversion factor
Mass	$1 \, \mathrm{kg} = 5.61 \times 10^{26} \, \mathrm{GeV}$
Length	$1 \, \mathrm{m} = 5.07 \times 10^{15} \, \mathrm{GeV}^{-1}$
Time	$1 \, \mathrm{s} = 1.52 \times 10^{24} \, \mathrm{GeV}^{-1}$
Electric field	$1 \, \mathrm{V/m} = 2.30 \times 10^{-24} \, \mathrm{GeV}^2$
Magnetic field	$1 \, \mathrm{T} = 6.94 \times 10^{-16} \, \mathrm{GeV}^2$

In the system of natural units, the electric charge is also dimensionless, $[e] = [\sqrt{\hbar c}] = [1]$. In addition, one can fix the dielectric constant and the magnetic permeability of the vacuum, $\epsilon_0 = \mu_0 = 1$, since they only relate mechanical units to electromagnetic ones. The fine structure constant α, which measures the strength of electromagnetic interactions, is then dimensionless. It is defined as the electrostatic energy of two electrons at a distance corresponding to their Compton wavelength, compared to the energy of an electron at rest:

$$\alpha = \frac{\frac{1}{4\pi\epsilon_0} \frac{e^2}{\hbar/mc}}{mc^2} = \frac{e^2}{4\pi} \simeq \frac{1}{137}$$

Focus Box 1.3: Natural units

- The Milky Way, our home galaxy, has a diameter of about 10^5 ly or 30 kpc.

- The nearest galaxies to the Milky Way are the dwarf galaxies Canis Major at 2.5×10^4 ly and Sagittarius at 7×10^4 ly. The closest full blown galaxy, the Large Magellanic Cloud is 1.6×10^5 ly or 50 kpc away.

- The Local Group, our cluster of galaxies, has a diameter of about 10^7 ly or 3 Mpc. It is part of the Virgo supercluster with a size of 1.1×10^8 ly or 33 Mpc.

- The farthest object observed to date is the newly discovered galaxy candidate HD1 [855] at a distance of 33×10^9 ly or 10.2 Gpc.

The second idiosyncrasy the reader will have to bear with is that in physics arguments we use the so-called natural system of units. It is explained in Focus Box 1.3. It uses the speed of light as the unit of velocity and Planck's constant as the unit of action. Our excuse is that its use greatly simplifies formulae by eliminating insignificant constants like c and $\hbar$ and others. The real reason for doing all this is of course that particle physicists are brought up this way; we cannot think in any other system of units. At the end, you will obviously want distances to come out in meters, times in seconds and energies in electronvolts or Joule. We therefore provide a conversion table for some useful quantities in Focus Box 1.3.

FURTHER READING

S. Clark, *The Sun Kings: The Unexpected Tragedy of Richard Carrington and the Tale of How Modern Astronomy Began*, Princeton University Press (2009)

G.V. Khazanov (Edt.), *Space Weather Fundamentals*, CRC Press (2016)

M. Marshall, *The Genesis Quest: The Geniuses and Eccentrics on a Journey to Uncover the Origin of Life on Earth*, Weidenfeld & Nicolson (2020)

J.C. McPhee and J.B. Charles (Edts.), *Human Health and Performance Risks of Space Exploration Missions*, NASA SP-2009-3405 (2009)

2 A Brief History

In this chapter, we briefly retrace the discovery of cosmic rays and their role in physics research over the last more than 100 years. This chapter largely draws on the collections of papers in Astroparticle Physics [531] and The European Physical Journal [469] commemorating the 100th anniversary of the balloon flights by Victor F. Hess.

2.1 AIR ELECTRICITY

In a 1785 paper published in the Mémoires of the French Académie Royale, Charles-Augustin de Coulomb reported extensive measurements he had made in May, June and July of that year concerning the spontaneous slow discharge of electrically charged bodies in contact with air [2]. For his measurements, he used the torsion balance he had already used to determine the force between electric charges as a function of distance, which we call Coulomb's law. He observed a slow relative drop in electric force between the charges as a function of time, independent of the material and the amount of charge. He found that the loss was faster in humid air, but still present with dry air.[1] He carefully assessed the charge losses by conduction through the supporting materials and found that they could not explain his findings: electric charges passed through the air. He concluded, that "the air is composed by an infinity of elements, partly insulating, partly conductive".

It took almost a hundred years of progress in constructing vacuum pumps [202] to prove this conclusion. In a paper read at the British Royal Academy in 1879 [3], William Crookes reported on what he called "less important offshoots" of his experiments on gas discharge tubes filled with rarefied gases. He charged an electroscope (see Focus Box 2.1) such that the gold leafs made an angle of about 110°. He sealed it into one of his glass tubes, evacuated to about a millionth of an atmosphere. The tube was immersed into water connected to ground and left untouched for 13 months. The electroscope still showed the same charge. It is thus indeed the air that discharges, as Coulomb had suggested. It remained to be detected how electrically neutral air turned into a (bad) conductor. The phenomenon was called "air electricity" and often confounded with other atmospheric phenomena like lightning.

The history of cosmic rays physics is full of prominent figures, but also physicists less well known to the public made important contributions. Excellent examples of the latter type are Julius Elster and Hans Geitel [620], two secondary school teachers, friends and collaborators from Wolfenbüttel in northern Germany. They were skilled experimenters and often built or modified measurement equipment for their purposes. After measuring the charge of raindrops and other electrical phenomena in the atmosphere they started to look into the hypothesis that free charge carriers in the atmosphere were responsible for "air electricity". They used a simple gold leaf electroscope (see Focus Box 2.1) converted to a crude voltmeter by a scale allowing to measure the opening angle of the leafs. It was connected to an open metal cup for charge collection from the surrounding air. They repeated Coulomb's measurements and confirmed his estimate of the charge loss rate [12], but found it independent of the applied voltage. They then covered the device by a wire mesh which could be charged positively or negatively to a potential of about 300 V. The observation was that the discharge was much faster when the mesh was negatively charged. This way they proved that indeed positive and negative free charges exist in the atmosphere, with negative ones (electrons) much more mobile than positive ones (ions). Their findings were promptly confirmed by the Scottish physicist and metrologist C.T.R. Wilson [18] in

---

[1]It is indeed true that moist air is more conductive than dry air, see e.g. [195].

DOI: 10.1201/9781003181385-2

a closed volume. He also confirmed the independence of the charge loss rate from voltage, i.e. "the continual production of ions throughout the air". Wilson carefully eliminated any spurious currents as well as humidity and dust, by repeating the experiment with "pure country air" outside London. Repeating the measurement in a railway tunnel showed no variation by the small overburden. Ten years later, Wilson invented the cloud chamber, an imaging particle detector which proved an immensely fruitful instrument in early cosmic ray research (see Section 2.4).

Experiments with cathode rays in evacuated tubes à la Crookes led J.J. Thomson to the discovery of the electron in 1897 [8]. He concluded that cathode rays, i.e. electrons, ionise the rarefied gas in a cathode ray tube [13, 16], and so do X-rays, i.e. photons, discovered at about the same time [5, 7]. Almost simultaneously, Henri Becquerel discovered the radioactive decay of uranium, and that the emitted radiation almost instantly discharges cathode ray tubes [6]. Marie and Pierre Curie discovered other radioactive elements like thorium and radium [9–11], which were contained in natural ores like uraninite and chalcocite. Ernest Rutherford [15] and Friedrich Ernst Dorn [14] identified the radioactive gas radon emitted from these ores. And Elster and Geitel indeed found radioactive ions in the atmosphere [17], by simply rubbing them off a negatively charged long wire exposed to air and putting the cloth on a photographic plate. Thus ionising radiation was known to be emitted from radioactive substances in the Earth's crust and in the atmosphere. The question remained whether the dominant source of the constant ionisation of the atmosphere was radiation from the soil, from air itself or from some source outside Earth. And whether the ionising radiation was made of charged particles or photons. The answer was difficult to find since all of these phenomena contribute to some extent to "air electricity".

2.2 EARLY PIONEERS

Thus not much progress was made during the first decade of the 20th century, even though extensive efforts were spent to understand the nature and origin of the ionising radiation. Systematic investigations were carried out by Elster, Geitel, Wulf and Linke in Germany, Wilson in the United Kingdom, many scientists in Canada and in the U.S. including Ernest Rutherford, Albert Gockel in Switzerland, and Domenico Pacini in Italy. An active group in Austria was formed for the research into natural radioactivity, since the only European uranium mine at the time was in St. Joachimstal in Bohemia, part of the Austro-Hungarian empire. In October 1910, the Austrian Academy of Sciences founded the Institute for Radium Research in Vienna. The International Committee for the Radium Standard involved all leading scientists working on the subject, under the leadership of Stefan Meyer, the founding director of the Vienna Institute.

Singling out radiation from the Earth's crust required to measure the ionisation rate as a function of altitude. The first measurements of "air electricity" on a manned balloon were a byproduct of meteorological research by Franz Linke. He carried a gold-foil electrometer from Elster and Geitel on six (out of twelve) balloon flights up to an altitude of 5500 m [21]. In parallel, ionisation rates were measured on the ground by his colleagues in Wolfenbüttel, Munich and Potsdam. He concluded that at an altitude of 1000 m, the ionisation rate was smaller than on ground. Between 1 and 3 km it stayed the same, but kept growing above, by a factor of four at 5500 m altitude. His findings were not really taken seriously and never quoted by his contemporaries, an astonishing fact since he was apparently well networked with the rest of the "air electricians". It may be due to the fact that he hid his findings in 90 pages of meteorological detail and quoted the voltage loss as a percentage per unit time. But this was only the beginning of a long series of efforts to measure the intensity at ever increasing altitude, as schematically shown in Figure 2.1 for some prominent protagonists.[2]

---

[2]For an account of early scientific ballooning see [504].

You may remember gold-foil electroscopes from you physics course at school. The one shown on the right is by William Ayrton from 1890. The pair of loosely supported gold leafs can be charged through the top plate and will move apart by Coulomb repulsion. The device is sensitive to very small charges. Measuring the angle between the leafs gives a crude measurement of the applied voltage. Instead of the top plate, an ionisation chamber can be fit on top of the device.

Credit: Science Museum Group

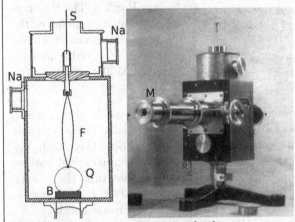

An ionisation chamber is a closed volume filled with gas, normally air. Its interior forms a capacitor, the voltage between its electrodes can be read from the outside using an electrometer, like the bifilar one invented by Theodor Wulf shown on the left [477]. Wulf patented the device in 1906 (DRP 181284). A cut through the instrument is also shown.

Credit: R.G.A. Fricke and K. Schlegel [477]

Wulf succeeded to make the device more sturdy and transportable without compromising its sensitivity. The lower end of thin metal filaments (F) was fixed to a quartz string (Q), using its elasticity as a restoring force. The string was mounted on an amber insulator (B). An external source can be connected to S and the case grounded. Two sodium dryers (Na) reduced the humidity on the inside of the device. The distance of the two filaments was measured through a microscope (M). A photograph of the instrument is shown on the right. It was manufactured by the Günter & Tegetmeyer (G&T) company. It was sensitive to voltage changes of the order of 0.01 V. A similar device [477], modified according to suggestions by Victor Hess, was used by him on his balloon flights. Kolhörster later had the device improved by an internal charger operated from the outside by a magnet. Millikan and Regener used electroscopes where the microscope was replaced by a camera. The distance of the filaments and the positions of barometer and thermometer needles could thus be automatically recorded.

The rate of voltage decrease dU/dt in the ionisation chamber measures density of created ions per unit time $d\rho/dt$, where ρ is the number of ions per unit volume V:

$$\frac{dU}{dt} = \frac{1}{C}\frac{dQ}{dt} = \frac{V}{C}e\frac{d\rho}{dt}$$

with the elementary charge e and the capacitance C. For capacitance and volume of a typical set-up, a voltage decrease of 1 V/h corresponds to an ionisation rate of a few ions/cm^3/s.

Focus Box 2.1: Electrometers

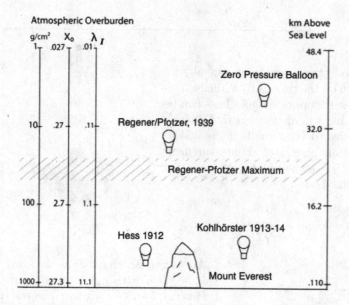

Figure 2.1 Float altitude of balloon experiments, and atmospheric overburden in units of mass surface density g/cm^2, radiation length X_0 as defined in Focus Box 2.7, and hadronic interaction length λ_I as defined in Focus Box 2.9. The Regener-Pfotzer maximum is the region of maximum ionisation rate. (Credit: adapted from Müller [484])

Progress was accelerated by the construction of a sturdy transportable electrometer, since the electroscope with its flimsy gold foils was delicate. The Catholic priest Theodor Wulf SJ invented the sensitive dual string electrometer, explained in Focus Box 2.1. It was used well into the 1920s to measure ionisation rates. Wulf's own measurements at various places [24, 28], including the top of the Eiffel tower in Paris – at that time the world's highest building – did not reveal a conclusive altitude dependence. In 1909 Karl Bergwitz, a student of Elster and Geitel and himself a secondary school teacher made another early balloon flight measuring the ionisation rate at 1300 m altitude [25]. He still found 25% of the ground rate, in contradiction to the assumption of gamma rays emitted from the ground. His advisers discouraged him from pursuing the subject further. Domenico Pacini and collaborators [32] made similar measurements on the mediterranean sea and below the surface of Lake Bracciano and also concluded that the radiation was probably not emitted by the Earth's crust.

Further balloon flights were made in 1910 by Albert Gockel [26], a German working at the newly founded University of Fribourg, Switzerland. He used a balloon on loan from the Swiss Aero-Club in Zürich and a rather unreliable electrometer he had also borrowed. These limitations were due to the fact that his research on "air electricity" was not at all supported by his boss, who thought that he should rather concentrate on solid state physics for the benefit of Swiss industry [467, 482], an argument against fundamental physics still heard today. The data collected during a short flight over Lake Zürich showed strong statistical and systematic variations. Nevertheless they confirmed that the radiation did not emerge from the Earth's surface. Gockel was first to call it "cosmic radiation," but missed a fundamental discovery. Two lessons are to be learned from his failure: if discovery is what you are after, use only first class instruments; and do not listen to your boss.

2.3 UP AND AWAY

The experimental proof of the extraterrestrial origin of cosmic rays is thus rightly credited to the Austrian Victor Francis Hess [31]. In a series of long balloon flights up to 5300 m altitude and with a first class electrometer manufactured to his standards by the Günther & Tegetmeyer company [460], he measured the intensity of the ionising

A device for counting the passage of ionising particles was invented by Rutherford's collaborator Hans Geiger in 1908 [23]. Its principle of operation and that of other gaseous detectors is explained in Focus Box 2.3. In its original version it worked only for the heavily ionising α particles with charge $+2e$. When Geiger moved back to Germany, he and Walter Müller at University of Kiel perfected the device in 1928 [47] such that it was also sensitive to β particles with a single charge $-e$, thus four times less ionisation (see Focus Box 1.2). It consists of a gas-filled metallic tube, serving as an ionisation chamber and cathode, with an axial anode wire strung in its centre and put under high voltage. When an ionising particle liberates electrons from the gas atoms, they are accelerated towards the anode. Since the field becomes very strong close to the thin anode wire, the drifting electrons gain momentum and can themselves ionise gas atoms. Thus an electron avalanche causes an appreciable current surge between anode and cathode. It can be made audible with a loudspeaker as a sharp snapping sound, and can be fed into an electronic circuit. The operation is stable over a wide plateau of high voltages, between ~ 1000 and $1500\,\text{V}$.

An early example is shown in the figure below [147]. A wire W is strung in the middle of a sealed metallic tube (T), insulated by the end-pieces. High voltage, typically produced by a dry pile (B) like a Zamboni, is fed to the wire as anode and the tube as cathode. When ionising radiation hits the tube, an avalanche is started by the primary ionisation causing a discharge. It is detected here by the electrometer on the right.

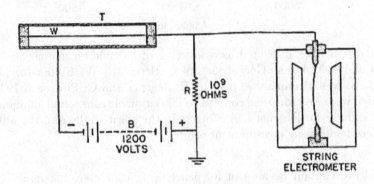

(Credit: US National Bureau of Standards)

The discharge can also be recorded electronically and its frequency counted. Geiger-Müller counters are sensitive to protons (H^+), alpha (He^{++}) and beta particles ($e^{\pm}$) through direct ionisation of the gas in the tube. They are sensitive to gamma rays (photons γ) indirectly through photoelectric effect or Compton scattering off atomic electrons (see Focus Box 2.7). As the first particle counting device and the first with electronic readout allowing coincidence measurements [49, 74], Geiger-Müller counters caused a break-through in cosmic ray physics [80].

Focus Box 2.2: Geiger-Müller counter

radiation systematically. His rather precise data showed that starting at an altitude of about 800m the intensity increases more or less quadratically with altitude. His measurements were repeated with an even higher accuracy by Werner Kolhörster [35] for altitudes reaching 9300m, again using electrometers by Günther & Tegetmeyer, which he had improved concerning air tightness and sensitivity. The results for the ionisation rate as a function of altitude obtained in manned balloon flights by Gockel, Hess and Kolhörster are shown in Figure 2.2. The extraterrestrial origin of the radiation was thus established, but many – like Hess himself – preferred to call it "altitude radiation" (Höhenstrahlung in

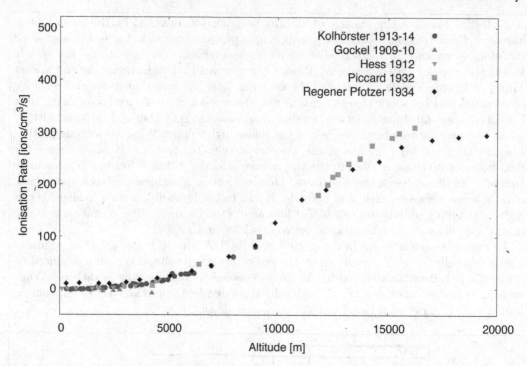

Figure 2.2 Ionisation rates measured as a function of altitude on manned balloon flights between 1909 and 1932 by A. Gockel [26], V.F. Hess [31], W. Kolhörster [34, 36] and A. Piccard [61], as well as unmanned ones by E. Regener and G. Pfotzer in 1934 [81]. The instruments used were electrometers coupled to ionisation chambers and an omnidirectional Geiger-Müller counter for Regener and Pfotzer. In the spirit of the day, the authors made no attempt to estimate their measurement errors.

German) or ultra-radiation because of its penetrating character. Its nature in terms of particle content remained unclear well into the 1930s.

The first world war (1914–1918) brutally interrupted physics research as a whole and cosmic ray research in particular. Moreover, after the war German physicists were excluded from international conferences for almost ten years [534]. In 1926, Robert A. Millikan from Caltech was already famous for his measurement of the elementary charge, recognised by the Nobel Prize in Physics of 1923. He got interested in the results of Hess and Kolhörster and decided to challenge them with his own measurements.[3] He and his collaborators invented an automatic recording of electrometer readings on a photographic film, moved by a clockwork [43]. It also recorded readings of a thermometer and a barometer. This device allowed him to use an unmanned sounding balloon, recovering instruments from two flights which ascended to altitudes of 11.2 and 15.5 km. His measurements confirmed that the ionisation rate grows at high altitudes, but his data showed only about a quarter of the increase previously measured by Hess and Kolhörster. He concluded that his data "constitute definite proof that there exists no radiation of cosmic origin having such characteristics as we had assumed" [43, p. 360f]. He nevertheless continued his research on mountain peaks and airplanes [45], trying to prove his belief that the ionisation was due to high-energy cosmic gamma rays. He intended to prove this by measuring the absorption coefficient and showing it to be inconsistent with the known properties of alpha and beta rays from radioactive

---

[3]The history of the transatlantic controversy concerning the discovery of cosmic rays is reviewed e.g. by De Maria, Ianniello and Russo [245].

elements. Indeed his recording electroscope was also submersible. His doubts about the cosmic origin of the radiation were dispelled by experiments he performed at two mountain lakes in southern California [44]. The principle of the measurement is shown in Figure 2.3. Muir Lake is situated at 3600 m altitude, Arrowhead Lake at 1550 m. Millikan found that the ionisation rate of an instrument was the same when it was sunk 1.8 m deeper in Muir Lake than in Arrowhead lake, making up for the absorption power of the two kilometres of air between the two. The absorption coefficient he found was low, so it reinforced his belief that cosmic rays were high-energy photons. They were supposed to originate from "...the creation of atoms the signals of whose birth constitute the cosmic rays" [48], especially the fusion of hydrogen into helium. Since he was one of the most influential physicists of his time, he was publicly credited as the discoverer of the extraterrestrial origin of cosmic rays, e.g. by a New York Times article reprinted in Science where they were called "Millikan rays" [40]. Victor Hess [41] and Kolhörster [42] promptly corrected this misconception. Despite all the progress in the field, Millikan never parted with his "birth-cry" photon hypothesis. At the symposium for his 80th birthday in 1949, the attendees thus had the privilege to listen to the same old story [130], suitably modified to include photons from matter-antimatter annihilation. Of course, high-energy cosmic gamma rays exist (see Chapter 7), but their rate is a minor contribution. The message to take away is clear: with all due respect, challenging leading authorities is part of the job of an experimentalist.

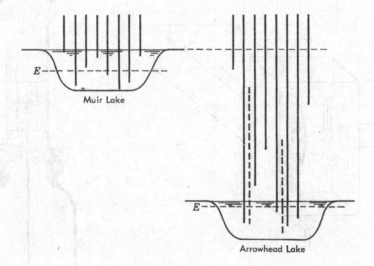

Figure 2.3 Principle of the Millikan measurement of cosmic ray absorption in mountain lakes as depicted by Rossi [169]. The absorption of cosmic rays (straight lines) by the air between the two lakes is compensated by immersing the electrometer about six feet deeper in the upper lake. Additional contributions from the air in between the two altitudes (dashed lines) do not exist.

And ample challenge there was, together with progress in instrumentation for cosmic rays research. On the ground, the invention of the Geiger-Müller counter (see Focus Box 2.2) allowed to count cosmic rays one-by-one. The coincidence technique (see Focus Box 2.4), applied to pairs of counters with an absorber in between, led Walter Bothe and Kolhörster to conclude in 1929 that cosmic rays were of a corpuscular nature [49]. Two very different balloon campaigns in 1932 – one manned, one unmanned – put an end to the debate on ionisation rate as a function of altitude. The manned one was initiated by the Swiss Auguste Piccard, then teaching at the Université Libre de Bruxelles, and realised together with the

Belgian engineer Max Cosyns. They constructed a pressurised payload capable of housing two people and their instruments. The spherical capsule was made of aluminium and carried by a hydrogen balloon with close to $15000\,m^3$ capacity. The first flight on September 14, 1930, starting from Augsburg in Germany, suffered from unfavourable winds and the capsule was damaged during the launch. The passengers had to tighten leaks and were unable to do the extensive experiments they had planned. The second flight on August 18, 1932, from the Swiss military airport in Dübendorf close to Zürich, was much more successful. While being inflated, the balloon was held by a detachment of the Swiss Army. Despite the early hours of the launch, more than 30,000 people attended the event [65]. The balloon reached a maximum altitude of more than $16\,km$ and many measurements with a Kolhörster-type electrometer as well as a pressurised one were completed [60, 61]. The ionisation rates are reported in Figure 2.2 together with those of his predecessors. It is clear that the ionisation rate continues to grow until stratospheric altitudes. The public interest in the adventure aspect of the experiment was enormous, Piccard became a national hero both in Switzerland and in Belgium. He was immortalised by Hervé as Professor Calculus (professeur Tournesol in the original) in the famous "Adventures of Tintin" comics.

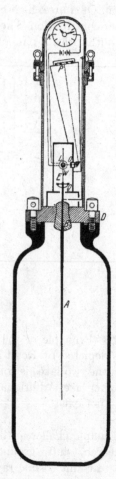

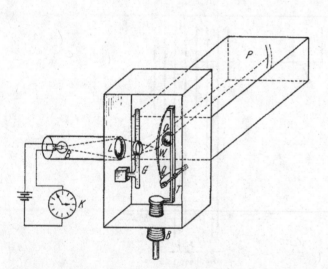

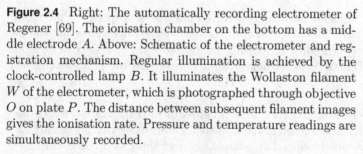

Figure 2.4 Right: The automatically recording electrometer of Regener [69]. The ionisation chamber on the bottom has a middle electrode A. Above: Schematic of the electrometer and registration mechanism. Regular illumination is achieved by the clock-controlled lamp B. It illuminates the Wollaston filament W of the electrometer, which is photographed through objective O on plate P. The distance between subsequent filament images gives the ionisation rate. Pressure and temperature readings are simultaneously recorded.

There could not be a greater contrast to the unmanned balloon campaign led by Erich Regener, then teaching at the Stuttgart Institute of Technology. In 1928, he had constructed the electrometer with automatic recording [50, 69] shown in Figure. 2.4. It was simpler and more accurate that Millikan's. It was also more sturdy since it had no moving parts except the clockwork regularly illuminating the single-filament electrometer. He submerged it in

Lake Constance [69] to depths of 230 m. The low absorption coefficient in water made him speculate about short wavelength photons as a source of the ionisation via Compton scattering off atomic electrons (see Focus Box 2.7). More interesting were his measurements during about 30 flights with sounding balloons in 1934–36, published together with his graduate student Georg Pfotzer. Without any public attention, they were launched from the courtyard of their institute. No adventure, no "conquest of the stratosphere", no public, no newspaper coverage. As a first result they verified the values obtained by Kolhörster [68] and extended them to stratospheric altitudes. They then compared the measurements with the ionisation chamber to simultaneous count rates of a single short Geiger-Müller counter (see Focus Box 2.2) sensitive to particles from all directions. The count rate was recorded with an electromechanical counter, photographed in regular intervals. The result is shown in Figure 2.6 on the left. The ionisation rate and count rate varied in exactly the same way with altitude [80, 81]. This is an important step, since it links an integral measurement of the ionisation to a measurement counting the impacts of ionising particles one-by-one. They further found that above an altitude of 18 km the rate of ionisation no longer increased. There is thus an altitude corresponding to a maximum of the ionising radiation, primary and secondary summed up. In follow-up experiments [83], Regener and Pfotzer measured the vertical flux of cosmic rays with three vertically stacked Geiger-Müller counters in coincidence as shown in Figure 2.5. The result shown in Figure 2.6 on the right clearly exhibits the Regener-Pfotzer maximum at a height in the vicinity of 20 km. Above, the ionisation rate and the counter hit frequency decrease. To see the levelling off of the rate at altitudes above 50 km – corresponding to primary cosmic rays, plus a little backsplash from the atmosphere – and reduced by about 50% compared to the Regener-Pfotzer maximum, required instruments on a rocket (see Section 2.6).

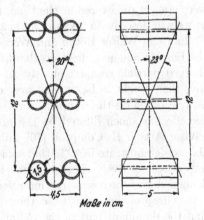

Figure 2.5 The Geiger-Müller hodoscope used by Regener and Pfotzer for stratospheric flights [87, 88] in 1936. The Geiger-Müller tubes in each layer were connected in parallel to a threefold coincidence circuit à la Rossi (see Focus Box 2.4).

Encouraged by the coincidence measurements of Bothe and Kolhörster [49], Bruno Rossi experimented with triple coincidences in various geometries, including ones which cannot be crossed by a single particle [74] (see Section 2.5). To look into the energy spectrum, he set up a tower of lead bricks with three imbedded counters as shown in Figure 2.7 [74]. He found that while a soft component of cosmic rays was easily absorbed in a few cm of lead shielding, some 50% of a hard component could traverse as much as one meter of lead. This was a completely unexpected result – no particles with such a penetration had yet been observed – and it met some scepticism. It strongly contradicted the gamma ray hypothesis, even though Rossi's argument that secondaries cannot be more penetrating than primaries is flawed in hindsight. We of course now know that the primary radiation is dominated by protons and helium nuclei which never reach the Earth's surface. The penetrating secondaries of their

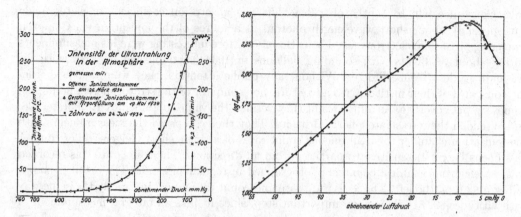

Figure 2.6 Left: Ionisation rate (left scale) and count rate (right scale) as a function of decreasing atmospheric pressure automatically measured on sounding balloon flights in 1934 by E. Regener and G. Pfotzer [81]. The instruments were two ionisation chambers and a single short Geiger-Müller counter sensitive to all directions. Right: Number of counts per 4 minutes as a function of decreasing atmospheric pressure from Pfotzer's thesis of 1936 [87, 88] with the vertical coincidence set-up shown in Figure 2.5, subtending a solid angle of about 20° around the zenith. The count rate is plotted with a logarithmic scale. It increases exponentially with decreasing overburden, except for a small hump at about 3 km attributed to geomagnetic effects. Above the Regener-Pfotzer maximum the rate decreases.

air showers, which do reach down to sea level, are mostly muons (see Chapter 3), which were not yet discovered at the time. But it would not be long before they were discovered, as we will describe in Section 2.4 below.

In 1929, Walter Bothe and Werner Kolhörster [49] had demonstrated the influence of the Earth's magnetic field on the flux of cosmic rays. It was thus known that they had a charged particle component. My all-time cosmic ray hero Bruno Rossi [55, 56] predicted that there ought to be an east-west effect if there were a preferred sign to the cosmic ray charge, based on the complex work by the Norwegian Carl Störmer [58, 59, 67]. Although the effect is much diluted by the production of secondaries, Luis Alvarez and his thesis advisor Arthur H. Compton [70], Johnson and Stevenson [73] and Rossi himself [74] were able to demonstrate in 1933 that more cosmic rays come from the west than from the east, indicating positively charged primaries. In 1937, Compton and Turner [96] measured the dependence of the cosmic ray intensity at sea level on the geomagnetic latitude. They found a 10% drop of the intensity at the equator compared to 50° latitude, which demonstrated that the dominant part of the radiation consisted of charged particles. Thus the primaries are predominantly particles, positively charged particles. The state of the art in cosmic ray research was documented in great detail and with an extensive bibliography by Erwin Miehlnickel in 1937/38 [102], just before the outbreak of World War II.

In 1937, Regener was removed from the University by the Nazi regime but found refuge in a research institute of the Kaiser-Wilhelm-Gesellschaft. He constructed instruments which were intended for spaceflight with one of Wernher von Braun's sounding rockets, but never used. Today, he is mainly recognised as one of the founding fathers of geophysics [189]. Victor Hess received the 1936 Nobel prize in physics together with Carl Anderson (see Section 2.4). He and his Jewish wife emigrated to the U.S. in 1938. The more opportunistic Kolhörster joined the Nazi party and continued his career in war-time Germany. Bruno Rossi was forced to emigrate in 1938 due to the Italian racial laws. After stays in Denmark

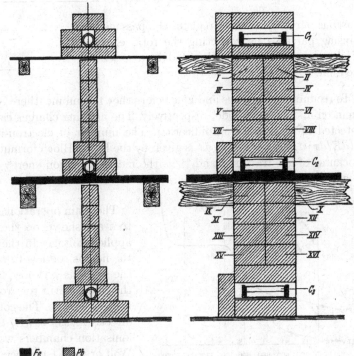

Figure 2.7 Rossi's 1933 set-up to search for highly penetrating particles in cosmic rays [74]. There is a preselection by a few centimeters of lead on top of the first tube. The tubes are separated by up to 50 cm of lead each. Their hits are registered in three-fold coincidence.

Fe   *Pb*

and Britain he ended up in the U.S., where he contributed to Project Manhattan during World War II. After the war, he continued cosmic ray research at MIT.

2.4   IMAGES OF COSMIC RAYS

During about four decades after their discovery, cosmic rays stayed the dominating source of high-energy particles. They cause particle showers by interaction with the nuclei of the atmosphere, such that a variety of long-lived particles reaches the Earth's surface. Imaging particle detectors allow these to be visualised and identified. The first such detector was the cloud chamber, developed by the Scotsman Charles Thomson Rees Wilson [30, 33] at the Cavendish laboratory. Wilson originally invented the cloud chamber to study the formation of clouds in the atmosphere. It was known at the time that the presence of dust particles favoured the formation of droplets in a supersaturated vapour. Wilson constructed a set-up to supersaturate a vapour by rapid expansion. He observed that indeed condensation also occurred in a dust-free environment. Ionising droplets by X-rays, he made a gross measurement of the ion charge [18, 20] later improved by J.J. Thomson [19]. Much to his surprise, ionising particles left tracks of condensation, even whole showers were caused by converted X-rays. Perfecting the construction of the chamber, ably doing most of the glassblowing himself [530], Wilson arrived at the set-up shown in Figure 2.8 on the left [30, 33]. Its operational principles are discussed in Focus Box 2.5.

The first to publish an image of a cosmic ray track was Dmitry Vladimirovich Skobeltsyn [46] from Leningrad, but without identifying it as such. In 1927, he was doing research on Compton scattering using a cloud chamber in a 1 kG magnetic field. He noticed at least one "track of unknown origin with such a large energy that its curvature cannot be measured" [46, Fig. 12 and p. 371–372]. He estimated the magnetic rigidity (see Focus Boxes 2.5 and 3.2) to more than 20 MV, so the momentum to at least 20 MeV for a particle of unit

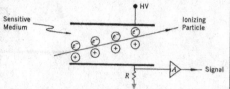

Gaseous particle detectors register the passage of
ionising particles by collecting the total charge
of electrons and/or ions liberated in a gaseous
medium.

   To recuperate electrons and ions before they recombine, there is an electric field collecting
them on anode or cathode, respectively. The arriving charges cause a short current surge,
detected after suitable amplification. The number of electron-ion pairs created is $N_i = -(dE/dx)\,d/W$, where dE/dx is given by the Bethe-Bloch formula (Focus Box 1.1), d is the
thickness of gas traversed and W is the mean ionisation energy required for a single pair;
in a gas $W \simeq 30$ eV.

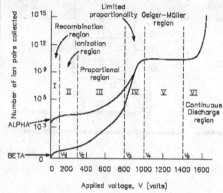

The main operational regions of gaseous detectors are shown on the left as a function of the
applied voltage. In the **recombination region**,
the field is too low to prevent recombination and
the signal is very low. In the **ionisation region**,
the charges drift towards the electrodes, but there
is no avalanche. The collected charge is small, but
directly measures dE/dx. This is the region where
ionisation chambers work, like the ones used by
Wolf or Hess (see Focus Box 2.1).

   In the **proportional region**, the signal is amplified by an ionisation avalanche close to
the anode, still proportional to the number of ionisation pairs but amplified by a factor
of 10^4 to 10^8. The signal is thus large, but subject to variations in amplification due to
high voltage, temperature and other environmental factors. This region is used in multiwire
proportional chambers (MWPC) and drift chambers.

   When the high voltage is further increased, in the **Geiger-Müller region** (see Focus
Box 2.2), the acceleration of primary electrons increases rapidly and they excite an avalanche
early on. In addition, a large number of photons is produced in the de-excitation of ionised
atoms. These also contribute to the avalanche by photoelectric effect. A discharge is caused
giving a large electrical signal on the anode. It can be made audible as a sharp snap.
Following the discharge, the counter is insensitive during what is called its dead time.

Focus Box 2.3: Gaseous particle detectors

electric charge. In the following years, the cloud chamber became one of the two standard
imaging particle detectors of the time, together with photographic emulsions discussed below. Wilson shared the 1927 Nobel prize for this invention with Arthur H. Compton, the
American physicist who discovered elastic photon-electron scattering named after him (see
Focus Box 2.7).

   In 1932, Carl D. Anderson at Caltech [62], using a cloud chamber of the type shown in
Figure 2.8 on the right, combined with a magnet to form a spectrometer (see Focus Box 3.2),
discovered the positron, the anti-electron predicted by Paul Dirac's equation of motion for
relativistic particles. Figure 2.9 shows the spectrometer as well as the cloud chamber photo
of a positron. Its charge $+e$ is determined through the curvature of its track, its direction
of flight by the energy lost when traversing a lead plate. Its energy loss per unit path
length, measured through the density of droplets, also indicates a light, minimum ionising
particle (see Focus Box 1.2). The year after, Giuseppe Occhialini and Patrick Blackett [72]

A coincidence is the appearance of two or more signals from particle detectors within a given time window. Walter Bothe and Werner Kolhörster were the first ones to apply this technique to cosmic rays research, using two vertically stacked Geiger-Müller counters with an absorber in between [49] as shown on the right.

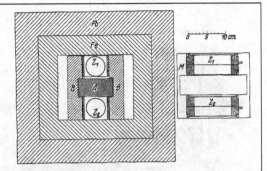

The two counters were connected to electroscopes like in Focus Box 2.2. Hits were recorded photographically on separate moving films. Coincidences were tediously identified by hand comparing the two films.

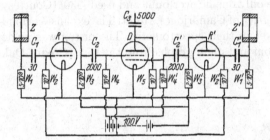

In 1930, Bothe invented the first electronic coincidence circuit shown on the left [52], using vacuum tubes. The signal from two Geiger-Müller tubes is clipped and amplified by the two circuits with triodes R and R'. The outputs are fed to the bi-grid tetrode D.

When there is a signal on both R and R', the tube D fires and an output pulse is presented to the coupling capacitor C_3, where the signals are counted. The time resolution of the set-up was of the order of a millisecond. This circuit was limited to a twofold coincidence, since there are no tubes with more than two grids.

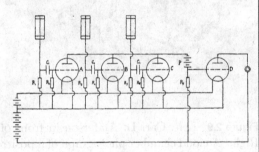

Bruno Rossi extended this idea to multiple coincidences by putting the inputs in parallel as show on the right [53]. Negative pulses from the tubes are fed to the grids of the triodes such that they block the anode current.

Only if all tubes are blocked simultaneously will there be a positive pulse fed to the discriminator tube D. His circuit improved the time resolution to $350\,\mu s$.

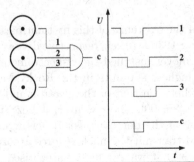

A modern implementation of a triple coincidence is shown on the left. Hits from detectors 1, 2 and 3 are shaped to logical signals and fed to an AND gate which delivers the output signal c. Typical time resolutions are of the order of nanoseconds, limited by time jitter of the detector input.

Focus Box 2.4: Coincidences

Figure 2.8 Left: The cloud chamber constructed by C.T.R. Wilson at the Cavendish laboratory in its final design. It was reportedly the only one he ever built and used [530] (Courtesy and copyright: Cavendish Laboratory, University of Cambridge). Right: The expansion cloud chamber used by Anderson and Neddermeyer in their spectrometer. The piston was made of glass to allow illumination from the bottom and photographs from the top. (Credit: Oak Ridge Associated Universities)

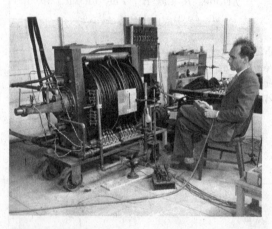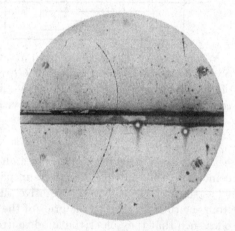

Figure 2.9 Left: Carl D. Anderson in front of his magnetic spectrometer for cosmic rays. Right: A cloud chamber photograph of a positron with kinetic energy 63 MeV, entering from below, traversing a 6 mm thick lead plate losing energy, and continuing with a curvature corresponding to 23 MeV [71]. (Credit: Archives, California Institute of Technology)

at the Cavendish Laboratory of Cambridge confirmed the existence of this first antiparticle. Anderson and Hess shared the Nobel prize in physics of 1936 for their respective discoveries.

In the years 1936/37, Anderson and Seth Neddermeyer installed their spectrometer on Pike's Peak, with an altitude of 4300 m one of the highest summits in the Rocky Mountains. They identified a charged particle with a mass between that of the electron and that of the proton and suggestively called it "mesotron" [85, 90]. Shortly after, Jabez Street and Edward Stevenson [98] confirmed their discovery of the new particle. It was wrongly conjectured at the time to be the particle that should transmit the nuclear force according to Hideki Yukawa's trial field theory [84]. Consequently it was wrongly categorised as a strongly interacting meson [203]. In the group of Marcello Conversi in Rome, as well as by

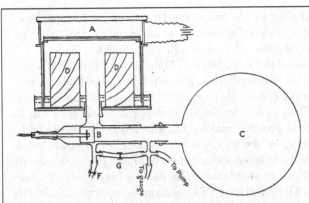

A schematic of Wilson's cloud chamber, seen in Figure 2.8(left), is shown on the left [33]. The cloud chamber A on top can be observed through its glass cover, it is illuminated obliquely through its cylindrical enclosure. On the bottom it is closed by a metal piston, with cylindrical walls partially immersed in water. The wooden hollow cylinder D reduces the air volume under the piston.

At the beginning of an observational cycle, the glass bottle C on the right is evacuated. The chamber A contains a mixture of air and water vapour in equilibrium. When plug B is pulled, the piston is abruptly pulled down, so that the saturated vapour in A is adiabatically expanded. In adiabatic expansion, the pressure P drops faster than the volume V ($PV^\gamma =$ const, with $\gamma \simeq 1.33$ for water vapour under normal conditions), such that the temperature also drops. The water in the mixture undergoes a phase transition. Condensation starts around condensation nuclei, i.e. ions (or dust particles if the air is not carefully cleaned). Thus a track of droplets forms along the path of an ionising particle. The chamber remains sensitive and droplets suspended for about a second afterwards. When they are charged, they move under the influence of the electric field created by the battery.

Anderson's construction of the chamber (see Figure 2.8 right) differs from Wilson's in that the piston is transparent so that illumination comes from the bottom. The chamber is immersed in a homogeneous magnetic field B in the direction of the chamber axis, so that the momentum p of particles with charge q can be measured by their radius of curvature $\rho = p/(qB)$. The often used quantity magnetic rigidity, p/q, is thus the direct output of a spectrometer measurement as discussed in Focus Box 3.2. Geiger-Müller tubes trigger the expansion of the chamber as well as a camera photographing the chamber volume. Absorbing plates can be used inside the chamber to measure the energy loss and thus the square of the particle charge. An example of the spectrometer set-up and the photo of a positron track are shown in Figure 2.9. The positron enters from the bottom, since its momentum is diminished when traversing the lead plate. Its positive charge is indicated by the sense of the curvature. The ionisation above the plate is slightly larger than below, since its velocity β is diminished, $-\langle dE/dx \rangle \propto 1/\beta$. The energy loss allows to infer the mass of the particle.

Focus Box 2.5: Cloud chamber operation

Evan J. Williams and George E. Roberts at University College of Wales, the decay of the new particle was first observed [275]. Under the most difficult conditions of World War II, Conversi, Oreste Piccioni and Ettore Pancini determined its lifetime of about $2\,\mu$s. However, their findings were not generally known until after the war [118, 120]. After Enrico Fermi's emigration to the U.S., Franco Rasetti, a close collaborator of his, went to Canada in 1938. He measured, in parallel to Conversi's group, the lifetime of "mesotrons" stopped in matter [114, 480]. After the war, Conversi and his colleagues showed that the muon had nothing to do with nuclear forces [120]. It was thus recognised as a heavy "brother" of the electron, with similar properties but about 200 times heavier.[4] The name mesotron disappeared in favour of the less suggestive muon (μ).

---

[4]However, it is not at all an excited state of the electron, as the absence of the radiative decay $\mu^\pm \to e^\pm \gamma$ demonstrates.

The second important imaging detector of the time was the photographic emulsion, silver halide grains suspended in gelatin. Henri Becquerel had observed in 1896 that radioactive radiation left foggy images on photographic plates. In 1910, Suekichi Kinoshita, a Japanese physicist of the Meiji era educated in Göttingen and Manchester, found that single alpha particles were able to produce a detectable photographic effect [27]. His findings were promptly confirmed by Max Reinganum [29], who also suggested that at slanting incidence the grains formed sparse sequences. Thus molecules are excited permanently and can be converted to metallic silver by a chemical process, making the affected grain visible under a microscope. The emulsion is then fixed by washing away the undeveloped silver halide grains. Despite the exceedingly thin emulsion layer on early photographic plates, Kinoshita and Ikeuti [37] succeeded shortly afterwards to show that alpha particles leave a track of developable grains along their trajectory. The sensitivity to ionisation depends on the density and size of silver halide grains. The density of developed grains is then proportional to the specific energy loss by ionisation, itself proportional to the square of the particle charge (see Focus Box 1.2). The emulsion can thus record the passage of ionising particles by tracks. It is a 3D visualisation and measurement device just like the cloud chamber. And it has a built-in memory. However, its usage is much more tedious due to the many steps involved.

Major improvements in the application of this technique to subatomic physics in general and cosmic rays in particular are due to the work of the Austrian physicist Marietta Blau,[5] another example of an almost forgotten pioneer. She was one of a bunch of (mostly freelance) female scientists attracted to the Vienna Institute of Radium Research by its director Stefan Meyer and by the role models of Marie Curie and Lise Meitner [512]. Blau had an education in X-rays and medical applications, so she was well prepared to apply imaging techniques to other research problems, which she successfully did at Meyer's institute from 1923 to 1938. She collaborated with industry to develop thicker and more sensitive emulsions, sensitive not only to alpha particles but also protons [39], which ionise four times less. In collaboration with her student Hertha Wambacher she succeeded to measure recoil protons from the scattering of neutrons [63, 64], which had just been discovered by James Chadwick. Comparing energy distributions measured in emulsions, cloud chambers and in scintillators (see Focus Box 2.6), they succeeded to make nuclear emulsions a quantitative tool. The two were improbable partners in research [280]. Blau came from a well-to-do Jewish family, Wambacher was an ardent Nazi who secretly joined the Nazi party in 1934, when it was still illegal in Austria. In 1936, Blau and Wambacher exposed a stack of emulsions on the Hafelekar near Innsbruck at 2300 m altitude for five months. Besides the usual cosmic ray tracks they also observed what they called "stars" [92–94, 101], groups of tracks originating from a common interaction vertex. These come from the blow-up of heavy nuclei excited by the absorption of a hadron. The 1938 incorporation of Austria into the III. Reich forced Blau to emigrate at the height of her academic career. This also ended her plans for balloon flights with emulsions. Via Oslo and with help from Albert Einstein, she and her mother found temporary refuge in Mexico, where funds for research were in short supply. In May 1944, Blau moved to the U.S., where she worked for industry and academia after the war. In 1960, she went back to Vienna, but found no support except free-lance work for the Radium Institute. She was nominated for the Nobel prize three times, including two nominations by Arnold Sommerfeld. Due to a devastating report by Axel E. Lindh [512] which clearly did not do justice to her contribution, the nominations were not retained.

Ilford company in Knutsford (Cheshire, Great Britain) had an important role in the development of nuclear emulsions [223]. Starting in 1939, they marketed the half tone emulsions used by Blau and Wambacher. After the war in 1949, in collaboration with the group of

---

[5]For a short biography, see:
https://lise.univie.ac.at/physikerinnen/historisch/marietta-blau.htm.

Atoms and molecules emit light when excited by an electromagnetic stimulus or by ionising radiation. This phenomenon is called **luminescence**, when the emitted spectrum has a component in the visible range. The decay constant of the light emission further characterises the phenomenon. When the emission is fast, compared to the typical lifetime of excited energy levels of $\simeq 100\,\mu s$, it is called **fluorescence**. If a meta-stable energy level is excited and the emission is delayed by more than a microsecond, it is called **phosphorescence** and may last from seconds to hours. **Scintillation** is the basically immediate emission of a short flash of light, it thus belongs to the fluorescence family. Decay times for the flash are of the order of nanoseconds and allow to accurately time the passage of ionising particles.

The decay time of the light emission and the spectrum of the emitted light depend on material properties. In order to be useful, the material must obviously be as transparent as possible. It can be of an organic or inorganic nature. **Inorganic scintillators** are single crystals grown at high temperature, usually alkali halides like NaI or CsI, or oxides like BGO or PWO. Their crystalline structure creates the energy bands which can be excited and emit light in the visible range when electrons fall back to the ground state. Some need activator impurities to enhance scintillation. The best known example is thallium, used e.g. in NaI(Tl) scintillation crystals. Inorganic scintillators have high nuclear charge, high density and a short radiation length. Important characteristics are shown in the table below: ratio of electric charge to atomic number Z/A, mass density ρ, hadronic interaction length λ_I (see Focus Box 2.9), radiation length X_0 and Molière radius R_M (see Focus Boxes 2.8 and 3.4) as well as light yield compared to sodium iodide.

Material	Z/A	λ_I [cm]	X_0 [cm]	ρ [g/cm^3]	R_M [cm]	Light Yield [% of NaI(Tl)]
NaI(Tl)	0.427	154.6	9.46	3.67	4.5	100
CsI(Tl)	0.416	171.5	8.39	4.51	3.8	165
BGO	0.421	22.8	1.12	7.13	2.2	20
PWO	0.413	20.7	0.89	8.30	2.0	<1
Plastic Scintillator Saint-Gobain BC-400	0.541	78.9	42.6	1.03	9.1	25

In **organic scintillators** individual molecules scintillate. They consist of fluorescent aromatic hydrocarbons, dissolved either in fluid or in a solidified polymer matrix like polyvinyltoluene. In the latter form, they can be manufactured in almost any shape and size. Since they are very transparent with absorption lengths of several meters, single units can be made large. However, their light output is small compared to most inorganic scintillators, so thicknesses of order centimetres are required to obtain a measurable signal from minimum ionising particles.

Nitrogen molecules in the **atmosphere** fluoresce when excited by extensive air showers. They emit light omnidirectionally in the ultraviolet range of frequencies. The high intensity of ionising particles at the poles causes aurorae, which tend to be dominated by emissions from atomic oxygen, resulting in a greenish glow. It comes from a forbidden transition of electrons in atomic oxygen which persists for a long time and accounts for the slow brightening and fading of auroral glow. These thus belong to the phosphorescence family of phenomena, with time scales of seconds.

Focus Box 2.6: Fluorescence and scintillation

Cecil Powell in Bristol, they succeeded to improve the sensitivity of their emulsions sixteen-fold. The Ilford G5 emulsion was thus sensitive to minimum ionising particles and remained a standard in the field. Powell and his group developed the usage of this tool in nuclear and cosmic rays physics to an almost industrial level. He recruited collaborators from all over Europe for the tedious analysis of emulsions, setting the stage for the large experimental collaborations in particle physics we are used to today.

Cecil Powell, César Lattes and Giuseppe Occhialini in 1947 discovered the charged pion, the lightest hadron.[6] Close in mass to the muon but strongly interacting, it explained the discrepancy between high production rates in the upper atmosphere and high penetration power, which had inspired Marshak and Bethe to form the essentially correct "two-meson hypothesis" [121]. Strong interactions of incident cosmic rays produce a large number of pions, which decay into muons via $\pi^{\pm} \to \mu^{\pm}\nu_{\mu}$ with a short lifetime of only 26 ns. It is the muons which penetrate the atmosphere, since they do not interact strongly. Their long lifetime of $2\,\mu$s, sufficiently prolonged by relativistic time dilatation, allows them to reach the Earth's surface. The soft component of cosmic rays is made of electrons and photons from electromagnetic showers, mostly initiated by the decay of neutral pions, $\pi^0 \to \gamma\gamma$. Basic features of electromagnetic and hadronic showers are discussed in Focus Boxes 2.8 and 2.9.

In the same year 1947, Clifford Butler und George Rochester, extending on work they had done during the war [224], discovered the kaons K^0 and $K^{\pm}$ [122], the first members of the second generation of hadrons, which were called "strange particles". They used a cloud chamber and magnet from Blackett's group in Manchester, placed under a lead target to locally cause hadronic showers and controlled by a hodoscope of Geiger-Müller counters to select high multiplicity showers. The observed signature was V-shaped pairs of opposite sign decay products, $K^0 \to \pi^+\pi^-$, as well as tracks with large-angle kinks without visible recoil, as from $K^{\pm} \to \mu^{\pm}\nu_{\mu}$. The discovery was confirmed in 1950 by Anderson's group [134] using the same technology, but much higher statistics due to a high altitude location on White Mountain (3200 m).

In the early 1950's, the strange baryons Λ, Σ and Ξ were identified in cosmic ray interactions using cloud chambers [138, 139] and emulsions [143–145]. But the 1950's also saw the rapid replacement of cosmic rays by particle accelerators as the primary source of high-energy particles. A good example of this transition is the discovery of the neutral pion in $\pi^0 \to \gamma\gamma$, since it was seen at roughly the same time in cosmic rays using emulsions [133] and at the 184-inch synchro-cyclotron at Berkeley [135].

Accelerators thus took over as far as the study of particles and their interactions was concerned. The cloud chamber was soon replaced by the bubble chamber as the major imaging particle detector; cloud chambers are only still used as demonstration devices. Electronic devices and digital technology allow today to register, identify and analyse high rates and a large variety of particles. These technologies of course also found their way into cosmic ray research. Emulsions, however, are still used when experimenting with rare events and very short-lived particles. A recent example [461] is the OPERA experiment, which saw tracks of τ leptons with a lifetime of about 3×10^{-13} s [439, 561, 666] in a CERN neutrino beam. Emulsions will also serve in the SHiP experiment [721] under construction at CERN, searching for new states of matter and the scattering of dark matter particles off electrons.

As far as cosmic rays are concerned, the emphasis shifted from using their secondaries as a tool for particle production towards understanding the composition and origin of primary cosmic rays. This will be the subject of the rest of this book.

---

[6]Excellent reproductions of the original emulsion images are found in [223].

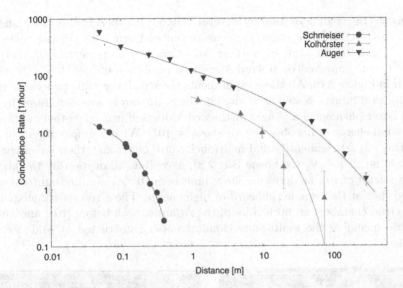

Figure 2.10 Rates for triple coincidences with the third counter off-axis, as a function of its distance. The counter surface was $81\,cm^2$ for Schmeiser and Bothe [103], $430\,cm^2$ for Kolhörster [104] and $200\,cm^2$ for Auger and collaborators [111]. The first two took data in the plane, Auger on Jungfraujoch and Pic du Midi. The curves are a fit to guide the eye, using a power law with soft exponential cut-off. The qualitative behaviour is similar, normalisation and extent depend on surface and altitude.

2.5 EXTENSIVE AIR SHOWERS

Dmitry Skobeltsyn had seen in 1929 that groups of almost parallel energetic electron tracks appeared in some of his randomly triggered cloud chamber pictures [51]. He attributed energetic electrons to Compton scattering of cosmic photons, multiple occurrences were thus "curious". The pioneer of extensive air shower detection is again Rossi – who else? – thus opening the way to detect very rare, high-energy cosmic rays. Directly after inventing his triple coincidence circuit (see Focus Box 2.4), he experimented with various geometries of their arrangement and with the thicknesses of absorbers above and in between the counters [74]. One of them was a triangular arrangement, with one counter on top and two arranged horizontally on the bottom, at a distance such that no single particle could hit all three tubes, and with a lead sheet on top. When varying its thickness, the so-called "Rossi curve" of coincidence rates resulted. It showed astonishing features. The coincidence rate first increases with lead thickness, then slowly decreases. And it is not zero when lead is completely removed. This means that the lead serves at the same time as an absorber for soft particles and a target for energetic ones. It causes showers of soft particles which are more and more absorbed when the target increases beyond an "optimum" thickness. But astonishingly enough, the rate does not go to zero when the lead target/absorber is completely removed. When Rossi first tried to publish his findings, his paper was rejected [215, p.72]. Only after an intervention by Heisenberg it was accepted by Physikalische Zeitschrift [66].

The result triggered activities by several others. Walther Bothe [103] and Werner Kolhörster [104] verified the effect in measurements at sea level. Pierre Auger took to Jungfraujoch and Pic du Midi for further coincidence measurements [107, 111], using a coincidence circuit improved by Roland Maze to a resolution of $5\,\mu s$ [109]. The results are summarised in Figure 2.10. All these experiments showed that cosmic rays cause showers of particles, in metal targets as well as in air. The name air shower was introduced by Bothe to denote the latter phenomenon. Auger distinguished himself from the others in that he estimated the total energy of his observed air showers [107]. With a simple argument based on his estimation of particle identity and multiplicity (10^6 electrons), their mean energy (critical energy in air $\simeq 10^8\,\mathrm{eV}$, see Focus Box 2.8), as well as absorption by the atmospheric overburden (factor of 10), he derived a lower limit of $\simeq 10^{15}\,\mathrm{eV}$. In hindsight a conservative limit indeed, but at the time an unheard-of high energy. These results stimulated imaging work with cloud chambers on both sides of the Atlantic, both before [108] and after World War II [136], including the multi-plate cloud chamber constructed at MIT, as shown in Figure 2.11.

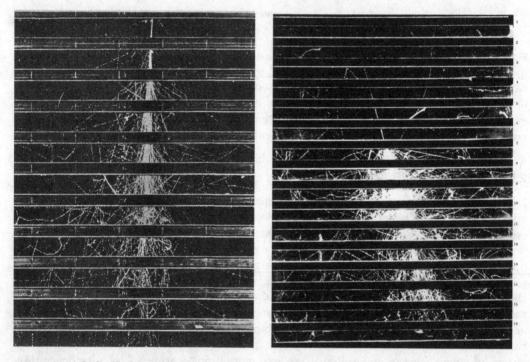

Figure 2.11 Left: High-energy electromagnetic shower [169, Fig. 7-5, p. 96] in an MIT cloud chamber with 1.25 cm thick brass plates. An electron or photon enters from above and interacts in the first plate. Electromagnetic showers start early; they are slim and regular. Right: High-energy hadronic shower [129] observed in an MIT cloud chamber with 1.3 cm thick lead plates. A proton of 10 GeV enters the chamber from above and interacts in metal plate 7. Hadrons are more penetrating, their showers are wide and irregular and contain even more penetrating muons.

It became qualitatively clear from these images that electromagnetic showers have a hard core of energetic electrons and photons, accompanied by softer particles at the fringes, like Rossi had concluded. Due to the large number of particles in the shower and the rather short radiation length, they are compact, symmetric and regular. Hadronic showers also have an electromagnetic component, mainly coming from $\pi^0 \to \gamma\gamma$ decay. In addition, light hadrons

The elastic electromagnetic interaction of an **electron** with atomic electrons in a material, Møller scattering ($e^-e^- \to e^-e^-$), is different from that of other particles since it involves four identical particles. This leads to a slightly different **ionisation energy loss** $\langle -dE/dx \rangle_{ion}$ as shown on the right for a lead target [765]. **Positrons** undergo Bhabha scattering ($e^+e^- \to e^+e^-$) and electron-positron annihilation ($e^+e^- \to \gamma\gamma$) which make additional small contributions to their energy loss at modest energies.

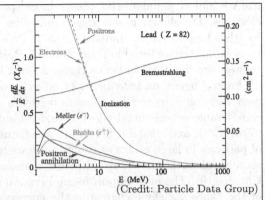

(Credit: Particle Data Group)

More important for light particles is the **bremsstrahlung** process in the electromagnetic field of a nucleus, via the Feynman diagram on the left. A nucleus (or electron), represented by the thick dot, is crucial to preserve energy-momentum conservation. The mean energy loss per unit path length for $e^\pm$ is determined by the **radiation length** X_0, $\langle -dE/dx \rangle_{brem} = E/X_0$. Radiation lengths for all common materials are listed e.g. by the Particle Data Group [765, Tab. 6].

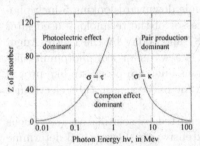

The energy, where the ionisation energy loss per radiation length is equal to the electron energy itself, is called the **critical energy** E_C. It has an important role in the understanding of electromagnetic showers, as described in Focus Box 2.8. Measured critical energies for electrons and muons have been parametrised by the Particle Data Group as a function of the nuclear charge [765, Sec. 34].

The passage of photons through matter involves three dominant processes: the photoelectric effect at energies below 100 keV; Compton scattering at energies between 1 and 10 MeV and pair production at high energies. The **photoelectric effect** is the absorption of a photon by an electron bound in an atom. The electron will be emitted in the process with the energy difference between its binding energy and the photon energy.

The remaining ionised atom takes up the small recoil. Since electrons at any energy level can be involved, electrons from the photoelectric effect have a wide range of energies. For metals, electrons with the maximum energy are emitted from the Fermi level of the material.

In **Compton scattering** $\gamma e^- \to \gamma e^-$, the incoming photon is elastically scattered off an atomic electron. Since the process is only important at energies above 1 MeV, the binding energy can be neglected. The cross section at lowest order in QED is given by the Klein-Nishina formula. At low energies the angular distribution is isotropic, at high energies it develops a forward peak.

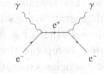

Pair production $\gamma(Z, A) \to e^+e^-(Z, A)$ is the conversion of a photon into an electron-positron pair in the electromagnetic field of a particle with mass M_A. Its threshold energy is small, $E_\gamma \geq 2m_e c^2 (1 + \frac{m_e}{M_A})$. The recoil energy of the nucleus, of the order of 1 MeV, is again negligible. The cross section is inversely proportional to the radiation length X_0 which summarises the material properties.

Focus Box 2.7: Electrons, positrons and photons in matter

like $\pi^{\pm}$ and K populate the shower more sparsely due to their higher mass compared to electrons. The shower is also longer and wider due to the fact that the hadronic interaction length is larger than the radiation length. Processes which electrons, positrons and photons incur in matter are listed in Focus Box 2.7. Basic properties of electromagnetic and hadronic showers are discussed in Focus Boxes 2.8 and 2.9, respectively.

From Auger's measurements of off-axis coincidences it was clear that further experimentation on extensive air showers needed a coverage of large surfaces on the ground. A considerable experimental effort to understand air showers took place in the Soviet Union in the 1950s and 60s [481, Sec. 4]. It was found that the lateral distribution of the number of particles in large arrays of counters allows to estimate the total number of particles in a shower, called the shower size, which is a measure of the energy of the primary particle (see Chapter 10). The most important observation at the time was probably the first indication of the "knee", a slope change in the energy spectrum around a few times 10^6 GeV (see Section 3.2) seen by Kulikov and Khristiansen [153].

An important step forward in the detection of extensive air showers was enabled by the invention of the photomultiplier tube [380], described in Focus Box 2.10. It allows to detect the feeble light signal emitted by gases, liquids and solids as a reaction to excitation and ionisation of atoms by passing radiation. The emitted light, in the region of visible and UV frequencies, is classified as scintillation or fluorescence, depending on details of the mechanism (see Focus Box 2.6). Examples visible by the naked eye are the aurorae borealis and australis, due to the occasionally extreme fluxes of charged particles close to the poles. Photomultipliers also had an important role in verifying the first indications of the "knee" in the cosmic ray spectrum. The so-called Culham array in the U.K. pioneered water Cherenkov counters [162] in the early 1960s. Cherenkov counters, described in Focus Box 2.11, register light emitted by relativistic particles in transparent media. The array had four water tanks in plastic-lined containers of $1.3\,\mathrm{m}^3$, each viewed by a single downward-looking 5-inch photomultiplier tube. Their set-up was a small scale predecessor of the technology used later by the Haverah Park experiment [174] from 1967 to 1987 and today by the Pierre Auger Observatory (see Chapter 10). Immersed in pure water, sea water or ice, photomultiplier tubes also equip Cherenkov detectors for neutrino reactions and rare processes like proton decay (see Chapter 7). In fact, photomultiplier tubes are the only vacuum tube electronics components still in widespread use today.

Rossi's group at MIT realised that the short pulses emitted by scintillators (see Focus Box 2.6) allow to analyse the arrival time structure of extensive air showers and determine the arrival direction (see Focus Box 10.3). Figure 2.12 shows the principle of the measurement [146]. In its first implementation, tanks of liquid scintillator each had a photomultiplier tube converting the scintillation signal into a short electrical pulse. An array of such counters triggered recoding of an event when exceeding a minimum pulse-height threshold. The signals from the array were simultaneously recorded by an oscilloscope which was photographed. As the sketch shows, the particles in an extensive air shower form a barely curved shell of a few meters thickness. The relative timing of the counters can thus be used to reconstruct the arrival direction. The pulse height gives a measure of the number of particles crossing the counters. Rossi's first array was destroyed by lightning igniting the flammable liquid scintillator. A larger array was thus set up using plastic scintillator disks [150] at the Agassiz Astronomical Station of Harvard University. The results after four years of data taking were summarised in a comprehensive paper in 1961 [161]. To analyse the pulse heights measured in the scintillators, the group developed the so-called density method. If one assumes that all particles crossing a scintillator slab are minimum ionising and none is absorbed, the measured pulse height is proportional to the sum of their total track length (see Focus Box 2.8). It is then independent of the incident angle since the length increases with increasing zenith angle, but the effective surface decreases in the same

When energetic electrons or photons ($E \geq$ 1 GeV) enter a dense material of sufficient thickness, an electromagnetic shower results. It is caused by repeated bremsstrahlung processes for electrons, and pair production processes for photons (see Focus Box 2.7). There we argued that both processes have the same characteristic length, the radiation length X_0.

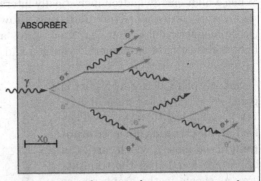

A simple model due to Bruno Rossi and Kenneth Greisen [115, 141] can serve to understand the **longitudinal development** of a shower. Consider an electron of initial energy $E_0 > E_c$. Let it lose half its energy in a bremsstrahlung process after a traversed thickness $\Delta x = X_0 \ln 2$, neglecting ionisation losses. The photon of $E_\gamma = E_0/2$ converts into a symmetric e^+e^- pair after another X_0. This process continues until all $e^\pm$ have slowed down below the critical energy. These lose all their remaining energy by ionisation, neglecting bremsstrahlung. This branching process will thus lead to 2^t particles after tX_0 penetration depths, with energies $E(t) = E_0/2^t$. The shower development will stop when $E(t) < E_c$, after $t_{max} = \ln(E_0/E_c)/\ln 2$. At that point, the number of ionising particles will be $N_{max} = E_0/E_c$. If we take t_0 as the average range of $e^\pm$ in units of X_0, the total track length T of all particles in the shower will be

$$T = t_0 X_0 N_{max} + X_0 \sum_{\mu=0}^{t_{max}-1} 2^\mu = (1 + t_0) \frac{E_0}{E_c} X_0 \propto E_0$$

Given a roughly constant ionisation energy loss $\langle -dE/dx \rangle_{ion}$ (see Focus Box 1.2), T will be proportional to the total ionisation energy of the shower. Thus both counting the number of charged particles in a shower and integrating their ionisation energy loss will yield a measure of the primary energy. A more elaborate analytical treatment of electromagnetic showers was first given by Bhabha and Heitler [91], Carlson and Oppenheimer [95], and Greisen [149].

A relevant quantity when using the Earth's atmosphere as a calorimeter is the penetration depth X_{max} at which the shower reaches its maximum number of particles. Our simple model gives $X_{\mathrm{max}} = X_0 \ln E_0/E_c$. The so-called elongation rate, $\Lambda = dX_{\mathrm{max}}/d\log_{10} E_0$ denotes the rate of increase of X_{max} per decade of energy. For dry air at normal pressure, the above quantities are $X_0 = 37\,\mathrm{g/cm}^2$, $E_c = 85\,\mathrm{MeV}$ and $\Lambda = 85\,\mathrm{g/cm}^2$ per decade in energy.

The **transverse development** of an electromagnetic shower is dominated by multiple Coulomb scattering of its low-energy electrons and positrons, since the emission angle of bremsstrahlung and the opening angle of $e^\pm$ pairs are both small. The angular distribution of particles is given by Molière's theory [126]. After t radiation lengths it has a Gaussian core with wider tails. The r.m.s. opening angle of this Gaussian core is $\theta_{rms} = tz \frac{19.2\,\mathrm{MeV}}{\beta pc}$ in 3D for particles with charge ze, velocity βc and momentum p [244]. If we again assume that the range of electrons is approximately X_0, this leads to a spatial width of $R_M \simeq \frac{21\,\mathrm{MeV}}{E_c} X_0$; this quantity is called the **Molière radius**. One Molière radius of a shower contains roughly 90% of its energy, two such radii contain 95%. A small Molière radius means slim showers. This quantity is thus important in defining the granularity of an electromagnetic calorimeter (see Focus Box 3.4).

Focus Box 2.8: Electromagnetic showers

When a high-energy hadron inelastically interacts with a nucleus, a large number of different final states can result. Thus **hadronic showers** in matter are much more complex than electromagnetic ones. However, final states at modest energies are largely dominated by light hadrons, most prominently charged and neutral pions.

Neutral pions decay into photon pairs, $\pi^0 \to \gamma\gamma$ almost immediately, $\tau(\pi^0) = (8.30 \pm 0.19) \times 10^{-17}$ s. They feed an **electromagnetic component** of the shower, which carries an important fraction of the total energy as shown in the graph on the right [438], of the order of one half. The electromagnetic fraction as a function of primary energy is expected to follow a power law [262], which is indeed observed.

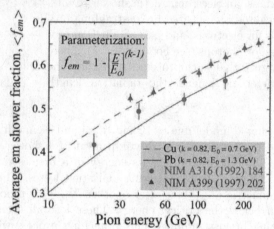

Other light mesons like $\pi^\pm$ and K have enough time to interact before they decay, increasing the number of strongly interacting particles. The longitudinal development of the shower component thus depends on the **nuclear interaction length** λ_I, which is inversely proportional to the probability of the average hadron to interact inelastically:

$$\lambda_I = \frac{A}{N_A \rho \sigma_{inel}}$$

The combination of the nuclear mass number A (mol weight in grams), Avogradro's number N_A (number of atoms per mol) and mass density ρ indicates the number of nuclear targets per volume. The inelastic cross section σ_{inel} measures the probability for an inelastic interaction. Nuclear interaction lengths for common materials are listed by the Particle Data Group [765, Tab. 6]. For typical dense absorber materials, the nuclear interaction length (in g/cm^2) is between ~ 10 times (Fe) and ~ 35 times (Pb) larger than the radiation length. For dry air the relevant characteristics are $X_0 = 36.62\,\mathrm{g/cm^2}$ and $\lambda_I = 90.1\,\mathrm{g/cm^2}$ for protons and neutrons, $\lambda_I = 122.0\,\mathrm{g/cm^2}$ for charged pions.

As there is a characteristic length for the development of the hadronic part of the shower, one can construct a simple longitudinal model [367] in analogy to the Heitler-Rossi-Greisen model for electromagnetic showers (see Focus Box 2.8). Since it is mainly relevant for air showers, we sketch the main ingredients in Focus Box 10.1.

Since hadrons are much heavier than electrons, kinematics involves much larger scattering angles. Thus also the lateral extent of the hadronic components depends on λ_I. For typical materials, 95% of the shower energy is longitudinally contained in about $5\lambda_I$, laterally in about $1.5\lambda_I$ at 100 GeV [491].

Focus Box 2.9: Hadronic showers

proportion. The total pulse height is thus proportional to the number of particles crossing and can be cross-calibrated with a plane of Geiger-Müller counters. The fluctuations around the mean are governed by Poisson statistics.

The detection of photons in a **photomultiplier tube (PMT)** is based on two physical processes, the photoelectric effect and secondary electron emission. The **photoelectric effect** was first observed by Heinrich Hertz [4] and explained later by Albert Einstein [22] as the emission of an electron when an atom absorbs a photon. This marks the beginning of the perception of light as consisting of individual quanta. Single electrons must be multiplied to create a measurable current. This is done via the process of **secondary emission**. When an electron is accelerated to sufficient energy and hits a metal surface, it ionises the material and more than one electron is scattered back. When these are again accelerated and hit further subsequent surfaces, an exponentially growing avalanche of electrons is created.

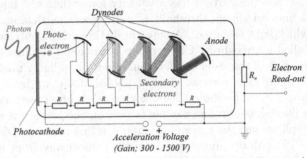

The combination of both processes is shown in the sketch on the right. A photon enters an evacuated photo-multiplier tube through a transparent window and hits the photocathode. A single electron per photon is emitted, accelerated and focussed onto a first dynode.

(Credit: Wikimedia Commons)

Several electrons result from secondary emission and are in turn accelerated towards further dynodes. The multiplication factor can thus typically reach several 10^6 and be controlled by the dynode voltage. Modern implementations of the PMT often use a more compact dynode set-up, realised e.g. by metal meshes, metal channels or micro-channel plates [387]. Photocathode and anode can be structured to achieve sensitivity to the photon impact point (Position Sensitive Photomultiplier, PSPM).

Semiconductor devices can also detect photons. Examples are the classical photodiode and its amplifying version, the avalanche photodiode. Photodiodes function like the familiar solar cells, by a PIN structure or a p-n junction (see Focus Box 3.5). When a photon is absorbed in the depletion region, a current surge results. The gain is typically small, so the devices are sensitive to high photon fluxes only. This changes, when the reverse bias voltage is increased towards the breakdown voltage, where a short circuit would occur. Upon absorption of a photon, an avalanche then occurs in the high voltage region and gains of the order of 10^2 can be reached. Since each diode can be much reduced in size, arrays with position sensitivity can be constructed.

Focus Box 2.10: Photomultiplier tubes and other photodetectors

The method allowed to locate the shower core position, and determine arrival direction and size at the same time [161]. It was found that the lateral distribution of particles around the shower axis follows a universal distribution with few parameters, basically independent of the energy and the direction of the shower. A Monte-Carlo sample of computer-generated showers was used to estimate errors in these quantities, which were found to be 5° for the incident angle, about 10 m for the core location and about 10% for the size.

To reconstruct properties of the primary from a sparse measurement of its atmospheric shower tail is far from trivial. Extrapolating to the shower size is but the first step. A description of an extended air shower can be done top-down if one knows the identity and energy of the primary particle. The primary hadronic interaction with air molecules must be described including properties of the secondaries, mainly light mesons like π and K as

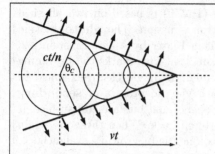

Light in a homogeneous medium of refractive index $n > 1$ propagates at a reduced speed c/n compared to the speed of light in vacuum c. When a charged particle moves through such a medium with a speed $\beta = v/c > 1/n$, it causes an electromagnetic shock wave of coherent light as sketched on the left. This phenomenon is called **Cherenkov radiation** [79, 395].

When the particle has a velocity lower than the limit of $1/n$, no light is radiated. This defines an energy threshold E_s for a given particle mass m, $E_s = mc^2/\sqrt{1 - 1/n^2}$, above which light is emitted. A detector distinguishing only between "light" and "no light" is thus called a **threshold Cherenkov counter**. The number of emitted photons in the visible range is of the order of a few hundred for a radiator thickness of $1\,\mathrm{cm}$ [263, p. 37]. Cherenkov counters thus need an efficient photon detection by e.g. photomultiplier tubes (see Focus Box 2.10). Threshold Cherenkov counters often use a conical mirror to focus the light onto a photodetector; an example is seen on the top of the set-up in Figure 2.15. In cosmic ray applications, the radiator is often a silicon aerogel [405].

The half opening angle θ_C of the conical wavefront has $\cos\theta_C = (\beta n)^{-1}$. In a dispersive medium (thus for all media except vacuum), the index of refraction depends slightly on wavelength, $n = n(\omega)$. As a consequence, the real wavefront is a superposition of slightly different ones. When the medium has a finite thickness in the direction of the particle motion, the emitted wave packets have a finite duration. They impact an intercepting detector plane as a thin ring of photons. Such a detector is thus called a **Ring Imaging CHerenkov counter** (RICH). The radius of the ring determines the Cherenkov angle θ_C and thus the particle velocity. The number of photons is proportional to the square of the particle charge. A RICH thus contributes to particle identification and velocity measurement. As an example, see the proximity imaging RICH located just below the spectrometer of the AMS-02 detector shown in Figure 9.1.

When a highly relativistic particle crosses the boundary between a medium and vacuum, it emits **transition radiation** [251]. The intensity is proportional to the particle's Lorentz factor $\gamma = 1/\sqrt{1 - \beta^2}$ and the square of its charge. This makes transition radiation suitable for discrimination between light and heavy particles. However, the radiated intensity per crossing is very small. Detectors thus use many layers of radiator or a fleece to multiply transitions. The emitted wave length is in the X-ray range and can be detected e.g. by proportional tubes (see Focus Box 2.3). The AMS-02 detector set-up features a transition radiation detector above the spectrometer, as shown in Figure 9.1.

Focus Box 2.11: Cherenkov light and transition radiation

well as light baryons. The development of the shower can be approximated in terms of basic properties like the radiation length (for the electromagnetic part) and the nuclear interaction and the decay length (for the hadronic part), as well as the inelastic cross sections of shower particles with air nuclei, as discussed in Focus Box 2.9. Analytically, this can be done by solving an integro-differential equation called the cascade equation. This will then give the extent, density and composition of the shower at sea level. A crude approximation of the problem can be solved analytically as shown by members of the Rossi group [151]. One can thus relate the shower size to the energy of the primary. Rossi found that the primary flux follows a power law with an index of $\gamma = -(2.17 \pm 0.1)$ between $3 \times 10^{15}\,\mathrm{eV}$ and $10^{18}\,\mathrm{eV}$. The reconstruction of events and the Monte Carlo calculations were done by a computer, the IBM-704 installed at MIT in 1956.[7] To my knowledge, this was the first time a digital computer was used for cosmic ray research.

Similar methods, gradually improved by more knowledge about hadronic showers, were and still are used by all subsequent ground arrays of counters to measure direction and energy of extensive air showers (see Chapter 10). The relation between primary and shower properties is mainly established numerically using Monte-Carlo methods. They were pioneered by Anthony M. Hillas, Peter K.F. Grieder and Jean Noël Capdeville [682].

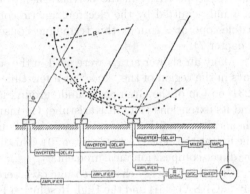

Figure 2.12 The principle of operation of Rossi's scintillator arrays [146]. Each scintillator was viewed by a single photomultiplier tube. The schematics of the electronics shows the trigger of an oscilloscope sweep by the coincidence of three counter signals. All counter signals were simultaneously displayed and photographed.

Bruno Rossi's group at MIT had a world-wide influence on the further development and methodology of cosmic ray research using extensive air showers. From 1950 to 1953, Minoru Oda [317] from the Institute of Nuclear Studies in Tokyo visited and collaborated with the set-up of the Agassiz array. On his return to Japan, he initiated an array of considerable extent and complexity [155]. Its major innovation was an underground muon detector, called the transition chamber, a sandwich of lead and iron plates with plastic scintillator planes. It demonstrated the correlation between the number of muons and the shower size, as well as the large fluctuations in muon number. Since the shower maximum is situated high in the atmosphere around the Regener-Pfotzer limit, air shower arrays were also constructed at high altitude. The first one, called BASJE, was situated on Mount Chacaltaya in Bolivia at 5200 m altitude, established by the MIT group with Japanese and Bolivian colleagues. With its shielded muon detectors it was equipped to distinguish between air showers caused by hadrons (muon rich) and photons (muon poor). It operated until 2019 [312]. At present it is being replaced by the ALPACA array [684]. A similar array in Tibet called LHASSO [595] recorded the highest energy photon shower seen to date, a 450 TeV gamma ray from the Crab nebula [686].

The way to the second major method of detecting and measuring air showers was opened by Kenneth I. Greisen, one of the first graduate students of Rossi in the U.S. [465]. He helped understanding electromagnetic air showers (see Focus Box 2.8) and co-invented the mechanism for the cut-off in the cosmic ray spectrum at ultra-high energies (see Chapter 10). In the early 1960s, he proposed to use the light emitted by the particle cascade in an air shower. It comes in two variants, Cherenkov light (see Focus Box 2.11) emitted by relativistic particles, in a narrow cone around the direction of the shower; and fluorescence light (see Focus Box 2.6), emitted omnidirectionally when N_2 molecules excited by the particle passage de-excite. Cherenkov light arrives first, about 10 μs before the first particles, in the direction of the shower and with a wide spectrum. Fluorescence has frequencies in the near-UV, but also a component in the visible range, which can be admired in polar lights.

In 1962, Greisen proposed to his graduate student Alan Bunner to try and catch fluorescence light from air showers using large photomultiplier tubes on Mt. Pleasant and Bald Hill

---

[7]See https://www.nercomp.org/about-us/our-history.

close to Ithaca N.Y. [172]. In an effective running time of 1000 hours, no serious candidate air shower was detected; only three events had been expected anyway. It seems 'that the first observation of air showers by the fluorescence technique was reported in 1969 by the Japanese group of Goro Tanahashi and Koichi Suga [177, 448]. This pioneering work was the basis of the very successful Fly's Eye project in Utah [214], where the weather conditions are more favourable than in the Southern Tier of New York State. Today, fluorescence cameras combined with air shower arrays equip the Pierre Auger Observatory in the southern and the Telescope Array in the northern hemisphere (see Chapter 10). Since Cherenkov light is mainly emitted by the electromagnetic component of air showers and requires pointing a telescope at a source, this technique is used to detect high-energy photon showers (see Chapter 7).

Many air shower arrays were used in the second half of the 20th century to study cosmic rays in the region of the "knee" at a few times 10^6 GeV. Examples are the EAS-TOP array [231] on top of the Gran Sasso underground laboratory in Italy, the KASCADE array [337] and its extensions [436] in Karlsruhe, Germany. They found indications of a change in the chemical composition of cosmic rays which could explain the change in energy dependence of the flux at the "knee" (see Chapter 10). This indirect evidence still needs to be substantiated by direct composition measurements.

Ultra-high-energy charged cosmic rays were observed by the Akeno Giant Air Shower Array (AGASA) [261] and the High Resolution Fly's Eye (HiRes) [409] in the 1990s by the two technologies introduced above. Results from these experiments did not agree on the existence of a high-energy cut-off, the Greisen-Zatsepin-Kuzmin limit introduced in Chapter 10. The two groups merged to create the Telescope Array (see `http://www.telescopearray.org`), combining a ground air shower array with fluorescence detection in the northern hemisphere, just like the Pierre Auger Observatory does in the southern hemisphere (see `https://www.auger.org`). Both experiments [408, 513] have since observed the cut-off at about 50 EeV.

2.6 THE COSMIC LABORATORY

With the advent of accelerators in the 1950s the role of cosmic rays as sources of new particles all but disappeared. Instead, as discussed above, the study of extended air showers gave the possibility to study the spectrum and nature of very-high and ultra-high-energy protons, electrons and nuclei of cosmic origin. The focus thus shifted from cosmic rays as a tool, to cosmic rays as a subject of study.

However, the interpretation of shower data has intrinsic limitations: none of the components of cosmic radiation can penetrate the atmosphere without interaction. There was thus – and still is – a renaissance of direct measurements at high altitudes with balloons and space craft, which enable ab initio measurements. Origins, acceleration and transport mechanisms are studied using the ever increasing precision of the available data.

The exploration of primary cosmic radiation in Low Earth Orbit started together with the space age itself. Pioneering measurements were made by James A. Van Allen in 1947 [123] using a V2 rocket, spoils of World War II recuperated from Germany together with its inventor Wernher von Braun. The rate of a single Geiger-Müller counter installed at the tip of the ballistic rocket is shown in Figure 2.13. It shows how the count rate falls beyond the Regener-Pfotzer maximum to reach a plateau above about 60 km altitude. Although these data are not corrected for back-splash by the atmosphere, this is arguably the first time that the direct observation of genuine primary cosmic rays has been reported. In the 1940s, Marcel Schein [116] and Rossi [125] showed in balloon missions that the primary cosmic ray flux was not dominated by electrons or positrons, but by nuclei. An important step ·was the result of high-altitude emulsion exposure by Phyllis Freier and others [124,

131]. It showed that primary cosmic rays contained heavy nuclei; an origin solely from the Big Bang was thus excluded. The chemical composition of cosmic rays stays an important subject of study today (see Chapter 8 to 10).

In the 1960s and 1970s, the Russian PROTON and SOKOL satellite detectors made first attempts of calorimetric energy measurements (see Focus Box 3.4) combined with rudimentary particle identifications [179, 181–183, 351]. The devices used in the PROTON space flights were sampling calorimeters of gradually increasing thickness preceded by a carbon target [166], as described in Focus Box 3.4. The particle identification by proportional chambers and a Cherenkov counter was compromised by back-splash from the carbon target. However, the energy measurements stay a reference for low-energy all-particle spectra [351]. The SOKOL mission also used a sampling calorimeter with scintillators as active elements, but preceded it by a dual set of Cherenkov counters to fight backsplash [230] (see Focus Box 2.11).

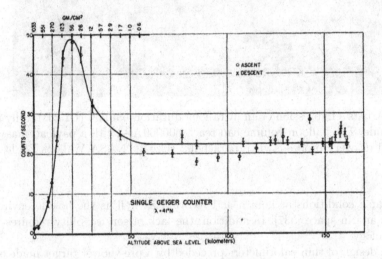

Figure 2.13 Count rate of a single unshielded Geiger-Müller counter installed at the tip of a V2 ballistic rocket as a function of altitude above sea level [123, Fig.4].

However, the start of the space age did not mean the end of scientific ballooning. Pressurised balloons are limited to about 30 km altitude, where the lifting balloon usually bursts. A second balloon or a parachute then takes the payload back, more or less gently. In the late 1940s, a new type of balloon open on the bottom – like a Mongolfière – was developed with a very thin polyethylene skin. These so-called zero-pressure balloons filled with hydrogen or helium, reached altitudes up to 45 km (see Figure 2.1) for short exposures, limited in duration by chemical attack on the balloon skin from components of the upper atmosphere. In the 1990s, NASA developed much more resistant composite balloon skins [329] strengthened by tendons to carry the payload. An artist's impression of a fully inflated one is shown in Figures 2.14. The resistance of the balloon material warrants longer flight times, which reached weeks in the 1990s when circling the Antarctic continent. In these ultra-long duration balloon (ULDB) flights, the permanent sunshine in southern summer and its circular stratospheric winds can be used to reach several weeks of exposure, with power supplied by solar panels.

Progress in balloon and satellite technology also encouraged progress towards more and more sophisticated energy measurement and particle identification. It followed the progress in detector technology used at accelerators with a delay of several decades. This delay is

Figure 2.14 Artists's impression of an ultra-long duration balloon (ULDB), fully inflated at its float altitude. The balloon volume can reach $600000 \, m^3$, with a payload mass of several tons. Payload recovery proceeds via a parachute. (Credit: NASA Wallops Flight Facility)

due to the harsh conditions at launch and recovery as well as the hostile environment at high altitude and in space [537]. In addition, the lack of serviceability requires the use of proven technology.

The basic design of thin calorimeters preceded by a pre-shower target made of material with low nuclear charge has persisted (see Focus Box 3.4), since it reduces the total weight of a calorimeter at the expense of energy resolution [293]. A conical shape of the pre-shower target also increases the solid angle coverage. This technology was thus used for example in the Advanced Thin Ionisation Calorimeter (ATIC) balloon detector [364], under the leadership of Eun Suk Seo from University of Maryland, with a thin homogeneous calorimeter made of bismuth germanate (BGO) crystals (see Focus Box 3.4). It measured the spectra of abundant primary nuclei in the energy range from 50 to about 200 GeV per nucleus [414]. The subsequent Cosmic Ray Energetics And Mass (CREAM) detector [384] added more sophisticated particle identification devices like silicon layers measuring particle charge (see Focus Box 3.5) and a transition radiation detector distinguishing between light and heavy particles (see Focus Box 2.11). An evolved version of the detector, ISS-CREAM, was installed on the International Space Station in 2017. It was prematurely discontinued in a unilateral action by NASA in 2020 following a disagreement about the management of the experiment [755].

Calorimeters aiming at total absorption (see Focus Box 3.4) of electromagnetic showers – and to some extent also of hadronic showers – also saw a transition towards space applications. Recent examples are the DAMPE set-up on a free-flying Chinese satellite [602] and the CALET experiment on the International Space Station [725]. Sampling and imaging techniques as well as homogeneous calorimetry are used in both, complemented by particle identification devices to distinguish between electromagnetic and hadronic showers and identify nuclei. Their results are covered in Chapter 9.

Magnetic spectrometers took somewhat longer to make it from balloon payloads into orbit, since tracking devices are vulnerable. The principle spectrometer components, i.e. magnets and tracking devices, are discussed in Focus Box 3.2. Balloon spectrometers started with the construction of a short superconducting solenoid by the NASA Lyndon B. Johnson Space Center, Goddard Space Flight Center and New Mexico State University, called the NASA/NMSU Balloon Borne Magnet Facility [196, 242]. The spectrometer had multi-wire proportional chambers (see Focus Box 2.3) tracking charged particles through the inhomogeneous field outside the solenoid, with a field strength varying between 0.15 and 2.2 T. It was surrounded by time-of-flight scintillators (see Focus Box 2.6), a gas Cherenkov counter (see Focus Box 2.11) and a calorimeter. Balloon flights in various configurations from the late 1970s to the 1990s were equipped and analysed by what was later called the WiZARD collaboration, involving a large Italian group led by Piergiorgio Picozza and Piero Spillantini. It yielded spectra of primary protons and helium nuclei with magnetic rigidities between a few hundred MV and a few hundred GV [323], the spectra of electrons [210] and light isotopes [243]. The spectrum of antiprotons was measured in the Low Energy AntiProton (LEAP) and Matter-Antimatter Superconducting Spectrometer (MASS) configurations [241, 246]. In the Isotope Matter Antimatter eXperiment (IMAX) set-up [256] using the same magnet platform, the proportional chambers were replaced by drift chambers (see Focus Box 3.2). It measured proton and helium fluxes [304] as well as the percentages of helium isotopes at low energies [285]. It evolved into the ISOtope MAgnet eXperiment (ISOMAX) [303] using a superconducting magnet in a Helmholtz configuration [294]. A successful flight in 1998 yielded light isotope abundances [352]. In the second flight in September 2000, the Kevlar pressure vessel ruptured and the orderly termination of the balloon flight failed, such that the instrument was destroyed when the payload hit the ground.

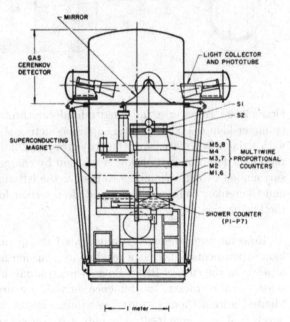

Figure 2.15 Schematic of a payload using the NASA/NMSU Balloon Borne Magnet Facility [210]. The short yokeless superconducting magnet generates an inhomogeneous field of up to 2.2 T. Outside the magnet, multi-wire proportional chambers track the trajectory of charged particles. A time-of-flight scintillator system, threshold Cherenkov counters, scintillators for charge measurement and a shallow electromagnetic calorimeter complete the set-up.

At the end of the 20th century – and just before the advent of a new detector generation inspired by accelerator experiments – Peter K.F. Grieder undertook to collect most of the available cosmic ray data in a thousand-page volume [318], like Miehlnickel had done before World War II [102].

A big step forward in balloon borne spectrometers was represented by the Japanese-led Balloon-borne Experiment with a Superconducting Spectrometer (BESS). Its set-up was built around a superconducting solenoid with an ultra-thin coil [298] providing a homogeneous magnetic field of 1 T. On the inside, a jet-type drift chamber with large rectangular drift cells measured the particle trajectory in the bending direction. It also provided multiple measurements of the specific energy loss dE/dx to assess the absolute electric charge of the particle. The spectrometer had a maximum determined rigidity of 200 GV and a geometrical acceptance of $0.35\,\mathrm{m^2 sr}$. The direction of the incoming particle was determined with a time-of-flight system. Distinction between light and heavy particles was helped by an aerogel Cherenkov counter. The instrument had seven successful flight from 1993 to 2002, with a successively upgraded detector. Physics results covered proton and helium spectra [348, 382], as well as measurements of muons from atmospheric showers [336, 348].

In the early 2000s, BESS was thoroughly revamped with an even thinner magnet of $2\,\mathrm{g/cm^2}$ thickness built at KEK [325], and a larger reservoir of liquid Helium in preparation for long balloon flights around the antarctic [349, 353]. This setup, shown in Figure 2.16 and called BESS-Polar, also had a larger aerogel Cherenkov counter contributed by NASA GSFC. It had two long-duration flights, 8 days long in 2004 and 30 days long in 2007/08. Its rich physics program [610] ranged from proton and helium spectra [580] to antiproton spectra [399, 472] and a search for anti-helium nuclei [473]. The combined BESS program yielded the lowest limit to-date on the anti-helium flux at modest energies (see Chapter 9).

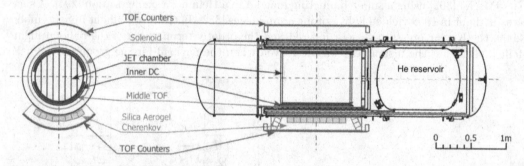

Figure 2.16 Transverse and longitudinal cut through the BESS-Polar balloon borne spectrometer [349]). The ultra-thin superconducting solenoid contains a jet-type drift chamber with axial wires measuring the trajectory in the bending direction. The non-bending direction is determined with low resolution by charge division on the 24 chamber wires. It is surrounded by drift chambers to resolve the left-right ambiguity, time-of-flight scintillators and Cherenkov counters. The liquid He reservoir for magnet cooling is seen on the right of the longitudinal cut.

However, even the most sophisticated set-up cannot solve the principle problem of balloon experiments. There is a residual overburden from the atmosphere, $5.73\,\mathrm{g/cm^2}$ at float altitude in the case of BESS-Polar, thus comparable to the instrument's own material upstream of the tracker. Its influence on the measured particle fluxes must be corrected by Monte-Carlo methods introducing residual systematics which cannot be reduced by the experimental technique itself. The only way to remove it is by experimenting in orbit outside the atmosphere.

Opportunities to do so came after the end of the Apollo programme in 1972. NASA launched a series of reviews discussing possible missions and their goals for science, economy and military applications. Among the outcomes were two projects concerning cosmic rays. The first was the recommendation by the U.S. National Research Council to launch a mission to the Solar system's magnetopause and beyond [228, 239]. It was realised by the two Voyager missions, launched in August and September 1977 from Cape Canaveral, Florida. Voyager 1 and 2 left the Solar system in 2012 and 2018, respectively. Both carry a Low Energy Charged Particle instrument (LECP) and a Cosmic Ray Instrument (CRS), which continue to send data to Earth (see Chapter 8).

The second outcome was NASA's plan to start an exploratory manned mission to Mars, with a space transport vehicle and a permanently manned space station as the first two steps. Under the Nixon administration, this project was reduced by budget constraints to the Space Transportation System (STS), vulgo Space Shuttle, for manned orbital flights. Its first crewed flight (STS-1) took place in 1981, the last one (STS-135) in 2011. Its design purpose was to transport crew and heavy loads to a permanently inhabited space station, originally called Freedom. Freedom was to surpass the capabilities of the Soviet (and later Russian) Mir space station, operating in Low Earth Orbit between 1986 and 2001. In 1984, President Reagan announced the construction of Freedom. The design process went through a great many iterations, first with a vertical truss, then with a horizontal truss. After many budget cuts, the design morphed into what was called the Alpha Space Station in 1993. Following negotiations in the early 1990s, it was joined with the design of a successor to Mir (Mir-2), the Japanese Kibō space laboratory and the ESA Columbus module, to form the International Space Station (ISS), officially created by a Memorandum of Understanding between NASA, the Canadian, Japanese and Russian Space Agencies as well as ESA in 1998.[8]

Part of the plans for the Freedom station was the Particle Astrophysics Magnet Facility, known as Astromag, defined by a NASA working group in 1988 [225] and selected a year later. Its design followed the basic idea of the balloon-borne superconducting magnet, this time with a strong magnet in the middle [226, 236] and experiments in the stray fields at both ends [240]. Proposals for the latter included the Large Isotope Spectrometer for Astromag (LISA)[9] experiment [234] and a follow-up experiment from the WiZARD collaboration [237], as well as an emulsion calorimeter to study high-energy hadronic interactions, called SCIN/MAGIC [238]. When plans for Freedom were shelved in 1992, the Astromag project was abandoned after an unsuccessful study for a free-flying version [249].

However, not only plans for a U.S.-led space station were still alive, the WiZARD collaboration also pursued their plans for a major cosmic rays spectrometer in space. Their project for the Freedom space station changed into a free-flying magnet spectrometer. In 1992, the collaboration found a flight opportunity as an attached payload on a Russian Earth observation satellite [428]. This gave rise to the PAMELA[10] project [270], with the specific task of studying cosmic antimatter. A sketch of the set-up is shown in Figure 2.17. Its heart was a small magnetic spectrometer of $13 \times 16 \, \text{cm}^2$ active surface inside a permanent magnet providing a uniform field of $\sim 0.48 \, \text{T}$. Ferromagnetic shielding outside the magnet reduced field leakage. Six layers of double-sided silicon micro-strip detectors (see Focus Box 3.5) tracked particles with high precision and measured their specific energy loss $dE/dx \propto Z^2$, for a total geometrical acceptance of $21.5 \, \text{cm}^2 \text{sr}$. The set-up was completed by scintillation detectors and an electromagnetic calorimeter, with tail catcher and neutron detector to help distinguish hadronic showers. A transition radiation detector [345] (see Focus Box 2.11) was also originally planned, but not included in the final set-up. PAMELA was launched in 2006 attached to the Russian Resurs-DK1 Earth observation satellite. Its orbit was first a quasi-polar (70°) elliptical one at altitudes between 355 and 584 km, changed in 2010 to a circular orbit at 550 km. The duration of the mission, originally planned to be three years, turned out to be about ten years, during which the efficiency of the device

---

[8]See https://www.nasa.gov/mission_pages/station/structure/elements/partners_agreement.html.

[9]This is not to be confused with the Laser Interferometer Space Antenna project for gravitational wave detection.

[10]The acronym stands for Payload for Antimatter-Matter Exploration and Light-nuclei Astrophysics.

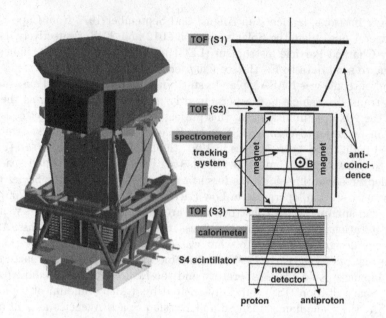

Figure 2.17 Schematic of the PAMELA satellite spectrometer and its components [391]. A time-of-flight (TOF) system with three layers determines the particle flight direction and triggers the data taking. A spectrometer with a permanent magnet and six layers of solid state tracking detectors measures the magnetic rigidity. A calorimeter with tail catcher and neutron detector completes the set-up.

changed from about 90% to 10% after 2010, due to radiation damage to the tracking detector front-end chips. A rich physics output resulted from the mission [631], which we will cover in Chapter 9.

The Italian component of the WiZARD collaboration obviously had ample experience inherited from their balloon experiments. Their members had also, in parallel, been involved in constructing collider detectors like L3 at LEP [258] and in analysing their precision data. Thus not only experience in modern particle detection but also the strict analysis techniques of modern particle physics migrated into the field of cosmic ray physics, which so far had suffered from little control of experimental systematics. As an example, Figure 2.18 shows a collection of results for the differential spectrum of the most abundant cosmic ray species, i.e. the number of protons per square metre, second and solid angle in a given rigidity interval. To make details visible, the gross energy dependence of the high-energy flux, which is roughly proportional to $R^{-2.7}$ has been taken out (see Chapter 3). The errors reported in the figure correspond to the quadratic sum of statistical and systematic errors given by the authors. Below about 30 GV, the results depend on the solar cycle and cannot be compared without correction (see Chapter 8). However, one also observes differences between results above 30 GV which are much larger than the quoted errors.

This dramatically changed with the advent of modern methodology in recent detectors, starting with PAMELA. Their sophisticated and redundant hardware, large acceptance and thorough calibration in accelerator particle beams yields better controlled systematics, together with excellent statistical accuracy of the data. The next such detector was the Alpha Magnetic Spectrometer, built by a collaboration of accelerator particle physicists led by Samuel C.C. Ting of MIT. A proposal was presented to NASA in 1994 and enthusiastically encouraged by Dan Goldin, the NASA administrator with the longest tenure so far, from

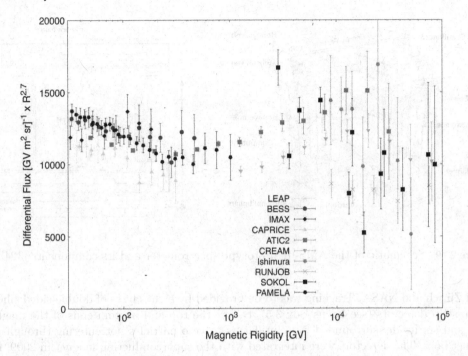

Figure 2.18 Differential proton flux, multiplied by $R^{2.7}$, from pioneering balloon and space experiments: LEAP [241], BESS-TeV [382], IMAX [304], CAPRICE [335], ATIC-2 [414], CREAM [637], Ishimura *et al.* [257], RUNJOB [365], SOKOL [255] and PAMELA [457]. For calorimetric results, magnetic rigidity is calculated assuming ^{1}H as the only isotope. Errors correspond to the quadratic sum of statistical and systematic errors. Data are extracted from CRDB [532, 739].

1992 to 2001. With little prior experience in space, NASA required the collaboration to first build a demonstrator prototype, called AMS-01, to travel on the Space Shuttle. The detector was constructed in record time, with about one half of the spectrometer equipped with silicon micro-strip tracking devices (see Focus Box 3.5) inside a permanent magnet. Simplified particle identification equipment surrounded the spectrometer with time-of-flight scintillators and aerogel Cherenkov counters. The organisational framework was cast as a U.S. inter-agency agreement, such that the Department of Energy was responsible for the detector construction, operation and analysis, NASA was responsible for transport to the ISS, installation and services and data transmission to ground. However, the internal organisation followed the traditional collaborative spirit inherited from particle physics experiments.

The prototype detector flew on Space Shuttle Discovery's flight STS-91 in June 1998, the last Space Shuttle mission to the Russian Mir station. It successfully collected 70 million cosmic rays in ten days. Details of the set-up, performance and physics result are summarised in a comprehensive report [324]. Based on the success of this precursor, the collaboration decided to construct a more ambitious detector for a long-term installation on the International Space Station, AMS-02. It was intended to exchange the permanent magnet for a superconducting one of 0.87 T [326], including the complex supplies by a cooling system with superfluid helium. From 2000 to 2008 the sub-detectors were constructed, mostly in Europe and financed by the international partners. The magnet was constructed by Space Cryomagnets Ldt., an offspring of Oxford Instruments plc., under the auspices of

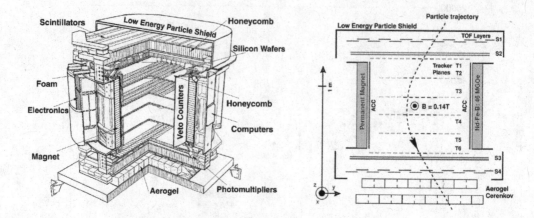

Figure 2.19 Schematic of the AMS-01 prototype spectrometer and its components [324].

ETH Zürich and NASA. Tracking was to be provided by eight layers of double-sided silicon micro-strip detectors (see Focus Box 3.5), two at the top and bottom ends of the magnet bore and six inside, surrounded by scintillators to veto particles not entering through the magnet bore. The detectors were integrated with the superconducting magnet in 2009, followed by an extensive calibration in CERN particle beams in 2009 and thermal vacuum tests at ESA-ESTEC in the Netherlands, which determined that the liquid helium supplies and cryocoolers would keep the magnet working for little more than two years. No resupply or service in orbit was foreseen.

However, for several years it had been uncertain if AMS-02 would ever be launched. After the 2003 Columbia accident leaving seven astronauts dead [394], NASA had decided to reduce shuttle flights and retire the remaining shuttles by 2010. A number of flights were removed from the flight manifest – the flight and cargo plan of NASA missions – including the flight for AMS-02 [390]. Studies to find other means of transport or convert AMS to a free-flying device turned out to be unsuccessful. Samuel C.C. Ting thus organised support by influential U.S. senators to keep the project alive [406]. In 2008, both chambers of Congress passed a bill compelling NASA to add another flight to the Space Shuttle program in order to deliver AMS-02 to the ISS. The bill was signed into law by president George W. Bush. In January 2009, three days after president Barack H. Obama took office, AMS-02 was back on the NASA manifest, together with an additional Space Shuttle mission to the ISS, STS-134.

Meanwhile it had become clear that the lifetime of the ISS would be much longer than originally planned. Originally scheduled to be de-orbited in the first quarter of 2016, its planned usage was successively extended to 2020, 2024 and currently 2028. The collaboration thus decided in 2010 – with the additional flight already scheduled – to abandon the short-lived superconducting magnet in favour of the weaker permanent one from AMS-01 [440]. In a crash program involving the whole collaboration [454], the permanent magnet was fitted into the vacuum case of the superconducting one. To rescue the spectrometer rigidity resolution, one tracking plane outside the magnet bore was moved to the very top of the set-up. An additional plane was equipped with spare parts and put in front on the calorimeter, lengthening the lever arm for bend-angle measurement to 3 m and thus achieving a maximum determined rigidity of 2 TV. At the expense of rigidity resolution at low rigidities, the set-up thus gained enormously in life expectancy and was fit to see rare and higher energy phenomena. The new detector, shown in Figure 9.1, was assembled in record time and exposed to beam calibration at CERN over again.

On August 26, 2010, AMS-02 was delivered from CERN to the Kennedy Space Center by a Lockheed C-5 Galaxy cargo plane. It was launched to the International Space Station on May 16, 2011 on shuttle Endeavour's flight STS-134, commanded by Mark Kelly. Two robot arms transported it from the shuttle cargo bay to its anchor point on the zenith side of the starboard truss of the ISS, where it was installed and connected to power and data lines on May 19. A few hours later, AMS was up and running and recorded its first helium nucleus.

The tracking system releases a constant power of about $140\,\mathrm{W}$ through its front-end electronics inside the magnet bore. This heat must be removed to keep the temperature constant. The tracker is thus equipped with a thermal control system [498] using bi-phase CO_2 as a coolant, with two redundant cooling circuits each equipped with two redundant pumps. By 2015, these were showing wear and it was clear that a replacement was required for long-term operation. In 2019, an improved replacement of the pumping system was sent up to the ISS including a set of special tools to replace the system in orbit and replenish the liquid CO_2. In a complex operation, this was done by the ISS crew in four space-walks. It involved cutting and welding a pressurised system in orbit for the first time and was described as the "most challenging since Hubble repairs" [712].

The final AMS-02 detector and its performance are discussed in Chapter 9. In its first ten years of operation on the International Space Station, AMS has collected more than 200 billion cosmic ray elementary particles and nuclei in the rigidity range from GV to several TV (see `https://ams02.space`). The percent precision of the AMS results is revealing unexpected features in cosmic ray spectra which we will discuss in great detail in Chapter 9. Steven Weinberg somewhat cruelly remarked[11]: "The only interesting science done on the ISS has been the study of cosmic rays by the Alpha Magnetic Spectrometer, but astronauts played no role in its operation". This statement has since been proven wrong. Astronauts' interventions have been crucial to ensure the experiment's longevity, just like for the Hubble space telescope.

The maximum determined rigidity for the most powerful spectrometer available today, AMS-02, is of the order of a few TV. At even higher energies, calorimetric detectors provide spectra ranging up to the PeV region, close to the "knee" feature of the all-particle spectrum (see Chapter 3). One of them, the CALorimetric Electron Telescope (CALET) [725, 728] discussed in Chapter 9, was attached to the Japanese Kibō module of the ISS in August 2015. It was constructed and is run by an international team notably involving members of the WiZARD collaboration. It has two calorimeters, an imaging one, $3X_0$ thick with Tungsten as absorber and scintillating fibres as active elements; and a homogeneous one made of lead-tungstate (PWO) crystals with $27X_0$ (see Focus Box 3.4). The AMS-02 and CALET detectors' positions on the ISS are shown in Figure 2.20.

The DArk Matter Particle Explorer (DAMPE) [602] on board of a Chinese free-flying satellite has a BGO crystal calorimeter of $31X_0$ thickness also preceded by a tracking device. It takes data in orbit since December 2015. The Russian NUCLEON experiment [550], a thin calorimetric detector with large acceptance from the PROTON/SOKOL family, carries a carbon target and two calorimeters, one based on the KLEM methodology [302] and a more conventional silicon-tungsten one, totalling $17X_0$. In all cases, the calorimeter is preceded by a charge-measuring detector to identify the incoming nucleus. Hadrons are separated from electrons by the shape of the shower. These modern calorimetric detectors and their results are also discussed in Chapter 9.

Progress is indeed remarkable, as seen when comparing recently published proton fluxes in Figure 2.21 with the wide-spread results from pioneering experiments shown in

---

[11]Robin McKie, *Twenty years of the International Space Station – but was it worth it?* The Guardian, Oct 25, 2020

Figure 2.20 Photograph of the International Space Station taken by the Expedition 66 crew in November 2021: AMS-02 is on the left on the main truss (circle), CALET is on the bottom right (ellipse) on the Exposed Facility of the Japanese Experiment Module Kibō (Source: NASA).

Figure 2.18. Experiments agree well on the shape of the spectrum, less well on the normalisation, although results are compatible given the quoted systematic errors. A possible reason for normalisation differences is systematics of the energy or rigidity scale, as discussed in Focus Box 9.9. Because of the high precision of the individual results, care must be taken when placing the data points inside their respective bins. This is done here using the prescription of reference [269], discussed in the beginning of Chapter 3. We assume a prior for the flux, $d\Phi/dR = aR^{-b}$, as a function of rigidity R, justified by the gross features of cosmic ray spectra. The data points in a rigidity bin ranging from R_1 to R_2 are then placed at $\tilde{R}$:

$$\tilde{R} = \left[\frac{R_2^{1-b} - R_1^{1-b}}{(1-b)(R_2 - R_1)} \right]^{-\frac{1}{b}}$$

We use priors with $b = 2.7$ for nuclei and $b = 3.0$ for electrons throughout the rest of this book and omit the tilde. It is seen that the shape and normalisation of the proton fluxes measured by calorimeters and spectrometers are in good agreement in the overlap region. Calorimetric results agree within errors among themselves, with the possible exception of the NUCLEON result when approaching PeV energies. However, for the latter, $\tilde{R}$ could not be calculated and bin centres given in [709] are used. This is potentially important in this representation since the measured fluxes $d\Phi(R)/dR$ are scaled-up by $d\Phi(R)/dR \times R^{2.7}$ so that details can be appreciated despite the rapid decrease of the flux itself.

These latest examples of modern cosmic ray experiments and results take us to the heart of our subject. To set the scene, we will discuss gross features of cosmic ray spectra and composition in the next chapter. In the Chapters 8 to 10, we will differentiate – by rough energy range – between solar, galactic and extragalactic cosmic rays and discuss experimental methods and results for each of these.

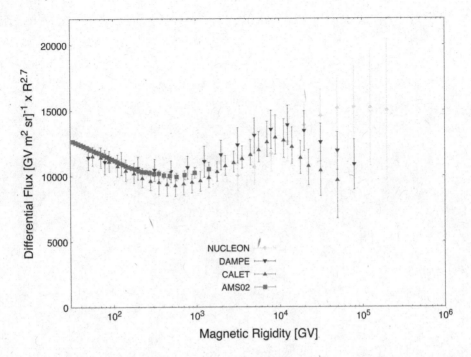

Figure 2.21 Differential proton flux, scaled by $R^{2.7}$, from recent space experiments: AMS-02 [560], DAMPE [693], CALET [840] and NUCLEON [709]. The data points are placed inside their bins according to the procedure given in [269]. For calorimetric results, magnetic rigidity is calculated assuming ^{1}H only. Errors correspond to the quadratic sum of statistical and systematic errors.

FURTHER READING

M. Bertolotti, *Celestial Messengers: Cosmic Rays, the Story of a Scientific Adventure*, Springer (2013)

W. Gentner, H. Maier-Leibnitz and W. Bothe, *Atlas typischer Nebelkammerbilder mit Einführung in die Wilson'sche Methode*, Springer (1940)

C.F. Powell, P.H. Fowler and D.H. Perkins, *Study of Elementary Particles by the Photographic Method*, Pergamon Press (1959)

G.D. Rochester and J.G. Wilson, *Cloud Chamber Photographs of Cosmic Radiation*, Pergamon Press (1952)

Ch. Spiering (Edt.), *Cosmic rays, gamma rays and neutrinos: a survey of 100 years of research*, Eur. Phys. J. H 37 (2012) 319–565

B. Strohmeier and R.W. Rosner, *Marietta Blau: Stars of Disintegration*, Ariadne Press (2006)

M. Walter, Ch. Spiering and J. Knapp (Edts.), *100 years of cosmic rays: The anniversary of their discovery by V.F. Hess*, Astropart. Phys. 53 (2014) 1–190

3 Gross Features

In this chapter, we give a preview of the observables and measurement techniques used for cosmic ray research. We also fly over the gross features of cosmic ray spectra, composition and anisotropy. The coverage in this chapter remains deliberately superficial, since we cover all subjects in more detail and more systematically in subsequent chapters. However, it seems useful to us to disclose to the reader from the beginning what we will be talking about in the rest of this book.

3.1 OBSERVABLES AND TECHNIQUES

The basic observables for cosmic ray particles are obviously momentum, mass and electric charge, as well as the direction from where they reach the detector and their time of arrival. Combinations of momentum, energy, mass and absolute charge[1] $|Z|$ can identify the particle. The sign of the charge Z distinguishes between particles and antiparticles. The direction is of limited usefulness for charged particles, since their trajectory is altered by interaction with magnetic fields at the level of the solar system, the galaxy and beyond (see Chapter 6). Their flux is thus in general found to be homogeneous, isotropic and constant in time, with notable deviations at low and maybe also at high energies. The former is caused by the Earth's magnetic field in conjunction with the dynamic influence of the Sun; this causes alterations of the incoming flux at low energies, which depend on time (see Chapter 8). The latter may come from the location and activity of sufficiently near-by sources. The stiffness of the trajectories at high momenta may result in a small residual anisotropy (see Chapters 9 and 10). Variations in time at high energies – if they exist – would indicate that the influence of violent cosmic events is not entirely washed out by the long storage time of charged particles in the galaxy.

The basic observable for isotropically arriving particles of a given species is thus the omnidirectional flux Φ, given by the number of particles of a given species impinging on a unit of surface $[m^2]$ per unit of time $[s]$, coming from a unit of solid angle[2] $[sr]$. The units given in square brackets refer to the SI system of units. The flux comes in two variants. The differential flux is determined separately in each interval of energy or rigidity. The integral flux counts particles above a minimum energy. Both typically follow a power law in terms of an energy-type variable, total energy, kinetic energy, kinetic energy per nucleon or magnetic rigidity. Details of how to determine the flux are given in Focus Box 3.1.

Momentum p – or rather magnetic rigidity $R = p/Z$ ($[R] = [GV]$) for a particle of electric charge Z – is the natural output of spectrometer measurements, as discussed in Focus Box 3.2. These quantities are also the right choice when discussing plasma acceleration, as shown in Chapter 5. Calorimeters rather measure kinetic energy $E_{kin} = p^2/(2M)$ for particles of mass M, as explained in Focus Box 3.4. When particles are correctly identified, momentum and energy measurements are of course equivalent. However, identification of nuclear isotopes requires high resolution measurements of both mass and charge. Calorimeters do not determine the sign of the particle charge, so do not distinguish between particles and antiparticles, even when M and $|Z|$ are known.

---

[1] We systematically express particle charge in units of the elementary charge $e = 1.602176634 \times 10^{-19}$ C, which is fixed in SI units.

[2] A solid angle is the three-dimensional analog of a planar angle. The latter (in radians [rad]) is defined as the arc length subtended by the angle on a circle around the apex with unit radius. The solid angle (in steradians [sr]) is the surface subtended by the three-dimensional angle on a unit radius sphere.

The differential flux $d\Phi/dE$ is given by the number of particles $\Delta N(E)$ counted in a time window Δt and an energy window between E and $E + \Delta E$. When it is measured by a detector covering an acceptance A, defined by its active surface, solid angle coverage and total detection efficiency, it is given by:

$$\frac{d\Phi(E)}{dE} = \frac{\Delta N(E) - B}{A \, \Delta t \, \Delta E}$$

Its units are usually chosen as $[d\Phi/dE] = [1/(\mathrm{m^2 sr\, s\, GeV})]$. The raw number of particle counts ΔN needs to be corrected for the background counts B. The acceptance A of a detector includes three main factors:

- A geometrical factor G ($[G] = [\mathrm{m^2 sr}]$) defined by the active surface and solid angle coverage as specified in Focus Box 3.3.

- The efficiency ϵ_T to trigger the data acquisition when a particle of the given species passes through G.

- The efficiency ϵ_S of the selection criteria used to assign particles to species and energy bin.

The selection efficiency ϵ_S may contain migration probabilities between species and between energy bins, in addition to a general inefficiency to accept qualifying particles. The integral flux $\Phi(E)$ is obtained integrating the differential flux above a threshold energy E:

$$\Phi(E) = \int_E^\infty \frac{d\Phi(E)}{dE} \, dE$$

For high-precision results, care must be taken when placing the data points inside their respective bins. This is done in this book wherever possible using the prescription of reference [269]. For a given distributions $f(x)$, the correct positioning $\tilde{x}$ of a point in an interval $x_1 < x < x_2$ is defined by:

$$f(\tilde{x}) = \frac{1}{\Delta x} \int_{x_1}^{x_2} f(x) \, dx$$

with $\Delta x = x_2 - x_1$. For rapidly varying distributions and/or large bins, this position is significantly different from the usual assumptions $\tilde{x} = (x_1 + x_2)/2$ or $\tilde{x} = \left(\int_{x_1}^{x_2} x f(x) \, dx \right) / \left(\int_{x_1}^{x_2} f(x) \, dx \right)$. Correct positioning obviously needs a hypothesis about the true distribution prior to the measurement. As we will see, spectra of cosmic rays systematically follow a power law $\propto a E^{-b}$. The data points in an energy bin ranging from E_1 to E_2 are then placed at:

$$\tilde{E} = \left[\frac{E_2^{1-b} - E_1^{1-b}}{(1-b)(E_2 - E_1)} \right]^{-\frac{1}{b}}$$

The same is of course true for other energy-type variables like kinetic energy (total or per nucleon) or magnetic rigidity (momentum per unit charge). For high-precision measurements, it is also necessary to account for the bin-to-bin migration of events caused by energy or rigidity resolution. This is usually done by an unfolding procedure [204, 208, 266, 388].

Focus Box 3.1: Cosmic ray flux

When all incident cosmic rays are lumped together, the flux as a function of energy indeed roughly follows a power law above $\sim 50\,\mathrm{GeV}$, as seen in Figure 3.1. This way of representing the overall spectrum of cosmic rays has an interesting history all by itself. To my knowledge, the first to compile such a collection of data was Simon P. Swordy from University of Chicago,[3] for a Scientific American article published with James W. Cronin and Thomas K. Gaisser in 1997 [277]. While the journal reduced his plot to an artist's impression of the spectrum, he published a better version in 2001 [321] and made the plot itself and its input data widely available. There are no published all-particle spectra at low energies, where experiments strive to measure the spectrum species-by-species (see Chapters 8 and 9). He thus scaled up a proton spectrum from CREAM by a factor of 2 to represent a rough estimate of an all-particle spectrum. The scale factor is determined in the following way. Assume that H and He are the dominant components of galactic cosmic rays. Assume further that their rigidity(!) spectrum is represented by a power law, $d\Phi/dR = aR^{-b}$, with respective coefficients a_H and a_{He} and roughly equal spectral index b. The sum of their energy(!) spectra is then $d\Phi/dE = (a_H + 2^{b-1}a_{He})E^{-b}$, neglecting masses. This corresponds to a ratio $\Phi(\mathrm{H}+\mathrm{He})/\Phi(\mathrm{H}) = 1.96$, almost exactly the factor of 2 used by Swordy in 1997. The only data from the original plot which are still contained in our present compilation are the ones extracted from [182], which Swordy painstakingly read off a graph showing an integral all-particle spectrum from the PROTON family of instruments, and converted into a differential spectrum. With this exception, Figure 3.1 presents an updated version of the "Swordy plot" using contemporary data.

The all-particle spectrum has few remarkable features, especially when presented in a log-log plot covering over 30 orders of magnitude in flux and close to 13 orders of magnitude in energy. The first feature marked in the plot is the so-called knee between 10^6 and $10^7\,\mathrm{GeV}$, where the fall-off steepens a bit. The second feature is called the ankle of the spectrum between 10^9 and $10^{10}\,\mathrm{GeV}$, where it flattens a bit. However, the spectral index is not constant between these features, nor is it the same for all species. For energies up to a few TeV, this is shown in Chapter 9 – element by element – with the high-precision data from recent space experiments. For energies above 10 TeV, it is shown for all particles on the right of Figure 3.1. The negative spectral index goes down more or less monotonously up to the knee, stays more or less constant up to the ankle, followed by a substantial rise. At a few times $10^{20}\,\mathrm{eV}$, the Greisen-Zatsepin-Kuzmin limit introduced in Chapter 10 then cuts off the spectrum rather sharply. The highest energy particle observed so far had of the order of 300 EeV [267], corresponding to about 50 J, seven orders of magnitude beyond the highest energies reached by man-made accelerators.

The all-particle spectrum emphasises that the acceptance and exposure time of an experiment is of key importance for its count rate and thus also for its energy reach. The acceptance is mainly determined by a geometrical factor depending on the surface and solid angle coverage of an experiment, as detailed in Focus Box 3.3.

Two main experimental concepts exist to observe cosmic rays on or near Earth.[4] *Direct* measurements on high altitude balloons or in space strive to count, identify and measure cosmic rays ab initio. *Indirect* measurements by air showers experiments use the atmosphere as a target and calorimeter, and rock or ice as an absorber if they have an underground component. They are called indirect since they do not observe the primary cosmic ray; they thus have a limited capability to identify the primary. However, they can cover enormous surfaces and thus be sensitive to the extreme energies and minute rates above the knee. Satellite and space station experiments are limited in acceptance, but can

---

[3]Thanks are due to William F. Hanlon from CFA, who made Swordy's original PAW KUMAC and data available to me.

[4]So far the only cosmic ray experiments outside the heliosphere are the instruments on board of the Voyager spacecrafts, as described in Chapters 2 and 9.

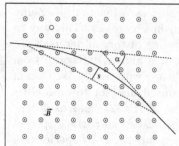

A particle of charge Ze and velocity $\vec{v}$ in a local magnetic field $\vec{B}$ experiences a force $\vec{F} = Ze\vec{v} \times \vec{B} = Ze(p_\perp/m)|\vec{B}|$, where $p_\perp$ is its momentum component orthogonal to the field. The longitudinal component $p_\parallel$ of its momentum remains unchanged. In an extended homogeneous field and with $\vec{p} \perp \vec{B}$, its trajectory will be a circle with radius $\rho = p/(0.3ZB)$ with p in GeV/c, B in Tesla. The radius of curvature thus directly measures the magnetic rigidity $R = p/Z$, its sign determines the sign of Z.

The curvature radius is determined by the sagitta $s = 2\rho \sin^2 \alpha/4$, where α is the angle subtended by the arc length l. The sagitta is thus $s = Zel^2 B/(8p_\perp)$, proportional to the square of the path length inside the field. The trajectory must be measured inside the field using at least three points. $N \gg 3$ equidistant measurements with spatial resolution σ give a rigidity resolution [164]:

$$\sigma_s(R) = \frac{\sigma R^2}{0.3Bl^2}\sqrt{\frac{720}{N+4}}$$

Multiple Coulomb scattering in the detector material also contributes to the error. For a total thickness d of material with radiation length X_0, one finds a contribution asymptotically independent of momentum:

$$\sigma_{ms}(R) = \frac{13.6\text{MV}/c}{\beta}\sqrt{d/X_0}$$

The total error $\sigma_{tot} = \sqrt{\sigma_s^2 + \sigma_{ms}^2}$ thus has a constant term from multiple Coulomb scattering dominating at low rigidities and a quadratic term from the sagitta measurement dominating at large rigidities. The bend angle α can also be used to measure the magnetic rigidity. It is given by $\alpha = l/\rho = lZeB/p_\perp$. It is thus proportional to the path length l, or in an inhomogeneous field to $\int B\,dl$. It can be measured outside the field, by determining the direction of the (straight) trajectories on entrance and exit. Given $n/2$ measurements on both sides with position errors σ one finds a resolution:

$$\sigma_\alpha(R) = \sqrt{\frac{8}{n}}\frac{\sigma}{l^2}\frac{R^2}{B}$$

The contribution of multiple scattering due to the material on and between entrance and exit must be added to this in quadrature. An often quoted figure of merit of spectrometers is the Maximum Determined Rigidity (MDR), where the relative error equals 100%, i.e. $\sigma_{tot}(\text{MDR}) = \text{MDR}$.

Superconducting electromagnets have been successfully used on balloon borne missions. In space, permanent magnets are preferred since they do not require consumables. Halbach arrays [444], made of permanent magnets in configurations like those sketched on the right, allow to construct a nearly homogeneous magnetic field inside, with minimum leakage outside the array. The latter is important since it avoids torque on the space craft by interaction with the Earth's magnetic field. PAMELA used a configuration similar to the lower sketch, AMS-02 a cylindrical one similar to the upper sketch.

In modern spectrometers, the trajectory of the particles is located with solid state tracking devices as introduced in Focus Box 3.5. These determine the passage of charged particles with micrometer accuracy. They also measure the specific energy loss $dE/dx \propto Z^2$ (see Focus Box 1.2).

Focus Box 3.2: Magnetic spectrometers

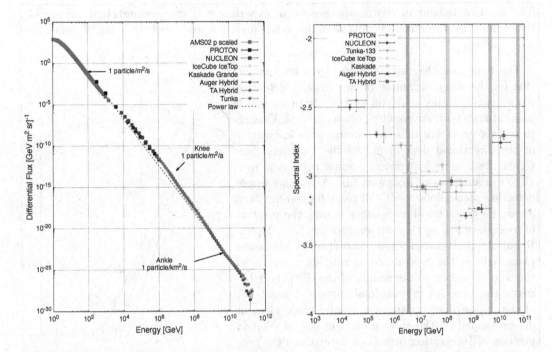

Figure 3.1 Left: All-particle spectrum of cosmic rays from selected satellite experiments and ground arrays. The AMS-02 proton spectrum [775] is scaled up as described in the text, to give a rough estimate of an all-particle spectrum at low energies. The spectrum from the PROTON family of instruments has been extracted by S. Swordy from [182]. All other spectra [539, 555, 558, 563, 566, 709] have been extracted using CRDB [739]. The dashed line simply connects the flux at $100\,\mathrm{GeV}$ to the flux at the ankle, $4 \times 10^9\,\mathrm{GeV}$; the slope corresponds to a constant spectral index of -2.84. Right: The spectral index as a function of energy above $10\,\mathrm{TeV}$, resulting from a fit to all-particle spectra in non-overlapping energy windows. The fit ranges are indicated by the horizontal error bars. The shaded bands indicate the position of spectral features known as first and second knee, ankle and GLZ cut-off (from left to right).

in fact almost reach knee energies despite obvious restrictions in size and mass for space payloads.

In Focus Box 3.2, we discuss the elements of magnetic spectrometers and some of their figures of merit. For each passage of a cosmic ray particle, the direct output of a spectrometer is its magnetic rigidity $R = p/Z$, i.e. the momentum per unit charge measured in Volts. When the particle is identified, i.e. its electric charge Ze is determined, rigidity can be converted into momentum. When the particle mass M is know, momentum can be converted into kinetic energy $E_{kin} = p^2/(2M)$ or total energy $E = \sqrt{p^2 + M^2}$. In principle, mass determination for nuclei identifies the isotope. However, a mass measurement with an accuracy distinguishing isotopes is only feasible at low energies. The conversion to energy is thus normally done supposing a certain isotopic composition.

The most important modern cosmic ray spectrometer is the AMS-02 observatory installed on the International Space Station since 2011. AMS-02 resembles a slice through a modern particle detector at a collider. Details of its set-up are discussed in Focus Box 9.4. Its heart is a magnetic spectrometer with a geometrical factor of about $0.5\,\mathrm{m^2 sr}$, delimited by two planes of plastic scintillator which provide the main trigger of the experiment. In its

The main ingredient in the acceptance of an experiment is the **geometrical factor** G, determined by the effective area and the solid angle covered, such that its SI units are $[G] = [\text{m}^2\text{sr}]$.

The figure on the right demonstrates its calculation for the simple example of two circular detection planes, planparallel and concentric. Particles which pass through both are counted, others are not. Evaluation of the geometrical factor requires prior knowledge of the directional dependence of the incoming flux. Given the general isotropy of cosmic rays, it is usually evaluated for an isotropic flux. A circular upper detection plane alone "sees" all particles coming from above, its geometrical acceptance is thus the product of its surface πr_1^2 and a solid angle of 2π, $G_1 = 2\pi^2 r_1^2$. Requiring that particles pass also through the lower plane reduces the acceptance by roughly the ratio of the second surface to the square of the distance z between the two. An infinitesimal surface element $d\vec{s}_1$ of the upper plane contributes a solid angle element $d\Omega$ subtended by the surface element $d\vec{s}_2$ at relative position $\vec{r}$. Integrating over both determines the geometrical factor G. For two circular planes one finds:

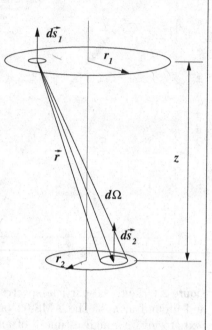

$$G = \frac{\pi^2}{2}\left[r_1^2 + r_2^2 + z^2 - \sqrt{(r_1^2 + r_2^2 + z^2)^2 - 4r_1^2 r_2^2} \right]$$

A good approximation for $z \gg r_i$ is $G \simeq 4\pi^2 r_1^2 r_2^2 / (r_1^2 + r_2^2 + z^2)$, the acceptance of a two-lens telescope. An analytic expression for the acceptance can obviously only be found for simple and regular geometries [184, 185]. More complex geometries require a numerical evaluation by Monte-Carlo techniques.

Focus Box 3.3: Geometrical factor in the acceptance

first nine years of operation, AMS-02 has recorded in excess of 200 billion cosmic rays in the rigidity range between a fraction of a GV (10^9 V) and several TV (10^{12} V). The current count rate can be followed on-line at the NASA web site https://ams02.space.

Kinetic energy, on the other hand, is the direct output of a calorimetric measurement, which we discuss in Focus Box 3.4. Since the original cosmic ray cannot be identified in the middle of shower particles, its charge must be measured upstream of the calorimeter. When the particle shower is completely absorbed in a sufficiently thick calorimeter, the energy measurement is straight forward. If not, leakage of energy in and transverse to the particle direction has to be corrected for. In any case, a calibration of the energy measurement is required to relate measured quantities to particle energy. For a balloon or space borne calorimeter, calibration is normally achieved in an accelerator particle beam. An excellent example of a modern cosmic ray calorimeter is the Calorimetric Electron Telescope (CALET), installed on the International Space Station since October 2015. It has a geometrical factor of about $1\,\text{m}^2\text{sr}$ for electrons and positrons. We describe the detector and its performance further in Focus Box 9.5.

For experiments using the atmosphere, sea water or ice as a calorimeter to detect ultra-high energy cosmic rays, calibration in an accelerator beam is obviously impossible, also because no accelerator reaches the required energies.[5] A conversion of measured quantities into energy then proceeds via the simulation of extensive showers, as further discussed in Chapter 10.

In addition to rigidity and energy, detectors determine the direction of the incoming particle. When the position and orientation of the detector is accurately known, the direction can be converted into an astrophysically relevant coordinate system like equatorial or galactic coordinates. However, except for extreme energies, the arrival direction of charged particles is essentially randomised by their collisionless interaction of with astrophysical plasmas. These exist at every scale, but especially in the solar system. The role of plasmas in acceleration and propagation of cosmic rays is discussed in Chapters 5 and 6. Despite the scrambling action of magnetic fields, there might be some information left in the arrival direction leading to a small anisotropy. This is especially interesting for cosmic rays coming from near-by sources at extreme energies where particle tracks are stiff even at cosmic scales. In Chapters 9 and 10, we will explore these possibilities further.

3.2 PARTICLE SPECTRA AND COMPOSITION

When cosmic rays are not separated into particle species, kinetic energy is the common denominator to measure their spectrum. In Figure 3.1, we already presented the differential flux for all cosmic ray particles lumped together. It generally follows a power law $aE^{-\gamma}$ with a spectral index γ which slowly varies between about 2.5 and 3.5. This obviously means that their number per energy interval drops by almost three orders of magnitude for every decade in energy. To get a feeling for counting rates let us consider the differential spectrum of protons alone, which is representative for the spectrum of all particles up to a factor of two as we have seen above. At $100\,\mathrm{GeV}$, the differential flux is about $4 \times 10^{-2}\,\left(\mathrm{m^2 sr\,s\,GeV}\right)^{-1}$. We assume a roughly constant spectral index of 2.7 as indicated by Figure 2.21 and integrate to get the count rate of protons above a certain energy, in units of $\left(\mathrm{m^2 sr\,s}\right)$, with an integral spectral index of 1.7. Assuming a detector which covers a whole hemisphere in solid angle, we can thus expect a count rate of roughly 2 protons per square meter and second above a threshold of $100\,\mathrm{GeV}$. Above $10^6\,\mathrm{GeV}$, we expect about 10 protons per square meter and year, and roughly 100 per square kilometre and year above $10^9\,\mathrm{GeV}$. Similar expected count rates as a function of threshold energy are indicated in Figure 3.1 for all particles lumped together.

Thus the acceptance of a detector – in terms of active surface and solid angle covered – determines the energy reach. For transportable devices, with acceptances below a $\mathrm{m^2 sr}$, and exposure times of a few years, this limits detailed studies to below the knee region at PeV energies. Above, square kilometre size surfaces are required, and ground based experiments use the Earth's atmosphere, ocean water or ice as calorimetric material.

It is thus small wonder that detailed knowledge about cosmic rays is concentrated below the knee region. Figure 3.2 shows the relative abundance of elements in low-energy cosmic rays, at an energy per nucleon between $160\,\mathrm{MeV}/n$ [370, 413] and $250\,\mathrm{MeV}/n$ [324]. At low energies, the cosmic ray flux varies significantly with solar activity, thus only data taken during the same period of the solar cycle are comparable. We thus use data collected close to the solar minimum of cycle 23 in 1997–98. The kinetic energy per nucleon, E_{kin}/A where E_{kin} is the kinetic energy of the nucleus and A its atomic number, is an often used quantity to compare properties of different nuclei. The reason is that an important inelastic reaction among nuclei, the spallation process, $(Z, A) + X \rightarrow (Z', A') + (Z - Z', A - A') + X$, sometimes also called nuclear fragmentation, leaves this quantity essentially unaffected. For example,

---

[5]The centre-of-mass energy of the most powerful accelerator, the CERN Large Hadron Collider, 14 TeV, corresponds to the impact of a $\simeq 10^{17}\mathrm{eV}$ proton on a nucleon at rest.

Electromagnetic and hadronic showers in matter differ substantially as explained in Focus Boxes 2.8 and 2.9. Thus also cosmic ray calorimetry needs to be adapted for cosmic $e^{\pm}$ on one hand, nuclei on the other hand. **Electromagnetic calorimeters** are characterised by the radiation length X_0 (see Focus Box 2.7) in both their longitudinal and transverse dimensions. Bremsstrahlung is emitted under a small average angle $\langle \theta^2 \rangle \simeq 1/\gamma_e^2$, where γ_e is the relativistic factor of the incoming electron. In the Molière theory of electromagnetic showers [126, 142] it is given as a function of the longitudinal depth x by:

$$\langle \theta^2 \rangle = \left(\frac{21.2\text{MeV}}{\beta pc} \right)^2 \frac{x}{X_0}$$

The transverse shower development is thus characterised by the Molière radius:

$$R_M = \sqrt{\langle \theta^2 \rangle_{x=X_0}} \simeq \frac{21.2\text{MeV}}{E_c} X_0$$

where E_c is the critical energy (see Focus Box 2.7). The two quantities X_0 and R_M define the longitudinal and transverse granularity of a calorimeter. 95% of the shower energy will be contained within a depth of $L(95\%) \simeq 17X_0$ and a radial extension of $R(95\%) \simeq 2.5R_M$ [191, 438, 491].

Hadronic calorimetry requires materials with higher absorption power, i.e. higher density, nuclear charge and mass. These material properties are summarised by the nuclear interaction length λ_I defined in Focus Box 2.9. Once the shower starts, it also has an electromagnetic component mainly coming from $\pi^0 \rightarrow \gamma\gamma$. The fraction of electromagnetic energy is of the order of 50% with a slow dependence on energy [466]. The hadrons in the shower are heavier than electrons and thus smaller in number; they spread wider and penetrate farther. Calorimeters aiming to fully absorb hadrons thus in general use a finer granularity electromagnetic compartment followed by a thicker one with coarser granularity. At 100 GeV primary energy, 95% of the energy are on average contained in about $5\lambda_I$ longitudinally and $1.5\lambda_I$ transversely; these coefficients scale roughly linearly with energy [491]. Unavoidable leakage is caused by neutrons, neutrinos and muons.

There are two different technologies used in calorimetry. **Sampling calorimeters** use dense, often metallic passive absorbers, sandwiched with lighter materials detecting and measuring ionisation. These are either gaseous wire chambers (see Focus Box 2.3), scintillators (see Focus Box 2.6) or solid state detectors (see Focus Box 3.5), in order of density. **Homogeneous calorimeters** use the same material as absorber and detector. Heavy scintillating crystals with a high-Z component like BGO, PWO, CsI or lead glass are thus a preferred choice.

The statistical part of the energy resolution is dominated by fluctuations in the number of particles, which obey Poisson statistics. Thus the relative energy resolution σ/E improves with energy like $\sigma/E \propto 1/\sqrt{E}$. For hadrons, fluctuations in the electromagnetic fraction add to the error. If only part of the shower is sampled, random fluctuation in the sampling fraction come on top. Systematic errors from leakage, signal processing and calibration have to be added to this in quadrature and dominate at high energies.

The thickness requirements to reach optimum containment and resolution are in obvious conflict with the weight capacity offered for balloon or satellite payloads. A compromise sometimes chosen is to precede the calorimeter with a low density pre-shower target like carbon [293]. Best resolution is obtained by combining solid state ionisation detection with tungsten absorbers [327]. This yields what is sometimes called an **imaging calorimeter** because of the rich shower detail provided; a rather costly solution limited to small dimensions.

Focus Box 3.4: Calorimeters

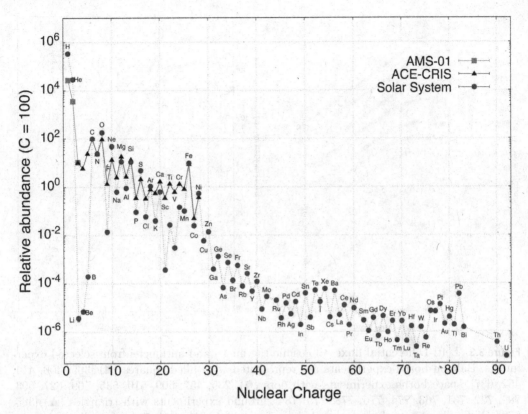

Figure 3.2 Chemical composition of cosmic ray nuclei during the solar minimum of cycle 23, measured by AMS-01 at a kinetic energy of 250 MeV/n [324] and by ACE-CRIS at 160 MeV/n [370, 413]. It is compared to data on the chemical composition of the solar system, compiled by Katherina Lodders [757]. Relative abundances are normalised to 100 for C.

a silicon nucleus interacting with a spectator nucleus X can be fragmented into lithium and sodium nuclei of almost equal velocity.

Figure 3.2 also compares the elemental composition in low-energy cosmic rays to that of the solar system in general [757]. The latter is extracted from the observation of solar spectra as well as the chemical analysis of rocky meteorites, like CI-group chondrites. The comparison indeed reveals systematic differences which are caused by the spallation process. Thus elements rarely produced in stellar nucleosynthesis, like the Li-Be-B group, are vastly overrepresented in cosmic rays. Lithium is even actively burned inside stars and thus about two orders of magnitude less abundant in the Sun than in meteorites. You may remember from your nuclear physics course that nuclei with even nuclear charge are more stable than those with odd charge and thus more abundantly produced at stellar sources. The abundance of odd charge nuclei is less depleted in cosmic rays since it again can be replenished by nuclear reactions on the way between the cosmic sources and the observer. This is especially obvious for the sub-iron group of elements, from scandium to chromium. Elements heavier than the iron group are produced in much smaller quantities than lighter ones. They have thus so far mostly escaped identification in cosmic rays, with notable exceptions (see Section 9.3.3). The production processes for cosmic rays are discussed in more detail in Chapter 4. The modifications cosmic rays undergo from source to detection are discussed in Chapter 6.

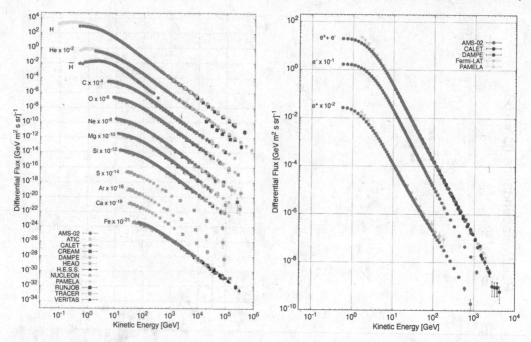

Figure 3.3 Left: Differential fluxes of cosmic ray nuclei and antinuclei from selected experiments. Balloon borne experiments are represented with filled squares (■) [365, 400, 414, 452, 637], space borne experiments with dots (●) [235, 457, 509, 510, 535, 569, 627, 693, 709, 722, 731, 763, 774, 775, 778, 779, 784], ground experiments with triangles (▲) [385, 678]. Right: Spectrum of electrons plus positrons [601, 619, 644, 719], as well as separate electron and positron spectra [456, 476, 508, 719, 720] from recent space borne experiments.

Figure 3.3 shows the kinetic energy spectra of representative nuclei (including the rare anti-nucleus $\overline{H}$) as well as leptons and anti-leptons in cosmic rays. We again observe the general power law behaviour we found for the all-particle spectra. Mind that in a representation plotting the logarithm of the flux as a function of the logarithm of the energy, quantitative detail is largely lost. However, differences between species and as a function of energy reveal important information about the sources, acceleration and transport mechanisms of cosmic rays. This will be discussed in the rest of this book. In addition, log-log representations give the simplistic impression that all measurements agree with each other and that statistics and systematics are unimportant. This is not always the case as we will see in the following chapters.

In the all-particle spectrum, hydrogen nuclei, i.e. free protons, represent about 75% of the total. About 17% of all nucleons are bound in helium nuclei and about 8% in heavier nuclei. About 13% of the total are bound neutrons. However, these fractions are not constant with energy. Indeed composition may change especially close to the knee and the ankle of the spectrum as discussed in Chapter 10.

The flux Φ in terms of particles per unit (area × time × solid angle) relates to the number density ρ of cosmic rays as $\Phi = \rho v/(4\pi)$. This assumes a flux characterised by a constant density of particles, a homogeneous average velocity v and isotropy on large scales. The energy density ρ_E can be calculated by integration, $\rho_E = 4\pi \int (E/v)\Phi \, dE$. To estimate the total cosmic ray energy density requires to extrapolate the flux Φ measured near Earth to outside the solar system. The data from the Voyager missions taken at more than 100 AU allow to safely do so [290]. As an example, Figure 3.4 [680] shows a comparison of the

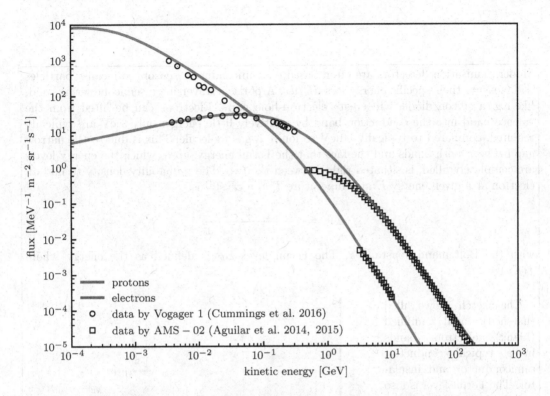

Figure 3.4 Simulated differential fluxes of cosmic ray protons and electrons (lines), adapted from [798], 5 Gyr after the begin of a three-dimensional MHD galaxy simulation, averaged over a ring centered at $r = (8 \pm 1)$ kpc with height 1 kpc. For comparison, data of Voyager 1 [575] and AMS-02 [515, 560] are shown. The AMS-02 data affected by solar modulation, i.e. below 10 GeV, are shown with light grey symbols. The simulated CR spectra are normalised to the observations at 10 GeV. (Credit: Courtesy of C. Pfrommer)

differential spectra from AMS-02 in near-Earth orbit to those measured by Voyager 1 [575] outside the heliosphere. When the part of the AMS-02 data affected by solar modulation are ignored, both spectra agree well with a simulation of galactic sources and propagation [798].

Integrating over energy one finds a total energy density ρ_E of the order of $1 \, \text{eV/cm}^3$. This is comparable to the energy density of typical magnetic fields on the galactic scale, with a strength of a few μG. It is thus clear that cosmic rays influence magnetic fields as much as the fields influence cosmic rays.

That all cosmic ray spectra follow a power law at high energies is indeed natural, since both acceleration by supernova remnants and known transport mechanisms favour such a behaviour. We will discuss this in Chapters 5 and 6. It is an interesting question, however, if non-standard sources like dark matter or other exotic forms of matter contribute in a quantifiable way. The best way to search for such phenomena is to concentrate on particles rarely produced by standard sources, especially antiparticles. We will come back to this question when we discuss galactic cosmic rays in detail in Chapter 9. But it is is equally questionable if known sources, acceleration and transport mechanisms truly saturate the high energy part of the spectrum [587]. We will come back to this question in Chapter 10. All of this evidently requires that one understands conventional astrophysical sources of cosmic rays. We will thus start from the beginning and consider how cosmic ray particles are synthesised in the Big Bang and inside stars.

Modern ionisation detectors are often based on semiconductor sensors to localise particles and measure their specific energy loss dE/dx. A particle traversing a semiconducting solid, like e.g. a silicon diode, will create electron-hole pairs. Electrons can be lifted from the valence band into the conduction band by relatively little energy: only 3 eV are typically required, compared to typically 30 eV to ionise a gas molecule. This is due to the narrow gap between both bands and the fact that the Fermi energy, up to which the energy levels are completely filled, is situated right between the two. The probability density to find an electron at a given energy E at temperature T in a crystal is:

$$f(E) = \frac{1}{e^{(E-E_F)/kT} + 1}$$

with the Boltzmann constant k. The Fermi level E_F is defined as the energy where $f(E_F) = 1/2$.

The sketch shows filled (hatched) and unfilled (blank) electron bands in a typical conductor, semiconductor and insulator. The Fermi level is also indicated.

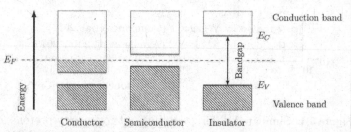

The gap between valence and conduction band is typically 1.12 eV wide for silicon at 300 K. The basis of a semiconductor diode are two layers, one of which is doped with positive (holes p), one with negative charges (electrons n).

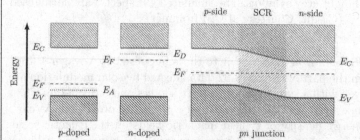

Where the two layers are in contact, a $p-n$ junction forms and charges migrate until an equilibrium is reached.

A depletion zone is formed around the contact, which works like a diode. When one applies an external voltage, a so-called reverse bias, the depletion zone can be enlarged. Electrons and holes liberated in this zone migrate to the two faces, within nanoseconds and very locally. Because of the low energy required to create mobile charges, the efficiency of such a detector is high, its response is linear in dE/dx and its energy resolution is very good. Only for very large ionisation densities, like those of heavy ions, saturation effects set in. To localise particles, the surfaces of semiconductor sensors are metallised with fine structures where charges are collected and fed into further electronic stages. Besides particle physics, semiconductor detectors are used in light detection, e.g. by CCD cameras.

Focus Box 3.5: Solid state particle detectors

FURTHER READING

V.S. Berezinskii, S.V. Bulanov, V.A. Dogiel and V.S. Ptuskin, *Astrophysics of Cosmic Rays*, North-Holland, 1990

Th. K. Gaisser, R. Engel and E. Resconi, *Cosmic Rays and Particle Physics* (2nd edition), Cambridge Univ. Press (2016)

A.M. Hillas, *Cosmic Rays*, Pergamon Press (1972)

T. Stanev, *High Energy Cosmic Rays* (3nd edition), Springer (2021)

M. Spurio, *Particles and Astrophysics: A Multi-Messenger Approach*, Springer (2015)

4 Particle Production

The analysis of cosmic ray composition and spectra permits us to peek into the cosmic nuclear and particle physics laboratory. Indeed, astrophysical sources range from the Big Bang itself to recent supernovae and thus cover the history of the Universe from a few moments after its beginning until today. The cosmic rays observed near Earth are however at the end of a long chain of events, which we will try to cover in the next three chapters. Fortunately photons and neutrinos allow us to look directly at the particle production processes. We thus gain access to the beginning as well as the end of the food-chain.

In this chapter, we describe the mechanisms of cosmic ray particle production at known and potential sources. As far as nuclei are concerned, this chapter heavily draws from Friederich-Karl Thielmann's reviews [833, 834]. Figure 4.1 gives a preview of the major processes responsible for the creation of nuclei as a function of their mass number. In the following sections, we will go through these processes one by one. A fascinating interplay between cosmology, nuclear physics and stellar evolution is thus exhibited.

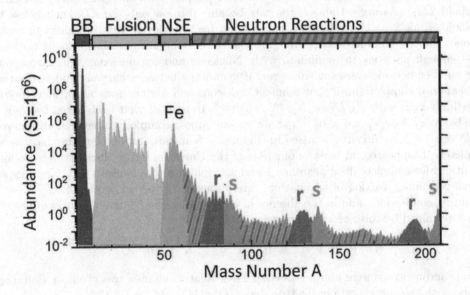

Figure 4.1 Solar abundances as a function of mass number A, showing the major processes contributing to their creation: Big Bang nucleosynthesis (BB), stellar fusion (Fusion), nuclear statistical equilibrium (NSE) and neutron-induced s- and r-processes (Neutron Reactions). (Credit: F. Käppeler, reproduced by permission)

4.1 BIG BANG NUCLEOSYNTHESIS

Shortly after the Big Bang the Universe was filled with an extremely hot soup of particles. By expansion, it cooled down ever after, until today's temperature of about 2.7 K. This growth is not the expansion of matter inside a fixed theatre of space-time. It is an expansion of space-time itself. Thus all lengths, including e.g. photon wavelengths, grow with time, and ambient temperature decreases accordingly. About one second after the Big Bang, quarks

DOI: 10.1201/9781003181385-4

and gluons had dressed into hadrons. Short lived particles had decayed and only electrons, neutrinos, photons, protons and neutrons were left.

A thermal equilibrium reigned, because there were sufficiently frequent scatters between particles to create a homogeneous temperature T. As explained in Focus Box 4.1, the scattering rate involves an average over the velocity distribution of the partners, which depends on their characteristic kinetic energy kT, with the Boltzmann constant $k \simeq 8.7 \times 10^{-5}$ eV/K. The condition for equilibrium is that the scattering rate is much larger than the inverse age of the Universe at that point in time, such that past scatters are abundant. Reactions of course only take place when they are kinematically allowed. Thus in a thermal bath of particles, only partners above the required centre-of-mass energy can ensure equilibrium. The required energy can be a reaction threshold or the Q-value[1] of a nuclear reaction. Thermal energy distributions, like the classical Maxwell-Boltzmann distribution, are very asymmetric and extend far beyond kT. As a rule of thumb, kT values above a few percent of the required energy are sufficient to maintain thermal equilibrium. Of course, equilibrium also requires that there are enough reaction partners to keep the rate high.

As the temperature dropped, equilibrium in the population of photons and baryons was maintained by the reversible reaction $\gamma\gamma \leftrightarrow N\bar{N}$, where N stands for a generic baryon. Equilibrium reigned as long as there were enough photons with energies above the reaction threshold $2M_N$. As argued above, the rate became too low only for kT much below this threshold. Finally at $kT \simeq 20$ MeV not enough fresh nucleons were produced to replace the ones lost by annihilation. At roughly the same temperature, antinucleons no longer found enough nucleons to annihilate with. Nucleons and antinucleons thus "froze out", their number became constant. However, if symmetry between baryons and antibaryons had remained complete until that moment, nucleons and antinucleons would have become exceedingly rare, with $N_N/N_\gamma = N_{\bar{N}}/N_\gamma \simeq 10^{-18}$. In reality, what is observed today is far more baryons, $N_N/N_\gamma \simeq 6 \times 10^{-10}$, as well as an almost complete absence of antibaryons, $N_{\bar{N}}/N_N \simeq 10^{-4}$, as already discussed in Chapter 3. Somehow antimatter must have almost completely disappeared, at least in our part of the Universe, before nuclei started to form. Conditions for complete disappearance – and astrophysical mechanisms to try and explain a small remaining fraction of antimatter – are discussed further in Section 4.5.

During equilibrium, and in the absence of antibaryons, the relative numbers of nucleons were determined by three reactions of weak interactions:

$$\nu_e\, n \leftrightarrow e^-\, p \ ; \quad \bar{\nu}_e\, p \leftrightarrow e^+\, n \ ; \quad n \rightarrow p\, e^-\, \bar{\nu}_e$$

Protons and neutrons were non-relativistic. Their relative number was given by Boltzmann factors, with the energy gain in neutron decay, $Q = M_n - M_p = 1.3$ MeV:

$$N_n/N_p = e^{-\frac{Q}{kT}}$$

At sufficiently low temperatures, $kT_n \simeq 0.87$ MeV, the characteristic time (i.e. the inverse rate) of the neutron-proton equilibrium reactions became longer than the age of the Universe. Thus, p and n froze out. At that point in time, their relative number was:

$$N_n(0)/N_p(0) = e^{-\frac{Q}{kT_n}} = 0.23$$

Free neutrons decay with a lifetime of $\tau_n = 879.4 \pm 0.6$ s. At a time t, the number of neutrons had thus fallen to:

$$N_n(t) = N_n(0)e^{-t/\tau_n}$$

---

[1] The Q-value is the amount of energy absorbed or released during a nuclear or particle reaction.

A central observable characterising quantum processes is their probability, defined as the ratio of the number of processes which take place to the number which could have taken place. In atomic physics, it is measured by the intensity of a spectral line. For particle reactions, it is the cross section or the decay width which measure the process probability.

The concept of the cross section is a geometric one, sketched on the right for a reaction $a + b \rightarrow c + d$. We associate a fictitious surface σ to each target particle b. Whenever a projectile a hits such a surface, the reaction takes place, otherwise it doesn't. The reaction rate is thus the product of the flux of incoming particles a, i.e. their number per unit surface and per unit time, the number of target particles exposed to this beam, and the cross section. This simple relation supposes that the surface density of target particles is small enough, such that the fictitious surfaces do not overlap and their total surface presented to the incoming beam is simply their sum. The cross section thus has the dimension of a surface, but of course no such surface exists. Instead, σ measures the probability for the process to take place.

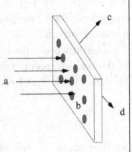

In accelerator based experiments, where the projectiles have a narrow energy distribution, the product between incoming projectile flux and the number of target particles is a characteristic of each experiment, called the luminosity. It is the maximum rate of reactions (per unit area) which would be observed if each projectile interacted with a target particle. The cross section is the proportionality factor between the luminosity and the actual reaction rate:

$$\frac{\#\text{reactions}}{\text{time}} = \underbrace{\frac{\#\text{projectiles}}{\text{time} \times \text{surface}} \times \#\text{targets}}_{\text{luminosity}} \times \text{cross section}$$

In cosmology and cosmic ray physics, projectiles usually follow an energy distribution with non-negligible width. The reaction rate Γ is then proportional to an integral over the incoming flux $d\Phi/dE$ and the energy dependent cross section $\sigma(E)$:

$$\Gamma = N_T \int \frac{d\Phi}{dE} \sigma(E) \, dE = N_T \langle \rho v \sigma \rangle$$

where N_T is the number of targets, ρ is the number density of the projectiles, v is their velocity and the average is over energy. That $\langle \rho v \rangle$ gives the incoming flux is easy to see for a homogeneous flow.

The cross section is measured in the large units of barn, $1\,\text{b} = 10^{-28}\,\text{m}^2$. Nuclear cross sections are of that size, but vary by several orders of magnitude. Cross sections measured for reactions among elementary particles are rather of the order of nanobarn, $1\,\text{nb} = 10^{-9}\,\text{b}$, or picobarn, $1\,\text{pb} = 10^{-12}\,\text{b} = 10^{-40}\,\text{m}^2$. For neutrinos they are even as small as $10^{-42}\,\text{m}^2$ (at 1 GeV).

Focus Box 4.1: Cross section and reaction rate

and the number of protons had risen to:

$$N_{\text{p}}(t) = N_{\text{p}}(0) + N_{\text{n}}(0) \left(1 - e^{-t/\tau_{\text{n}}} \right)$$

The relative number of protons and neutrons thus evolved like:

$$\frac{N_{\text{n}}(t)}{N_{\text{p}}(t)} = \frac{0.23 e^{-t/\tau_{\text{n}}}}{1.23 - 0.23 e^{-t/\tau_{\text{n}}}}$$

If nothing else had happened, neutrons would have disappeared after an hour or so, and we would have a Universe with only protons, electrons and neutrinos. But neutrons and protons close enough in phase-space coalesce to form stable nuclei. This nucleosynthesis starts by the formation of deuterium:

$$\text{n p} \leftrightarrow \gamma \, {}^2\text{H}$$

The binding energy is modest, $E_\gamma \simeq 2.2\,\text{MeV}$, but large enough to keep bound neutrons from decaying. The cross section for the formation process, $\sigma \simeq 0.1\,\text{mb}$, is much larger than for weak interactions.

The enormous density of ambient photons kept the reverse reaction, photo-desintegration of deuterium, competitive down to low temperatures of about $kT_D \simeq 0.05\,\text{MeV}$. Below, ${}^2\text{H}$ became stable and helium synthesis started:

$$\begin{array}{cc} {}^2\text{H n} \rightarrow {}^3\text{H } \gamma & {}^2\text{H p} \rightarrow {}^3\text{He } \gamma \\ {}^3\text{H p} \rightarrow {}^4\text{He } \gamma & {}^3\text{He n} \rightarrow {}^4\text{He } \gamma \end{array}$$

At that moment, the ratio of neutrons to protons was:

$$r = \frac{N_\text{n}(t)}{N_\text{p}(t)} = 0.14$$

We assume that all available neutrons were fused into helium nuclei. One needs two protons and two neutrons to form a ${}^4\text{He}$ nucleus and thus $N_\text{n} = 2N_\text{He}$, $N_\text{p} = 2N_\text{He} + N_\text{H}$. With $M_\text{He} \simeq 4M_\text{H}$, the mass fraction of helium at that point was:

$$Y \equiv \frac{M_\text{He}}{M_\text{He} + M_\text{H}} = \frac{4N_\text{He}}{4N_\text{He} + N_\text{H}} = \frac{2r}{1 + r} \simeq 0.25$$

The observed primordial ratio is $Y = 0.2449 \pm 0.0040$ [544], one of the triumphs of Big Bang theory. Details of the simple calculation presented above evidently depend on the baryon-to-photon ratio $\eta = N_\mathcal{N}/N_\gamma \simeq 6 \times 10^{-10}$, determined e.g. by the Planck satellite measuring the cosmic microwave background [536]. The fraction of ${}^4\text{He}$ should go up with η since there are more baryons available for fusion. Deuterium and ${}^3\text{He}$ should go down, since they are consumed in synthesising the more stable ${}^4\text{He}$. Figure 4.2 shows a recent compilation of measurements [544] compared with calculations in Big Bang theory [548]. The agreement for the lightest elements is impressive, except for the abundance of ${}^7\text{Li}$, which comes out a factor of three larger than observed. This lack of ${}^7\text{Li}$ is attributed to Li burning in old stars (see below).

4.2   UP TO IRON

Under the influence of gravity and driven by inhomogeneities in the distribution of matter and temperature, ingredients from Big Bang nucleosynthesis and the subsequent formation of atoms and molecules start forming large scale structures. Other ingredients of the Universe, such as dark matter may also play a role here. Gas clouds with matter in free fall start off the formation of stars. As a cloud contracts, its temperature increases according to the thermodynamics of ideal gases. Since the gas composition is dominated by hydrogen and helium, formation of more complex nuclei starts when the temperature allows two hydrogen nuclei to overcome the Coulomb barrier frequently enough and fuse. Hydrogen burning sets in. The main processes at this stage are called the pp-cycle and the CNO-cycle. The former is schematically depicted in Figure 4.3 on the left:

$$\begin{array}{rcl} {}^1\text{H} + {}^1\text{H} & \rightarrow & {}^2\text{H} + \text{e}^+ + \nu_e + 0.42\,\text{MeV} \\ {}^1\text{H} + {}^2\text{H} & \rightarrow & {}^3\text{He} + \gamma + 5.49\,\text{MeV} \\ {}^3\text{He} + {}^3\text{He} & \rightarrow & {}^4\text{He} + 2\,{}^1\text{H} + 12.86\,\text{MeV} \end{array}$$

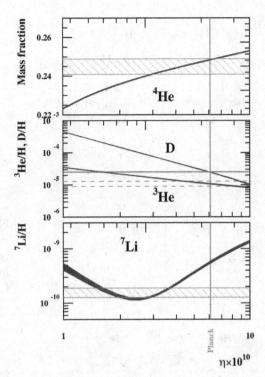

Figure 4.2 Primordial mass fraction of ^{4}He (top), ratios of ^{3}He and ^{2}H to the hydrogen abundance (middle) and ^{7}Li/H ratio (bottom) as a function of the the baryon-to-photon ratio $\eta = n_b/n_\gamma$. The curves are calculations of Big Bang nucleosynthesis [600], the horizontal bands come from a compilation of experimental determinations [544, 548], the vertical line marks the η determination by the Planck collaboration [536].

In the first step, deuterium is formed after fusion of two protons and a β^+-decay, p $\to$ n e$^+\nu_e$. Since the mass of the proton is less that that of the neutron, this inverse decay only takes place for bound protons, where the required energy comes from a gain in binding energy. In total, the cycle burns six hydrogen nuclei and produces two to restart the chain. The large amount of energy liberated in the last step comes from the fact that ^{4}He is a "doubly magic" nucleus where both protons and neutrons form a closed shell. It is thus strongly bound, as shown in Figure 4.6 and explained in Focus Box 4.2.

A competing hydrogen burning process, which uses ^{12}C as a catalyst, is the so-called CNO-cycle reaction first discussed by von Weizsäcker [99, 105] and Bethe [106]. It requires carbon as input and is thus only important for massive stars. In these, the temperature is high enough at an early stage, so that the triple-α reaction can produce carbon as sketched in Figure 4.4:

$$^4\text{He} + {}^4\text{He} \quad \to \quad {}^8\text{Be} - 0.1\,\text{MeV}$$
$$^8\text{Be} + {}^4\text{He} \quad \to \quad {}^{12}\text{C} + 7.4\,\text{MeV}$$

This then allows the CNO-cycle to burn hydrogen to helium as shown in Figure 4.3 on the right:

$$
\begin{aligned}
^{12}\text{C} + {}^1\text{H} &\to {}^{13}\text{N} + \gamma + 1.95\,\text{MeV} \\
^{13}\text{N} &\to {}^{13}\text{C} + \text{e}^+ + \nu_e + 1.20\,\text{MeV} \\
^{13}\text{C} + {}^1\text{H} &\to {}^{14}\text{N} + \gamma + 7.55\,\text{MeV} \\
^{14}\text{N} + {}^1\text{H} &\to {}^{15}\text{O} + \gamma + 7.34\,\text{MeV} \\
^{15}\text{O} &\to {}^{15}\text{N} + \text{e}^+ + \nu_e + 1.68\,\text{MeV} \\
^{15}\text{N} + {}^1\text{H} &\to {}^{12}\text{C} + {}^4\text{He} + 4.96\,\text{MeV}
\end{aligned}
$$

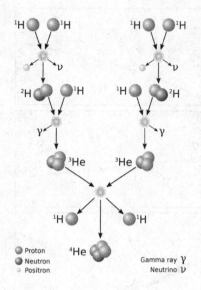

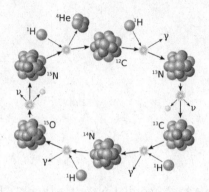

Figure 4.3 The pp-cycle (left) and the CNO-cycle (above) of nuclear fusion and decay reactions describing the formation of helium from hydrogen fuel. (Credit: Wikimedia Commons)

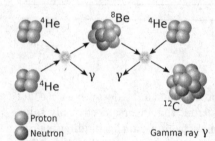

Figure 4.4 The triple-α fusion process, producing ^{12}C via an intermediate unstable ^{8}Be nucleus.(Credit: Wikimedia Commons)

The relative luminosity contribution of the pp- and CNO-processes as a function of temperature [453] is shown in Figure 4.5. For the Sun, the latter account for only a few percent of the He production. However, the electron neutrinos from the beta decays in the CNO-cycle have been observed [758], providing a rare look into the otherwise invisible processes in the Sun's core. In stars with a mass larger than about $1.3 M_\odot$, the CNO-cycle dominates the energy production.

These processes reduce the number of hydrogen nuclei by a large factor at every cycle. They are a formidable source of energy, building up an internal pressure around the centre of the star. Its inner $\sim 10\%$ are hot enough to maintain fusion, at a temperature of a few 10^7 K. The pressure thus balances the gravitational forces and a stable and rather long lived equilibrium is formed. This constitutes a so-called main sequence star, where the luminosity L is related to the mass M by an approximate power law:

$$\frac{L}{L_\odot} = \left(\frac{M}{M_\odot}\right)^{3.5}$$

where $L_\odot$ and $M_\odot$ are mass and luminosity of the Sun. The luminosity – thus also the life time – is given by the mass of the star; the higher the mass, the lower the lifetime. The sun has sufficient fuel for about 10^{10} years. The dependence of lifetime on mass can e.g. be observed in star clusters. Their members are formed at the same time. By observing luminosity and colour, the masses of the members can be estimated. As a function of age, less and less bright massive stars are observed.

The liquid drop model was the first nuclear model to describe the binding energy E_B and the radius R of nuclei. The latter is given by a compact packing of A incompressible nucleons, such that $R = r_0 A^{1/3}$ and the nuclear volume is roughly proportional to A, the atomic mass number. The radius parameter is $r_0 \simeq 1.25$ fm. The (negative) binding energy, $E_B = M(Z, A) - Z m_p - (A - Z) m_n$, with the nuclear charge Z and the nucleon masses $m_{p/n}$, is described by the Bethe-Weizsäcker formula:

$$E_B = \underbrace{-a_1 A}_{\text{Volume}} + \underbrace{a_2 A^{\frac{2}{3}}}_{\text{Surface}} + \underbrace{a_3 \frac{Z(Z-1)}{A^{\frac{1}{3}}}}_{\text{Coulomb}} + \underbrace{a_4 \frac{(N-Z)^2}{A}}_{\text{Asymmetry}} \pm \underbrace{a_5 A^{-\frac{1}{2}}}_{\text{Pairing}}$$

It is roughly proportional to the number of nuclei, thus the first term. However, there is a core where the nuclear force is saturated, surrounded by a surface layer where it is not, diminishing the binding energy. Coulomb repulsion between protons is taken into account by a term proportional to the number of proton-proton pairs. Up to here, the considerations are purely classical. In particular, they do not account for the fact that light nuclei with an equal number of protons and neutrons are particularly stable; others have a less negative binding energy by the asymmetry term. In addition, there are more stable nuclei with both proton and neutron numbers even and few with both odd; this is described by the pairing term. All parameters a_i are positive. The asymmetry term is positive and reduces the binding energy when $N \neq Z$. The pairing term has a positive sign for odd N and Z, negative when both are even, and zero otherwise. The coefficients are determined by adjusting to measured nuclear masses, e.g. in [373].

Understanding "magic" nuclei with high binding energy as seen in Figure 4.6 requires a quantum mechanical approach, the nuclear shell model. It is constructed in analogy to its atomic equivalent, but using a model potential. For a shell with principle quantum number n, there are n sub-shells with orbital quantum number $l = 0 \ldots (n-1)$. For each of them, there are $m_l = -l \ldots l$ projections of l on an arbitrary axis. These $2l + 1$ states are degenerate in energy. For each one, there are two spin orientations, with $m_s = \pm 1/2$. Without magnetic field, these are also degenerate in energy. A state is thus characterised by four quantum numbers, n, l, m_l, m_s. The degeneracy with respect to magnetic quantum numbers is removed by spin-orbit coupling. The notation used in nuclear physics for the principal quantum number is $n = N + l + 1$, where N is the number of nodes in the wave function. Energy levels are shown on the right [140].

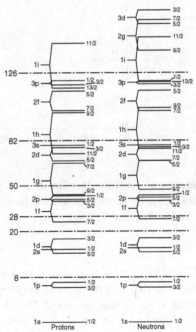

The spin-orbit coupling removes the energy degeneracy and the levels partially overlap. Large gaps in energy appear each time a shell is completed. This explains the "magic" numbers of protons and neutrons and in particular the "doubly magic" nuclei like ^{4}He, ^{16}O, ^{40}Ca, ^{48}Ca, ^{48}Ni, ^{56}Ni, ^{208}Pb etc., which are especially stable.

Focus Box 4.2: Nuclear structure and binding

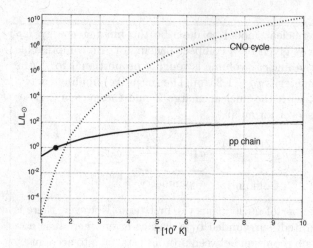

Figure 4.5 Stellar luminosity production relative to the Sun as a function of temperature for the pp chain and the CNO cycle, adapted from [453]. The dot marks conditions in the solar core.

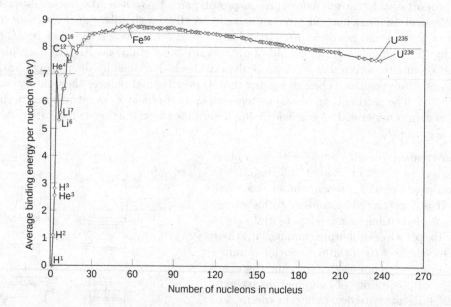

Figure 4.6 Average binding energy per nucleon as a function of the mass number A. The most tightly bound nuclei are found in the iron group and in the so-called magic nuclei. (Credit: Wikimedia Commons)

The layer structure of a hydrogen burning star is sketched on the left of Figure 4.7. The mass fraction of hydrogen as a function of radius varies from some 30% close to the centre of the core to the canonical 75% for the unburned fuel, but depends of course on time. The burning is steady, since fluctuations in temperature are automatically compensated by expansion or contraction of the core region. Energy is transported to the outside by convection and radiation.

At the end of the hydrogen burning phase, fuel runs out and the core consists only of helium, surrounded by hydrogen. The temperature of the core is about 1.5×10^7 K which is too cool for helium fusion. Hydrogen burning continues in an expanding shell outside the core, signalling the beginning of shell burning. Due to the higher temperature, more

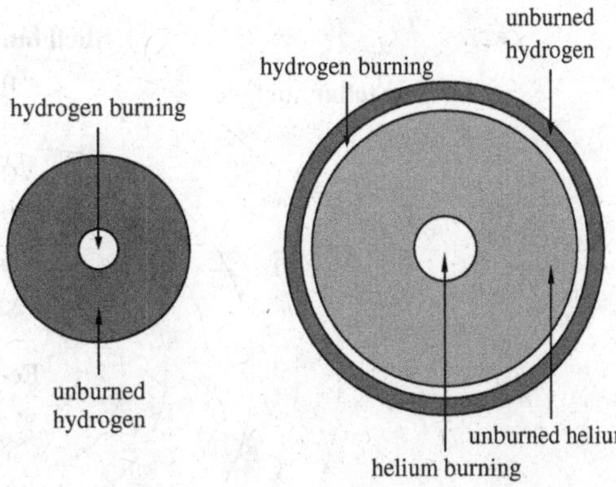

Figure 4.7 The left figure shows the schematic layer structure of a main sequence star in the hydrogen burning phase. On the right, the layer structure of a red giant in the helium burning phase is shown.

power is produced than in core hydrogen burning. This will drive an expansion of the outer unburned hydrogen and cool it down. The surface will be redder than during the main sequence phase, and the radius blows up by one or two orders of magnitude. A star in this phase is called a red giant. For a sun-like star this phase lasts about 2×10^9 years.

The core itself, without its own energy production, can no longer withstand the inward gravity. It will start to slowly contract, so that its temperature increases by the gain in gravitational energy. Heat is transferred outward by black body radiation. What happens from then on depends on mass. For high enough masses, core contraction will eventually pass the threshold for helium fusion and the core starts burning again, by the triple-α reaction depicted in Figure 4.4 and followed by a further α capture, $^{12}\text{C} + ^4\text{He} \rightarrow ^{16}\text{O} + \gamma + 7.16\,\text{MeV}$. This situation is depicted on the right of Figure 4.7. Low mass stars, below about two solar masses, may go directly to a degenerate core as described below, while others usually do that only after core helium is exhausted.

When a star's core is not producing thermal energy one would think that there is no mechanism to prevent total collapse. This is not true, since fermions – especially the electrons in the core plasma – cannot be indefinitely compressed. The Pauli principle prevents half-integer spin particles – called fermions – from occupying the same quantum state. This creates a repulsive potential between electrons and leads to the quantum mechanical electron degeneracy pressure, much stronger than Coulomb repulsion. The equation of state of such a degenerate Fermi gas is very different from that of an ideal gas. In ideal gases, pressure depends on density and temperature. In a degenerate Fermi gas, pressure is proportional to a power of the density alone, independent of temperature. So as long as the temperature is cool enough for the gas to be (close to) degenerate, temperature no longer matters. Only higher density increases the pressure.

When helium is depleted by burning, further shell burning processes are initiated by a sequence of core and shell states, as schematically listed in Figure 4.8. However, the main source of energy dissipation changes. While it was mainly due to photon radiation in H- and He-burning phases, neutrinos become the dominant loss factor in subsequent phases. This severely shortens their duration. The next in line is carbon burning:

$$
\begin{aligned}
^{12}\text{C} + ^{12}\text{C} \quad &\rightarrow \quad ^{20}\text{Ne} + ^4\text{He} + 4.6\,\text{MeV} \\
&\rightarrow \quad ^{23}\text{Na} + ^1\text{H} + 2.2\,\text{MeV} \\
&\rightarrow \quad ^{24}\text{Mg} + \gamma + 13.93\,\text{MeV}
\end{aligned}
$$

One would suspect that this phase be followed by oxygen burning, but there is an intermediate neon burning stage. It occurs because ^{20}Ne is a loosely bound nucleus, easily disintegrated by $^{20}\text{Ne} + \gamma \rightarrow ^{16}\text{O} + ^4\text{He}$ at this temperature. The alpha particle can recombine with another ^{20}Ne to form ^{24}Mg. The output is thus similar to the preceding carbon burning.

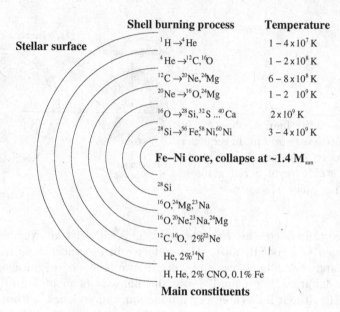

Figure 4.8 Schematic layer structure of a star at the end of shell burning. Major processes, their reaction products and temperature range, are indicated on the top right. Major fuels in each shell are indicated below.

Subsequently, successive fusion with α particles follows the diagonal in the N/Z-plane of Figure 4.9. But the rising temperature also allows fusion among oxygen nuclei themselves:

$$\begin{aligned}
^{16}\text{O} + ^{16}\text{O} \quad &\rightarrow \quad ^{28}\text{Si} + {}^4\text{He} + 9.6\,\text{MeV} \\
&\rightarrow \quad ^{31}\text{P} + {}^1\text{H} + 7.7\,\text{MeV} \\
&\rightarrow \quad ^{31}\text{S} + \text{n} + 1.5\,\text{MeV}
\end{aligned}$$

Major products of this stage are ^{28}Si and ^{32}S, but also nuclei up to ^{40}Ca are produced.

At this point of the development, the temperature has risen above 3 to 4×10^9 K, corresponding to photon energies of the order of 1 MeV. The high-energy tail of the photons can photo-desintegrate the products of oxygen burning and produce ample protons, neutrons and alpha particles. The temperature is high enough to pass Coulomb barriers for further fusion reactions. A statistical equilibrium between nuclei from calcium to nickel is thus established, with abundances given by thermal equilibrium. This phase is called Nuclear Statistical Equilibrium (NSE) and produces all nuclei around the iron group as shown in Figure 4.1. At this point, energy production by fusion stops, since the maximum binding energy per nucleon is reached and fusion becomes endothermic. A core made of strongly bound nuclei from the iron group results. Figure 4.9 schematically summarises major reaction paths in stellar nucleosynthesis.

In the preceding fly-over of stellar fusion, only major processes were quoted. In reality, there is a whole network of nuclear reactions contributing to the various phases of nucleosynthesis. Their relative importance is influenced by the density of input nuclei and cross sections as a function of temperature, as well as decay rates. Data bases and codes like NucNET [483], WinNet [514] and SkyNET [626] allow to follow the evolution in detail.

The life of all main sequence stars, as sketched above, is rather similar. The only difference is that higher mass stars are more luminous and their lifetime is shorter. What happens after the main sequence life has ended depends on the mass of the star. Low mass stars,

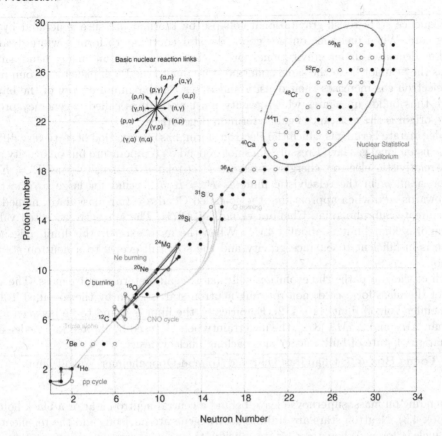

Figure 4.9 Schematic of stable isotopes (full dots) and unstable isotopes (empty dots) up to the iron group. Major fusion reactions are indicated by lines: H burning by the pp-cycle and the CNO cycle, He burning by the triple α process and subsequent α capture, carbon, neon and oxygen burning. Only major fusion paths are shown. The region of nuclear statistical equilibrium is also indicated. The abbreviations (a,b) in the insert denote transformation between nuclei A and B via the reaction $A + a \rightarrow B + b$.

with a mass less than a few times $M_\odot$, never get hot enough to trigger carbon burning. The outer H and He zones are expelled into a stellar wind, known – somewhat misleadingly – as planetary nebula. The star ends up with a small degenerate carbon/oxygen core, which is called a white dwarf, since its surface temperature can be as high as 2×10^6 K, thus of blue-white colour. Its typical mass is comparable to that of the Sun, $M < 1.4M_\odot$, but its radius is comparable to that of the Earth. The mass limit for white dwarfs, the so-called Chandrasekhar limit is explained in Focus Box 4.3.

Stars in the range $8M_\odot < M < 10M_\odot$ can collapse directly to a nuclear statistical equilibrium during carbon burning. Even heavier stars will first go through all core and shell burning phases up to silicon burning. When all fusion fuel is exhausted, their Fe/Ni core collapses to a neutron star via a supernova explosion. This expels the star's material in an explosive fashion, creating a plasma shock wave which is a formidable particle accelerator as we will see in Chapter 5. Core collapse supernovae of a single star are not the only type. Binary systems bring about other types of supernovae relevant for nucleosynthesis, as we will see below.

Main sequence stars resist gravitational collapse by thermal pressure generated by their burning core. When fuel runs out for good, the star contracts to form a white dwarf. As the radius decreases under gravitational forces, electrons occupy an energy band, since as fermions they cannot be in the same quantum state due to the Pauli principle. Compression of the electron gas increases their density and raises the maximum energy of the band. A pressure thus builds up agains which gravity must work. It is called degeneracy pressure since its origin is the resistance against fermion degeneracy.

The degenerate Fermi gas which the electrons form, has an equation of state very different from the usual ideal gas law. Pressure does not depend on temperature but on density alone. At non-relativistic electron energies, pressure P is related to density ρ as $P = K_1\rho^{5/3}$ valid for small ρ, in the relativistic limit as $P = K_1\rho^{4/3}$ valid for large ρ. The radius tends towards zero when approaching the so-called Chadrasekhar mass limit, named after Nobel laureate Subrahmanian Chandrasekhar [361, 371]. The currently accepted value of the Chandrasekhar limit is about $1.4M_\odot$. When the mass exceeds this limit, degeneracy pressure is insufficient to balance gravity and the star will evolve to a neutron star or a black hole.

Which of the two paths the evolution will take is again determined by mass. The upper bound of the mass for a cold, non-rotating neutron star is given by the so-called Tolman-Oppenheimer-Volkoff limit [110, 112]. Empirically, the limit is found to be between a little more than $2M_\odot$ and $2.5M_\odot$ [812], the uncertainty being caused by limited knowledge about the equation of state of bulk matter approaching nuclear desity.

Focus Box 4.3: Chandrasekhar and Tolman-Oppenheimer-Volkoff limits

Depending on mass, supernovae leave behind a central neutron star or a black hole (see Focus Box 4.3). Neutron stars are stabilised by degenerate nucleons and the repellent part of the nuclear force. When they rotate quickly, they can radiate electromagnetic energy like a beacon. Such sources are called pulsars (see Section 4.5). The maximum mass of a neutron star is about two solar masses (see Focus Box 4.3), from progenitors of about $30M_\odot$. However, the core collapse of progenitors which are this massive more often ends up as a black hole, as simulations show [741].

Most stars are born in binary systems, twin stars of diverse masses circling each other. When the twins are in different phases of evolution, or if they merge due to loss of rotational energy, this brings about an important mass increase for one of them. Prominent examples are binary systems, where one of the members has already shrunk to a white dwarf, the other evolves into a read giant. The enormous radius increase of the latter pushes its outer layers into a distance, where the gravity of the companion exceeds its own gravity. Thus matter is transferred from the red giant to the white dwarf. What is transferred is of course mainly hydrogen and helium, which, depending on the accretion rate, can quietly burn in a new outer shell of the white dwarf. At the same time, the oxygen/carbon core of the white dwarf will be compressed and heated. When the white dwarf exceeds the mass of the Chandrasekhar limit, the core is unstable against contraction. It will ignite and cause an explosive disruption of the whole system. Such supernovae are classified as type Ia. They have a narrow mass range and dissipate a known amount of energy. They thus qualify as a standard light source to measure the expansion velocity of the Universe [284, 297]. This observation is the prime evidence – originally not yet very firm [591] but supported by recent results from the Dark Energy Survey[2]– for an accelerating expansion of the Universe and the existence of Dark Energy (see e.g. [449]).

---

[2]See https://www.darkenergysurvey.org/des-year-3-cosmology-results-papers/

As far as nucleosynthesis is concerned, what matters is that mass transfer re-ignites fusion reactions in a white dwarf. The explosion expels nuclei from the iron group from its core as well as lighter elements from its outer regions. The exact mechanisms of ignition and explosion are rather complex and go beyond the level of our discussion here. Models say that supernovae of type Ia produce about $0.6M_\odot$ of ^{56}Ni, which undergoes subsequent β-decay to ^{56}Fe, about six times more than core collapse supernovae. Although the latter are more frequent, type Ia supernovae are thus comparable or even the dominant source of iron group nuclei.

4.3 BEYOND IRON

Nucleosynthesis by fusion reactions abruptly stops with iron group nuclei as output. This is due to the fact that for heavier nuclei, binding energy per nucleon slowly declines with mass as seen in Figure 4.6. In addition, Coulomb repulsion of course continuously increases and fusion would require excessive temperatures. Reaction rates for charged particles vary exponentially with temperature, and so do concurrent photo-disintegration rates so that the latter quickly win. Neutron capture reactions, on the other hand, suffer from no Coulomb barrier and rates have almost no temperature dependence. However, neutrons have a short lifetime of a few minutes. Thus, special astrophysical conditions are required for the production of heavy nuclei by neutron absorption. The dominant processes are labelled s (for slow) and r (for rapid), their paths are indicated in Figure 4.10.

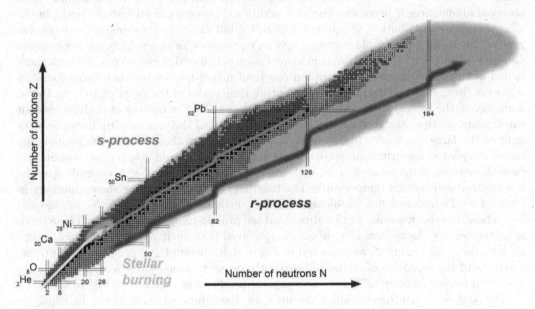

Figure 4.10 Nuclear chart showing stable isotopes in black, proton-rich β^+-unstable isotopes in dark shade and neutron-rich β^--unstable ones in light shade. Isotopes unstable against α-decay are highlighted. The grey fuzzy area indicates the border of unstable nuclei, which are still stable against proton or neutron emission. Its lower limit is called the neutron drip line. Also shown are the principle paths of nucleosynthesis for fission, s- and r-processes (arrows) explained in the text. (Credit: EMMI, GSI/Different Arts)

The s-process dominates under condition of neutron densities which are high enough to allow a single capture with subsequent β decay, with capture rates in the range of a few per decade. The so-called weak s-process produces elements up to Sr and Y from a seed of Fe nuclei. An example for the beginning of the chain is $^{56}\text{Fe} + \text{n} \rightarrow {}^{57}\text{Fe} \rightarrow {}^{57}\text{Co} + \text{e}^- + \bar{\nu}_e$. The dominant neutron source for this process is α-capture on neon, $^{22}\text{Ne} + {}^4\text{He} \rightarrow {}^{25}\text{Mg} + \text{n}$. The so-called main s-process produces even heavier elements up to Pb. Its main neutron source is α-capture on carbon, $^{13}\text{C} + {}^4\text{He} \rightarrow {}^{16}\text{O} + \text{n}$. The s-process progresses close to the valley of stability indicated by black squares in Figure 4.10, since it relies on isotopes sufficiently stable to survive until a β-decay occurs. The neutron densities produced during core and shell burning, ranging between $10^7/\text{cm}^3$ and $10^{10}/\text{cm}^3$, are sufficient to maintain s-processes and thus produce heavy elements up to about Pb. There it hits a region, where α-decay (i.e. spontaneous asymmetric fission) becomes dominant. The end-points of the s-process chain are thus circular processes like this:

$$
\begin{aligned}
{}^{209}\text{Bi} + \text{n} &\rightarrow {}^{210}\text{Bi} + \gamma \\
{}^{210}\text{Bi} &\rightarrow {}^{210}\text{Po} + \text{e}^- + \bar{\nu}_e \\
{}^{210}\text{Po} &\rightarrow {}^{206}\text{Pb} + {}^4\text{He} \\
{}^{206}\text{Pb} + 3\,\text{n} &\rightarrow {}^{209}\text{Pb} \\
{}^{209}\text{Pb} &\rightarrow {}^{209}\text{Bi} + \text{e}^- + \bar{\nu}_e
\end{aligned}
$$

Since the s-process involves stable isotopes, the relevant neutron capture cross sections can be measured with sufficient precision, e.g. at nuclear reactors or other intense neutron sources, like the CERN n_TOF facility. The network of processes can thus be reliably simulated, and resulting abundances can be predicted with precision. When comparing to observed abundances it becomes clear that additional processes contribute to peaks in the solar abundance distribution, as shown in Figure 4.1, but also to heavy element production in general. These are due to rapid neutron capture, r-processes for short. They involve neutron densities sufficient to sustain multiple neutron captures before β-decay occurs, during a short period of time. Neutron densities are ten orders of magnitude higher than those found in s-process sites, and thus lead to neutron capture time scales of the order of 10^{-4} to 10^{-6}s, much below the typical time scales of nuclear β-decay. Thus a nucleus of a given element can capture multiple neutrons taking it close to the limit of the neutron drip line, the lower limit of the large fuzzy area in Figure 4.10. Photodesintegration through the high-energy tail of the photon spectrum competes with neutron capture, leading to a quasi-equilibrium between isotopes of the same element, with a maximum neutron number N depending on the free neutron density and temperature. The maxima form a contour line shown in violet in Figure 4.10. They show kinks when closed neutron shells are reached, at $N = 50, 82, 126$ and 184. There, the capture cross section drops and the process temporarily pauses. These semi-magic nuclei (see Focus Box 4.2) are also longer lived than their neighbours. They "wait" for beta decay increasing Z, as indicated in Figure 4.10, before the r-process can continue. Products of the r-process of course subsequently decay to more stable isotopes, typically about ten nuclear mass units below the s-process products as sketched in Figure 4.1.

The very high neutron densities required for the r-process chain occur in explosive events rather than steady burning, since they do not require long neutron exposure in time. Extremely neutron-rich environments can be found in core collapse supernovae of the type Ib, Ic (hypernovae, leading to black holes) and type II.

A major source of r-process material is the merger of two neutron stars in a binary system. An example was signalled by the gravitational wave event GW170817, the first one observed by the LIGO and Virgo antennae associated with optical observations [624, 625]. Spectroscopic inspection of the remnant revealed two components, both of r-process origin [603–605, 621, 635]: a hot component consisting mainly of lower mass nuclei ($A < 140$), a cooler component of heavier nuclei ($A > 140$) rich in actinides like U, Th or Cf [692]. Further evidence for r-processes at work comes from the week-long afterglow of the event. Such long-term heating is only possible from radioactive r-process nuclei on their way to stability. Incidentally, models suggest that a single neutron star merger may produce between three and thirteen Earth masses of gold [651]. It is unfortunately hard to collect such riches.

There are thus at least two astrophysical sites where r-processes can occur. The relative importance of these is a subject of active research.

4.4 DARK MATTER

Now we need to come back to the level of particle physics, almost faultlessly described by the Standard Model, which is now complete with the discovery of the Higgs boson. It might well be, that there is no fundamentally new physics concerning particles and their interactions [443] until the Planck mass is reached, about 10^{19} GeV. However, there is compelling evidence that the matter and energy contents of the Standard Model is incomplete.

The best know example is the astrophysical evidence for the existence of dark matter, a form of matter that emits no light, neither reflects nor absorbs light. Speculation about elusive forms of matter is as old as philosophical thinking [642]. The first to use gravitational observation to find an object invisible to astronomical telescopes was the mathematician Friedrich Bessel. In the 1840s, he used accurate measurements of the stellar parallax, which is the apparent shift of the image of a near star with respect to a distant constellation as the Earth moves around the Sun. It can be used to triangulate the distance to the star. Bessel used it to conclude that Sirius had a dark companion, today known as Sirius B. His work also contributed to the discovery of Neptune a few years later. Thus even if they only manifest themselves through gravity, objects do not go undetected.

The American astronomer of Swiss origin Fritz Zwicky is often cited as the discoverer of dark matter. This is arguably also because he had a very strong personality;[3] Freeman Dyson characterises him as "intensely Swiss" [735]. In 1933, he studied the velocities of galaxies in the Coma cluster [75, 100]. He applied the virial theorem to link the mass of the cluster to the width of the velocity distribution measured by Edwin P. Hubble [54]. The virial theorem applied to astrophysics and Zwicky's argument are explained in Focus Box 4.4. From estimates of the number of galaxies in the cluster and their average mass, he arrived at a total mass corresponding to an expected velocity variance of about 80 km/s. The observed velocity spread is much larger, 1000 km/s along the line of light, estimated via Doppler shift measurements [54]. The visible mass is thus much too small to explain the variance of velocities. Zwicky concluded [75]: "If this should prove true, one would get the surprising result that dark matter is present in much larger density than luminous matter". In 1936, Sinclair Smith published a similar estimate [89] for the Virgo cluster. He also found a much larger mass than the value estimated from luminosity measurements by Hubble [78].

One of the most convincing proofs for the existence of dark matter are the rotation velocities of objects around their galaxies, thus one step in scale down from the cluster argument of Zwicky and Smith. They are also measured by the electromagnetic Doppler effect, which shifts the line spectra of abundant elements in young stars – like hydrogen and nitrogen– towards the red when the source recedes, towards the blue when it approaches. Spectroscopy for astronomy [505] works according to the principles of diffraction. Focus Box 4.5 shows how this is used to measure the radial mass distribution in spiral galaxies. Precision measurements of the velocity as a function of distance from the centre of a galaxy were pioneered in the spectroscopic survey of Andromeda by Vera Rubin and Kent Ford in the early 1970s [180]. In January 2020, it was announced that the Large Synoptic Survey Telescope (LSST), which is currently under construction in Chile, will be named the Vera C. Rubin Observatory.

---

[3]For a lively account of the debate around Zwicky's way of dealing with colleagues, and his daughter's defence of her father, see the article by Richard Panek in Discover Magazine [407].

The virial theorem states a relation between the mean values of kinetic and potential energies for a system in equilibrium [163]. The mean over time of the kinetic energy T for a system of N particles is:

$$2\langle T\rangle = -\sum_{i=1}^{N}\langle \vec{r}_i \vec{F}_i\rangle = \sum_{i=1}^{N}\langle \vec{r}_i \vec{\nabla} U\rangle$$

where $\vec{F}$ is the force at position $\vec{r}$. The second equation is valid for conservative forces arising from a potential U. If this potential is radially symmetric and follows a power law $U \propto r^k$, like the gravitational or Coulomb potentials, we get the simplified form:

$$2\langle T\rangle = k\langle U\rangle$$

On the other hand, the kinetic energy of a single particle of mass m and velocity v is $T = \frac{1}{2}mv^2$. If we assume an equal partition of the velocities over the three spatial directions, then $\langle v_x^2 + v_z^2 + v_z^2\rangle = 3\langle v^2\rangle$, where v is the mean velocity along any line of sight. For the total kinetic energy, we get $2\langle T\rangle = 3M\langle v^2\rangle$ with the total mass $M = \sum m$. The total potential energy is $U = -\alpha GM^2/R$ where R is the total radius and G is the gravitational constant. The constant α depends on the radial mass distribution. For an unrealistic radial equipartition one would e.g. get a coefficient of 3/5, more realistic coefficients are also of order unity. Using the standard deviation $\sigma^2 = \langle v^2\rangle - \langle v\rangle^2$ as a measure for the width of the velocity distribution and $\langle v\rangle = 0$ for an equilibrium system, we get the mass estimator which Fritz Zwicky used for the Coma cluster:

$$\frac{\alpha GM}{R} = 3\sigma^2 \quad ; \quad M = \frac{3}{\alpha}\frac{R}{G}\sigma^2$$

A wider velocity distribution of the galaxies in the cluster thus leads to a higher virial mass estimator. Zwicky estimated the total mass of the Coma cluster as 800 galaxies of a billion solar masses each, thus $M \simeq 1.6 \times 10^{42}$ kg. Its radius is about a million light years, i.e. $R \simeq 10^{22}$ m. The width of the velocity distribution predicted by the virial theorem is about $\sigma \simeq 100$ km/s, while the observed one is of the order of 1000 km/s. There is thus much more invisible mass present in the cluster than visible one.

Focus Box 4.4: Virial theorem and Zwicky's dark matter estimate

Gravitational lensing, suspected to exist by Newton[4] and by Einstein prior to the formulation of general relativity [282], is a formidable tool to measure the mass of large astronomical structures, even when this mass does not emit light. The principle is based on the fact that light rays follow straight lines in space-time distorted by the gravity of objects. In this manner the gravity of a heavy object in the foreground causes multiple deformed images of an object in the background. Measuring the deformation of the image of a galaxy behind a cluster, for example, one can calculate the mass of the cluster in the foreground. This technique is used in the compound image of Figure 4.11 showing the cluster of galaxies 1E 0657-56, better known as the bullet cluster. What is shown in white are bright objects from optical observation, thus normal matter. Overlaid in dense grey is an image of hot gas taken with X-rays, showing normal matter again. And a gravitational lensing image in lighter grey, showing the distribution of all matter.

---

[4] "Query 1. Do not Bodies act upon Light at a distance, and by their action bend its Rays; and is not this action (caeteris paribus) strongest at the least distance?" [1, 3rd book, part I, p. 313]

The Doppler effect causes the wavelength of light from a moving source to be shifted. When the source moves relative to the observer with velocity v, its wavelength λ is shifted from its value at rest by $\Delta\lambda$:

$$\frac{\Delta\lambda}{\lambda} = \sqrt{\frac{1+v/c}{1-v/c}} - 1 \equiv z$$

When the source is receding, $z > 0$, corresponding to a red shift for visible light. When it is approaching, $z < 0$ and visible light is blue shifted. Frequency standards like the bright ($n = 3 \rightarrow 2$) line of hydrogen (H-α, $\lambda = 656\,\text{nm}$) or the nitrogen line (NII, $\lambda = 658\,\text{nm}$) are typically emitted by hot gas around young stars. Their wavelength shift is often used to measure source velocities.

An example on how this is done is shown on the right. The photo shows an image of the spiral galaxy NGC3198 seen practically edge-on. Luminous matter right of the galactic centre is receding, approaching on the left. The velocity component along the line of sight is measured. Using the orientation of the rotation axis, indicated by the line, it is converted to rotational velocities.

Credit: Sloan Digital Sky Survey

Credit: T.S. van Albada *et al.* [212]

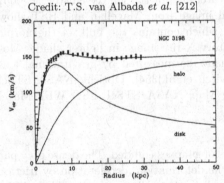

The result is shown on the left [212], the rotational velocity v_{cir} as a function of distance to the galactic centre. Matter is moving with a velocity almost independent of distance from the galactic centre, even at points far outside the luminous region. This is not what one would expect if only luminous matter were present.

Stars and other bright objects in a spiral galaxy are moving around its axis with a roughly circular motion. The velocity as a function of distance to the axis, R, is given by Kepler's third law: $v_{cir}^2(R) = \frac{GM(R)}{R}$. Here $M(R)$ is the total mass included in the path and G the gravitational constant. Up to the galaxy's outer limit, $M(R)$ grows with radius and so does the rotational velocity. Outside the visible limit of a few kpc, one should have $M = \text{const}$, and a decrease of the rotational velocity, $v_{cir} \sim 1/\sqrt{R}$. An example of a modern measurement of the rotational curve is shown above [212]. One observes $v_{cir}(R) \simeq \text{const}$ at large R. This means that there is an extended halo of invisible mass reaching far beyond the optical limit. The calculated velocity distribution caused by the galactic disk is shown by the curve, together with the deduced contribution by the halo of dark matter.

Focus Box 4.5: Doppler effect and rotation curves of spiral galaxies

Figure 4.11 A combined image, covering $7.5 \times 5.4\,\text{arcmin}^2$, of the galaxy cluster 1E 0657-56, also known as the bullet cluster. The optical image from Magellan and HST shows galaxies as bright spots. Hot gas in the cluster, which contains the bulk of the normal matter, is shown by the Chandra X-ray Observatory X-ray image in lighter shade. Most of the mass in the cluster is shown in darker shade, as measured by gravitational lensing. (Credit: X-ray: NASA/CXC/CfA/M.Markevitch et al.[354]; Optical: NASA/STScI; Magellan/U.Arizona/D.Clowe et al.[376]; Lensing Map: NASA/STScI; ESO WFI; Magellan/U.Arizona/D.Clowe et al.[376])

The image shows the situation after two clusters of galaxies collided. The dense grey part corresponds to the luminous hot gas, and shows the deformation, deceleration by friction, and the coalescence which is expected after such a collision for ordinary matter. The collision leaves a conical trail of heated gas behind like a bullet passing through matter. The light grey cloud shows that the bulk of the matter in both clusters is not luminous and extends farther than the luminous one. The smaller cluster now on the right has traversed the larger on the left with little disturbance for its dark component. Its bulk is in advance with respect to the much less abundant normal matter.

One concludes that dark matter accounts for about 85% of the mass of galaxies and their clusters, but this percentage can vary a lot. A recently discovered galaxy cluster, named Dragonfly 44, is even suspected to contain almost 100% of dark matter [583].

Since it can be observed by different techniques, it is thus clear that dark matter exists. It contributes to the gravitational confinement of galaxies and their clusters, and is probably also involved in the formation of large cosmic structures. Its gravitational interaction is the same as for normal matter, we are thus entitled to call it matter even though it is unconventional since it has not yet been seen to interact otherwise. If it consists of particles, and there is no reason to doubt that it does, they must have the following properties:

- They must be electrically neutral, otherwise they would shine light.

- They are moving at non-relativistic speeds, thus they are probably heavy. Dark matter of this kind is called cold dark matter.

- They are probably half-integer spin and subject to Pauli exclusion, otherwise dark matter would collapse onto itself.

- They interact very weakly with each other and with normal matter, otherwise reaction products would be abundant.

Particles which correspond to such a generic profile are called Weakly Interacting Massive Particles, WIMPS.

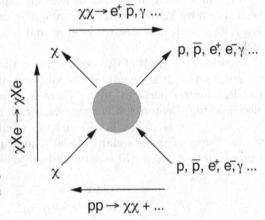

Figure 4.12 Schematic diagram of the three ways to hunt for dark matter particles χ. Top left to right: annihilation of dark matter particles into ordinary matter-antimatter pairs. Left bottom to top: interaction of dark matter with heavy nuclei like Xe. Bottom right to left: Production of dark matter particle pairs in proton-proton collisions.

The search for hypothetical dark matter constituents, which we generically denote with χ, is thus well targeted. They are attacked from three sides simultaneously, as schematically depicted in Figure 4.12:

- One can try to produce dark matter particles pair-wise at high-energy colliders like the LHC. Since they do not interact much with ordinary matter, their presence would be signalled by missing energy and momentum in the final state [658]. So far, the energy of the colliders and/or the sensitivity of experiments has been insufficient for an observation.

- One also searches for the rare interactions of dark matter particles with ordinary matter, in a scattering process. This is sometimes called direct detection. It requires large mass cryogenic detectors sensitive to the tiny recoil that a χ-nucleus scattering would cause [724]. Despite the ever increasing sensitivity of experiments, there has been no detection yet.

- Dark matter particles can annihilate pairwise. If χ is an ordinary fermion, it will have an antiparticle $\bar{\chi}$. If it is of the Majorana type, as is often assumed, the two will be identical. Annihilations into ordinary particles can then be observed as an extraordinary contribution to cosmic rays, neutrinos or photons [517]. Since dark matter is everywhere, observatories covering a large proportion of the sky are the most promising options. The Fermi satellite observatory and ground based observatories for high-energy gamma rays like H.E.S.S., Veritas and MAGIC look for photon signals for dark matter. The neutrino observatory IceCube at the South Pole looks for neutrinos. These pointing messengers are covered in Chapter 7. Space observatories covered in Chapter 9 look for dark matter signals in charged cosmic rays.

- Dark matter particles are not necessary stable, but must have an extraordinarily long lifetime, since Dark Matter is still abundant at the current age of the Universe. Signatures for χ decay are similar to annihilation signals.

Dark matter particles are thus fenced in from all sides and it seems that their discovery is only a matter of time and energy. One cannot ignore, however, that they may escape detection forever. A popular model for (almost) undetectable dark matter are the so-called sterile neutrinos [357, 358], right-handed companions of ordinary neutrinos with a modest mass of order 10 keV. Because of their "wrong" spin orientation, their interactions with matter are heavily suppressed, such that they may hide forever. But then again the existence of sterile neutrinos may be detectable by other means [410].

Dark matter may in fact not be the only explanation for the observed phenomena. As has been shown by Mordehai Milgrom [205–207], one can also modify Newtons law of universal gravitation such that it stays valid at and below solar system dimensions, but is modified at larger scales.

However, the spirit of this book requires that we take dark matter for granted, and that we assume that it is quantised like all other matter. Let us generically denote its quanta, the dark matter particles, by χ. There is thus at least one neutral fermion, χ^0 with zero electric charge, lepton and baryon number. It may be unstable, but with a very low decay rate. Even if it is stable, it will undergo annihilation reactions. If these were to proceed via the forces of the Standard Model – i.e. if they were mediated by W, Z or H bosons – leptonic final states would eventually lead to electron-positron pairs:

$$
\begin{aligned}
\chi\chi &\rightarrow e^+e^- \\
\chi\chi &\rightarrow \mu^+\mu^- \\
&\hookrightarrow e^+\nu_e\bar{\nu}_\mu\, e^-\bar{\nu}_e\nu_\mu \\
\chi\chi &\rightarrow \tau^+\tau^- \\
&\hookrightarrow e^+\nu_e\bar{\nu}_\tau\, e^-\bar{\nu}_e\nu_\tau
\end{aligned}
$$

If the mass of χ is of order TeV, final states $\chi\chi \rightarrow$ ZZ, W$^+$W$^-$ and HH will also occur, followed by subsequent decay of the final state bosons. Which final states can occur is only limited by phase space, with relative rates given by the couplings. Hadronic final states, $\chi\chi \rightarrow q\bar{q}$, will normally also show up with the same restriction. Quarks will pick up companions from the vacuum to form hadrons, mesons and to a lesser extent baryons and antibaryons. Leptonic decay chains of hadrons will again result in electrons and positrons. The kinematics for these reactions of course depend on the mass of the dark matter particle χ. If it is heavy enough, it will allow final states with most Standard Model particles. As an example for cosmic rays from dark matter with Standard Model interactions, Figure 4.13 gives the differential flux of positrons near Earth for a generic χ of mass 800 GeV [455, 551].[5]

For non-relativistic dark matter, a kinematic cut-off is provided by the first of the above reactions: the maximum electron/positron momentum is equal to the mass of the χ particle. If it were the only reaction, a single peak in the energy distribution would occur, broadened only by the kinetic energy distribution of the dark matter particles. Below this cut-off, the momentum distribution can be populated by electrons from the decay chains of heavier

---

[5]Halo option NFW, propagation option MED (plus MIN and MAX for the leptonic flux) and magnetic field option MF1 are used to obtain the plotted results from PPPC 4 DM ID.

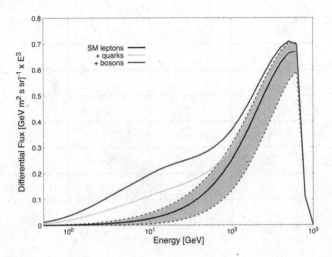

Figure 4.13 Expected differential flux $(\times E^3)$ for positrons calculated using PPPC 4 DM ID results [455], produced by a Navarro-Frank-White profile of dark matter [279] in the Milky Way, with $m_\chi = 800\,\text{GeV}$. Standard Model couplings and an annihilation rate of $\langle\sigma v\rangle = 10^{-26}\,\text{cm}^3/\text{s}$ are used to predict the flux coming from annihilation into leptons, quarks and bosons. Diffusive transport through the galaxy towards the solar system is included using a simplified description of galactic magnetic fields [551]. Uncertainties due to the diffusion through the galaxy are indicated by the shaded band for the flux of leptonic origin.

particles. Positrons close to the kinematic limit are preferentially produced by the leptonic final states listed above.

If dark matter particles had appreciable couplings to Standard Model forces, they ought to have been detected by collider experiments like the ones at the LHC or by direct detection in their interaction with nuclei. It thus appears more likely that the interactions of χ are mediated by a new force; let us generically denote its quanta be ϕ. This force would have very weak couplings to Standard Model particles. Annihilation could thus preferentially result in ϕ-pairs instead of Standard Model particle pairs. The decay properties of ϕ are unknown, but it is probably a light quantum with $m_\phi \ll m_\chi$, for practical reasons we will argue below. Hard leptons may thus preferentially result from its leptonic decay modes. Figure 4.14 shows again an example of a differential positron spectrum expected from $\chi\chi \to \phi\phi$, followed by leptonic decays of the light quantum $\phi \to e^+e^-$, $\mu^+\mu^-$ and $\tau^+\tau^-$, obtained by the same code with identical options. The spectrum is rather similar to the one produced by Standard Model lepton pairs (see Figure 4.13).

The rate of annihilation reactions is determined by three properties of dark matter particles: their number density ρ and velocity distribution $f(v)$, and of course the annihilation cross section σ (see Focus Box 4.1). The reaction rate Γ then scales as follows:

$$\Gamma \propto \rho^2 \int f(v)\sigma(v)v\,dv = \rho^2\langle\sigma v\rangle$$

In the simplest case, σv would be constant for non-relativistic dark matter. This is for example the case for isothermal dark matter with a density which decreases as the square of the distance r from the galactic center, $\rho \propto 1/r^2$. In more general cases suggested by cosmological simulation [265, 279, 342, 375, 433], the density profile is rather $\rho \propto 1/r$ and the velocity $\langle v\rangle \propto \sqrt{r}$. However, as long as the dark matter interaction properties are not known from particle physics, the cross section itself remains unknown. Fortunately, it can be

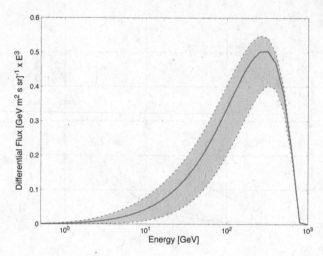

Figure 4.14 Expected differential flux ($\times E^3$) for positrons calculated using `PPPC 4 DM ID` results [455], produced by a Navarro-Frenk-White profile of dark matter [279] in the Milky Way, with $m_\chi = 800\,\text{GeV}$. A hypothetical light neutral boson ϕ is supposed to be pair produced, $\chi\chi \to \phi\phi$, followed by exclusively leptonic decays of ϕ. Uncertainties due to the diffusion through the galaxy are indicated by the shaded band.

estimated from the current abundance of dark matter. The argument goes as follows. In the early universe, thermal equilibrium reigned between dark matter and other particles. The number density ρ_{eq} of non-relativistic χ particles thus varied according to the Boltzmann factor:

$$\rho_{\text{eq}} = g_\phi \left(\frac{mkT}{2\pi} \right)^{3/2} e^{-\frac{m}{kT}}$$

where g_ϕ is the number of degrees of freedom for χ (2 for a neutral Majorana fermion), m its mass and T the ambient temperature. Due to the expansion of the Universe, m/T is also a measure of time. The number density thus decreased exponentially over many orders of magnitude, as shown by the black line in Figure 4.15 [485].

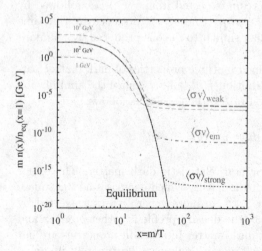

Figure 4.15 Evolution of the cosmological abundance for WIMPS [485] as a function of time, indicated by the ratio of WIMP mass and temperature, $x = m/T$. The WIMP number density is scaled by its value at $x = 1$. Mind that the abscissa covers 25 orders of magnitude. Deviations from thermal equilibrium (solid black curve) occur at freeze-out, with typical $\langle\sigma v\rangle = 2 \times 10^{-26}\,\text{cm}^3/\text{s}$ for weak (dashed), $2 \times 10^{-21}\,\text{cm}^3/\text{s}$ for electromagnetic (dashed-dotted) and $2 \times 10^{-15}\,\text{cm}^3/\text{s}$ for strong interactions (dotted). Three WIMP masses are indicated for the weak cross section case.

Just like for baryons and as explained in Section 4.1, at a certain point in time, this equilibrium was no longer maintained since the temperature fell below the required minimum. This freeze-out happened when the reaction rate $\rho_{\mathrm{eq}}^2 \langle \sigma v \rangle$ fell below the expansion rate of the Universe, the Hubble constant. Thus the rarified χ particles could not find enough partners to annihilate with and their number became constant. The higher the cross section for interactions among dark matter particles, the later such a freeze-out would occur and the lower the current abundance would be. This is indicated in Figure 4.15 with cross sections typical for weak, electromagnetic and strong interaction processes. Since the relic abundance of dark matter is precisely known from results of the Planck satellite, one can inversely infer $\langle \sigma v \rangle$ from the abundance. This requires to know the mass of the χ particle, since it relates the number density to the energy density determined by the CMB data. For TeV masses, one obtains $\langle \sigma v \rangle = \mathcal{O}(2 \times 10^{-26}\,\mathrm{cm}^3/\mathrm{s})$ [809] as used in Figures 4.13 and 4.14.

If dark matter interactions are mediated by a new light boson ϕ with $m_\phi \ll m_\chi$, the possibility of a Sommerfeld enhancement for the cross section arises. This comes from a kind of bound state between the two dark matter particles. The mechanism is discussed further in Focus Box 4.6. It may indeed lead to cross sections which are several orders of magnitude larger than the ones quoted above. In particular, if the mass is less than a few hundred MeV for a scalar ϕ, or about a GeV for a vector ϕ, kinematics may forbid its decay into heavier hadronic states. An enrichment of leptonic final states may thus result.

4.5 ASTROPHYSICAL ANTIMATTER

As noted in Section 4.1, antimatter produced from a symmetric Big Bang seems to have mysteriously disappeared – at least from our part of the Universe – before antinuclei could be formed. In 1967, Andrei Sakharov [173] formulated criteria for a dynamic way of making antimatter disappear for the profit of matter:

- Antinucleons, notably antiprotons, must be able to decay, thus violating the conservation of baryon number.

- There must be an objective way by which Nature distinguishes matter from antimatter. Thus the symmetries which connect the two, called charge conjugation C, and CP (together with the left-right symmetry parity P) must also be violated. Only this way antiprotons can disappear and protons stay.

- The disappearance mechanism must be at work outside of thermal equilibrium. The masses of protons and antiprotons are strictly the same, due to CPT symmetry involving the time reversal operation T, and converting at reaction back to itself. In thermal equilibrium, where the particle abundance is only a function of mass and temperature, their density would necessary be equal.

Baryon number conservation is one of the strictest known conservation law. The Super-Kamiokande experiment has been searching for proton decays for over 20 years in a detector of very large mass, 22.5 to 27.2 kilotons. None has been observed, leading to lifetime limits $\tau_{\mathrm{p}} > 10^{34}$ a for leptonic decays, 24 orders of magnitude larger than the age of the Universe [737]. Violation of the symmetry between particles and antiparticles, on the contrary, has been observed in the decays of mesons containing strange and bottom quarks.[6] The small violation is due to weak interactions and described in the Standard Model of particle physics by a complex phase in the quark mixing matrix. It might exist for leptons also, but has not been observed so far [769]. Opportunities when the Universe fell out of thermal equilibrium existed in the early phases of its development. Every phase transition presented

---

[6]See e.g.: https://pdg.lbl.gov/2019/reviews/rpp2019-rev-cp-violation.pdf.

At low energies, an attractive force between particles can lead to an enhancement of the cross section by the formation of a bound state. The enhancement was first discussed by Arnold Sommerfeld [57] and is named after him.

A mechanical analog exists from gravity. Suppose a particle with velocity v approaches a star with radius R and mass M from infinity. Neglecting gravity, the cross section for hitting the star is simply $\sigma_0 = \pi R^2$. However, the attractive gravitational force will suck in particles with an impact parameter $b_{max} > R$. Using conservation of energy and angular momentum gives b_{max} such that

$$\sigma = \pi b_{max}^2 = \sigma_0 \left(1 + \frac{v_{esc}^2}{v^2}\right)$$

where $v_{esc}^2 = 2G_N M/R$ is the square of the escape velocity from the star and G_N is Newton's gravitational constant.

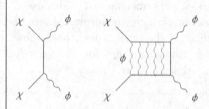

An analogous effect takes place in quantum theory when the attractive force carrier ϕ of a theory has a Compton wavelength smaller than αm_χ, where m_χ is the mass of the matter particle in question and $\alpha \ll 1$ is the fine structure constant characterising the strength of the force. As sketched on the left, bound states of χ with itself (or its antiparticle) can then form at low energies.

For the annihilation of dark matter particles, this can occur for sufficiently small mass ratios m_ϕ/m_χ. At low velocities of the dark matter particles, of the order of $100\,\text{km/s}$, the enhancement factor can reach several orders of magnitude [418].

Focus Box 4.6: Sommerfeld enhancement of the dark matter annihilation cross section

such an opportunity, but the best candidate might well be supraluminal expansion usually called the inflationary phase.[7] In conclusion, while we know that antimatter disappeared early after the Big Bang, we do not know how that happened.

4.5.1 SECONDARY POSITRONS

In absence of primordial antimatter, the conventional way to produce the observed small amount of antimatter is by interactions of primary cosmic rays with interstellar matter. Since both reaction partners are supposed to be matter particles, conservation of lepton and baryon number requires the production of particle-antiparticle pairs. Electromagnetic processes produce mainly lepton pairs, e.g. by photon conversion in the presence of a nucleus $\gamma + (A, Z) \to e^+ e^- + (A, Z)$ or by inelastic electron-nucleus scattering. Heavier leptons subsequently decay and eventually also leave an electron-positron pair plus neutrinos. Thus high-energy electrons and photons on their way through interstellar material can produce energetic positrons. Inelastic hadronic interactions, initiated mainly by the dominant proton and helium components, produce multiple mesons, which can also enrich energetic positrons via their decay chains, like $\pi^+ \to \mu^+ \nu_\mu$, $\mu^+ \to e^+ \nu_e \bar{\nu}_\mu$.

Positrons do not travel very far, since they loose energy quickly and are prone to annihilation reactions with interstellar matter. The expected positron fluxes are heavily influenced by the properties of interstellar matter and magnetic fields in the galaxy. We will analyse models of cosmic ray propagation and the production of secondaries in more detail in Chapter 6. Of major influence is the column density of matter traversed by particles on their

---

[7]See e.g.: https://pdg.lbl.gov/2019/reviews/rpp2019-rev-inflation.pdf.

voyage from source to detection as well as the properties of the interstellar plasma. These are not well known a priori, but bounds can be deduced from other cosmic ray data. The ratios of mainly secondary nuclei (like boron) to mainly primary ones (like carbon or oxygen) (see Section 9.3.2) provide strong limits on these parameters. Nevertheless important uncertainties remain for fluxes of secondary leptons. Figure 4.16 shows two predictions for the positron flux in the solar system [434, 458], which predate the precision measurements from AMS-02. Both predict a maximum of the positron spectrum below 10 GeV, followed by a monotonous decrease. More recent determinations [729, 749], taking into account constraints from AMS-02 for secondary to primary nuclei ratios, reduce the propagation uncertainty without changing the qualitative picture.

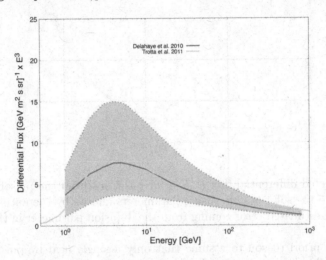

Figure 4.16 Expected differential flux ($\times E^3$) for secondary positrons from GALPROP [458] (grey line) and from Delahaye *et al.* [434] (black line) predating the availability of precision AMS-02 data. For the latter prediction, estimated uncertainties due to propagation parameters are indicated by the shaded band. The accuracy was limited by the cosmic rays data available at the time.

4.5.2 PULSAR POSITRONS

Stellar electromagnetic processes can also produce positrons. A prominent example are pulsars, rapidly rotating neutron stars (or to a lesser extent white dwarfs) where the rotational axis makes an angle with the direction of the magnetic field. This leads to a magnetic field variable in time surrounding the compact object, which in turn produces an electric field according to Maxwell's laws. Since neutron stars are small and rotate with frequencies up to kHz, the large induced electric field on their surface can rip out electrons from stellar material and accelerate them to high energies. A back-of-the-envelope calculation sketched in Focus Box 4.7 indicates that energies up to PeV can in principle be attained for a sufficiently strong magnetic dipole rotating in vacuo. Electromagnetic processes can then produce positrons as described above. However, the environment of a neutron star is filled with plasma, reducing the field strength. Moreover, electrons and positrons will cool by synchrotron radiation in the strong magnetic field. Both effects reduce the maximum electron or positron energy from direct pulsar emission by several orders of magnitude. As an example, Figure 4.17 shows the spectrum of positrons expected from the pulsar Geminga (J0633+1746), situated about 0.25 kpc from the solar system and about 3.4×10^5 y old, calculated analytically in [590] including diffusive propagation towards us. More details of the particle acceleration and radiation from pulsars are obtained by numerical simulation [645]. The results indicate that photon radiation occurs mainly in a cone around the polar region as well as a torus around the equator.

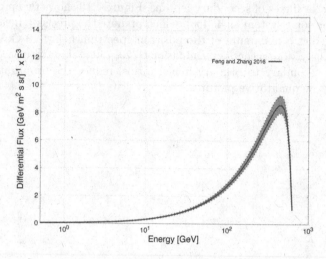

Figure 4.17 Expected differential flux ($\times E^3$) for positrons from the pulsar Geminga [590] and propagated to the solar system, using parameters quoted by the authors. The shaded area indicates the estimated error coming from the diffusion parameter in their calculation.

There is no a priori reason to assume that only a single near-by pulsar would dominate the positron flux observed at Earth. Contributions from several sources, with different distance and maximum energy, would thus cause the flux to be somewhat bumpy [646], especially at its peak and beyond. An example is shown in Figure 9.25. With the statistics expected from e.g. AMS-02 until its end-of-life it is not clear if such spectral features can be observed.

To obtain the TeV photon energies observed e.g. by the MAGIC telescope from the Crab pulsar [594] and by the H.E.S.S. telescope from the Vela pulsar and its surroundings [714], a more powerful acceleration mechanism is required. This may well be offered by the so-called magnetocentrifugal particle acceleration [629]. The basics of this mechanism are shown in Figure 4.18. Charged particles follow co-rotating magnetic field lines and are radially decelerated, but azimuthally accelerated. When their azimuthal velocity reaches the speed of light, they hit what is called the light cylinder surrounding the axis of rotation and radial motion stops. Particles then follow Archimedean spirals with high relativistic factors. A suitably aligned observer will periodically see their synchrotron radiation emitted tangentially.

4.5.3 SECONDARY ANTIPROTONS

Secondary cosmic ray antiprotons are produced in the inelastic interactions of ordinary cosmic rays, thus mainly protons and helium nuclei, with interstellar matter. Above the kinematic threshold, proton-antiproton pairs can be produced e.g. in pp $\rightarrow$ p$\bar{\text{p}}$X, where X denotes the remainder of the hadronic final state. Instead of antiprotons, antineutrons can be produced, which in due time convert to antiprotons via beta decay. Further indirect reactions include the production of hyperon pairs, with $\bar{\Lambda}$ and $\bar{\Sigma}$ decays then contributing to antiproton production. A very careful study of the cross sections for these processes has

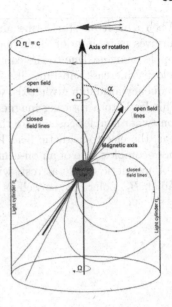

Figure 4.18 Schematic diagram showing the magnetocentrifugal mechanism of particle acceleration [629]. A neutron star rotates around the z-axis, with its magnetic field misaligned by an angle α. Particles travelling along the comoving magnetic field lines move towards the light cylinder with radius r_L, where their velocity approaches the speed of light. They then gain further energy along an Archimedean spiral. Synchrotron radiation is emitted tangentially to the light cylinder.

been conducted by Winkler and collaborators [526, 636, 672]. There, the total cross section σ is separated into antiproton and antineutron production, $\sigma_{\bar{p}}$ and $\sigma_{\bar{n}}$, which in turn are separated into prompt, σ^0, and hyperon contributions, σ^Λ:

$$\sigma = \sigma_{\bar{p}} + \sigma_{\bar{n}} \quad ; \quad \sigma_{\bar{p},\bar{n}} = \sigma^0_{\bar{p},\bar{n}} + \sigma^\Lambda_{\bar{p},\bar{n}}$$

Winkler *et al.* use latest cross section data from RHIC and LHC experiments and well-established input from strong interactions theory. The result, propagated to near Earth, is shown in Figure 4.19. The procedures used to describe the diffusive transport of antiprotons from their sources to the solar system are described in Chapter 6.

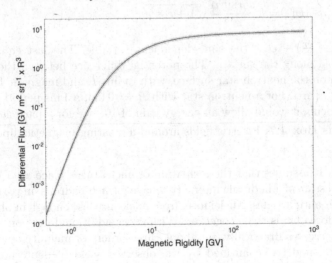

Figure 4.19 Expected differential flux ($\times R^3$) for antiprotons near Earth [636], multiplied by R^3. The band indicates the uncertainty coming from cross section data and the propagation model used.

The formation of complex nuclei in high-energy particle interactions, like pp $\rightarrow$ ^{2}H X or pp $\rightarrow$ ^{3}He X, is traditionally described by the coalescence model, which we quickly review

When a neutron star forms after the collapse of its progenitor, a strong magnetic field survives in its interior. This magnetic field is usually approximated by that of a magnetic dipole moment. Due to the contraction, the moment of inertia will reduce and the angular velocity increase substantially. The neutron star will thus normally be rotating with high frequency around a given axis, which is not necessarily aligned with the magnetic moment. This leads to a rotating magnetic dipole, a classical problem of electrodynamics. For the somewhat simpler case where the magnetic dipole moment is anchored at the center of the star, the electromagnetic field has been explicitly calculated e.g. by Sarytchev [426]. The more complex case of an off-center moment has been treated by PÃtri [634].

A magnetic dipole at rest, with magnetic moment μ, generates a magnetic field in the radial, B_r, polar, B_θ, as well as azimuthal direction, B_ϕ:

$$
\begin{aligned}
B_r &= 2\frac{\mu}{r^3}\cos\theta \\
B_\theta &= \frac{\mu}{r^3}\sin\theta \\
B_\phi &= 0 \\
|B| &= \frac{\mu}{r^3}\sqrt{1 + 3\cos^2\theta}
\end{aligned}
$$

When it rotates with angular velocity ω around an axis misaligned by an angle α with respect to the direction of the magnetic moment, an electric field is induced at a distance r in the θ- and ϕ-directions [426]:

$$
\begin{aligned}
E_r &= 0 \\
E_\theta &= -\frac{\omega\mu\sin\alpha}{cr^2}\sin\alpha\left(\cos\Phi - \frac{\omega r}{c}\sin\Phi\right) \\
E_\phi &= -\frac{\omega\mu\sin\alpha}{cr^2}\sin\alpha\cos\theta\left(\sin\Phi + \frac{\omega r}{c}\cos\Phi\right) \\
|E| &= \frac{\omega\mu\sin\alpha}{cr^3}\sqrt{1 + \cos^2\theta}
\end{aligned}
$$

where $\Phi = \omega\left(t - \frac{\omega r}{c}\right) - \phi$ is the time-dependent variable. The last equation gives the time-averaged field along the surface. The potential difference between the polar and the equatorial region on the neutron star surface, with radius R and magnetic field B_0, is thus of order $B_0(\omega R)/c\sin\alpha$. For a neutron star with $R = 10\,\mathrm{km}$, a $1\,\mathrm{ms}$ period of rotation and $B_0 \simeq 10^{11}\,\mathrm{T}$ in vacuo, it would allow an energy gain of $10^{15}\,\mathrm{eV}$ for a particle of unit charge.

Focus Box 4.7: Electric fields around a rotating magnetic dipole

in Focus Box 4.8. It assumes that the formation of nuclei takes place after the quarks and gluons in the final state of a hadronic interaction have formed hadrons, in particular protons, neutrons and their antiparticles. Nuclei form from nucleons close enough in phase space, such that the nuclear force binds them together. With every additional nucleon, the probability for coalescence decreases drastically, by about three orders of magnitude. This prediction of the coalescence model is confirmed by the observed yield of deuterium, tritium and ^{3}He and their antinuclei in proton-proton collisions at the LHC. Figure 4.20 shows results obtained by the ALICE experiment [656] for the yield of light antinuclei [656], measured in a limited region of phase space and extrapolated to full coverage. The observation is in good agreement with expectations from the coalescence model. If antihelium nuclei were observed in cosmic rays, it would thus be close to impossible that they are of secondary origin.

The coalescence model assumes that nucleons which are close enough to each other in phase space form a nucleus, i.e. A nucleons including Z protons, with a certain probability. The phase space distribution $d^3 N_A/dp_A^3$ of the nucleus is thus related to that of individual nucleons by the relation:

$$E_A \frac{d^3 N_A}{dp_A^3} = B_A \left(E_\mathrm{p} \frac{d^3 N_\mathrm{p}}{dp_\mathrm{p}^3} \right)^Z \left(E_\mathrm{n} \frac{d^3 N_\mathrm{n}}{dp_\mathrm{n}^3} \right)^{A-Z}$$

with $\vec{p}_\mathrm{p} = \vec{p}_\mathrm{n} = \vec{p}_A/A$ and the coalescence parameter B_A. This parameter can be associated to the volume of a sphere in momentum space, $4\pi p_0^3/3$, inside which the nucleons must be concentrated:

$$B_A = \frac{A}{m_\mathrm{p}^{A-1}} \left(\frac{4\pi}{3} p_0^3 \right)^{A-1}$$

The coalescence parameters for deuterium, tritium, ^{3}He and their antinuclei have been measured by the ALICE experiment in proton-proton interactions at the LHC [656]. They found that the coalescence parameters are the same for nuclei and antinuclei, as one would expect. They are roughly constant, independent of centre-of-mass energy up to 7 TeV, but show a mild dependence on the transverse momentum of the nucleus with respect to the initial proton direction.

If one assumes that proton and neutron distributions in a deeply inelastic final state are the same, the coalescence relation simplifies to:

$$E_A \frac{d^3 N_A}{dp_A^3} = B_A \left(E_\mathcal{N} \frac{d^3 N_\mathcal{N}}{dp_\mathcal{N}^3} \right)^A$$

where $\mathcal{N}$ denotes a generic nucleon. The total yield, integrated over phase space, is thus expected to follow a power law of atomic number, proportional to $c_1 A c_2^A$. This is indeed observed, as shown in Figure 4.20.

Focus Box 4.8: Coalescence model of nuclei formation

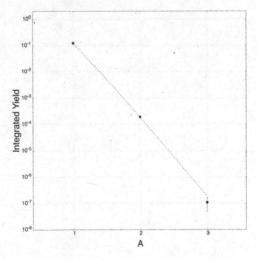

Figure 4.20 Measured yield of antihydrogen, antideuterium and $^3\overline{\mathrm{He}}$ nuclei by proton-proton interactions at the LHC at a center-of-mass energy of 7 TeV, measured by the ALICE experiment [656]. The dotted line corresponds to the A-dependence expected from the coalescence model, i.e. $c_1 A c_2^A$.

FURTHER READING

M. Bauer and T. Plehn, *Yet Another Introduction to Dark Matter: The Particle Physics Approach*, Springer (2019)

G. Bertone, *Particle Dark Matter*, Cambridge University Press (2010)

D.D. Clayton, *Principles of Stellar Evolution and Nucleosynthesis*, University of Chicago Press (1983)

J.J. Cowan, F.-K. Thielemann and J.W. Truran, *Stars, Stellar Explosions, and the Origin of the Elements*, Princeton University Press (2022)

R.A. Freedman, R.M. Geller and W.J. Kaufmann III, *Universe*, 10th Edition, W.H. Freeman and Company (2014)

E.W. Kolb and M.S. Turner, *The Early Universe*, CRC Press (1990)

5 Cosmic Accelerators

In this chapter, we discuss cosmic ray acceleration mechanisms, liable to carry charged particles from the energies typical for thermonuclear reactions inside stars to high energies. Energies up to $\mathcal{O}(10^{11})$ GeV are rarely, but significantly observed in cosmic ray spectra, as shown in Chapter 3. We concentrate on acceleration by the turbulent magnetic fields found in clouds of plasma populating interstellar space, a mechanism first proposed by Enrico Fermi [128, 148]. This extends on our discussion of bulk plasma acceleration by (regular) rotating magnetic fields surrounding pulsars in Focus Box 4.7. We then quickly go over astrophysical environments where favourable conditions for acceleration can be found.

5.1 FERMI ACCELERATION

By the end of the 1940's, many physicists, prominently including Hannes Alfvén and Edward Teller [127], thought that cosmic rays were mostly locally produced and confined to the solar system by magnetic fields. However, high-energy particles are hard to produce by the energy output of main sequence stars, which is predominantly in the form of low-energy photons and neutrinos. Enrico Fermi was thus led by energetic considerations [128] to postulate powerful sources in the whole galaxy.

The energy density of cosmic rays in interstellar space is about $\rho_E \simeq 1 \, \text{keV/m}^3$. The power required to generate such an energy in steady state is:

$$L_{CR} = \frac{V \rho_E}{\tau} \simeq 3 \times 10^{43} \, \frac{\text{GeV}}{\text{s}}$$

The volume of the galactic disk is $V = \pi R^2 d \simeq \pi (15 \, \text{kpc})^2 (200 \, \text{pc}) \simeq 4 \times 10^{63} \, \text{m}^3$. The residence time τ of cosmic rays in the galaxy is about 6×10^6 a (see Chapter 9). Stars cannot provide such a power in the right form, but supernovae can.

A single core collapse supernova of type II every 30 years in the Milky Way, which ejects 10 solar masses at a typical velocity of 5×10^6 m/s, will provide of the order of $L_{SN} \simeq 2 \times 10^{45}$ GeV/s, exceeding the required power hundredfold. Thus a relatively modest efficiency in transferring this energy to cosmic rays will suffice.

The total energy output of stars is comparable. The several hundred billion stars in the Milky Way produce of the order of 3.1×10^{46} GeV/s, but mainly in the form of photons with energies of the order of eV. One is thus hard pressed to imagine a mechanism to accelerate cosmic rays to TeV energies and beyond. These photons, however, are energetic enough to ionise interstellar matter. They and other ambient photons are also targets for the so-called inverse Compton interaction, by which energetic cosmic rays can promote them to high energies (see Chapter 7).

Inspired by such considerations, Enrico Fermi was the first to propose viable mechanisms of acceleration for charged cosmic rays [128, 148]. The density of charged particles in interstellar plasma is low, and so are the corresponding electromagnetic fields. But since the distances over which they act are truly astronomic, moving magnetic fields provide the necessary means of acceleration. Charged particles curl around the magnetic field lines on a helical path, a superposition of circular motion orthogonal to the local field and a streaming motion along the field direction, as shown in Focus Box 5.1. The radius of the circular motion is called the gyro- or Larmor radius. It is large, typically of the order of 10^7 m for an electron, 10^{10} m for a proton, but small on an astronomical scale. On that scale, charged particles are thus bound to follow magnetic field lines. When the field strengthens as a function of space or time, this can lead to a quasi-elastic reflection of the charged particle, as also

DOI: 10.1201/9781003181385-5

described in Focus Box 5.1. When the field strength increases, the gyroradius decreases and the gyration velocity increases. At the same time, the streaming velocity decreases to conserve energy. If the magnetic field is strong enough, this can eventually stop the streaming and reverse its direction. Such a magnetic field is called a magnetic mirror. Other reversals of direction can occur when the field direction itself reverses. Thus, elastic reflection can occur on sufficiently strong and variable magnetic fields, without direct contact to other particles. Such scatters are therefore sometimes called contactless, to distinguish them from direct elastic scattering off a material target.

Highly variable magnetic fields occur, when a violent event like a supernova throws out ionised matter in large quantities. Turbulent magnetic fields subsequently expand through space with the ejected plasma in the form of clouds or winds. In fact, external magnetic fields are "frozen" inside the plasma, i.e. they travel with the plasma material itself, as shown in Focus Box 5.2. Immediately following the explosion, shock waves are ripping through the interstellar material. This occurs, when the velocity of ejected material exceeds the speed of sound in the interstellar plasma (see Focus Box 5.5). The ejection velocity can be an appreciable fraction of the speed of light, such that shock waves are the normal case. Upstream of their passage, they leave a heated and turbulent plasma, the ideal environment for contactless reflection, thus accelerating charged particles efficiently.

A single such reflection will not normally convey a GeV or TeV energy to a proton. However, particles will be reflected to-and-fro through the shock front by reflection off magnetic mirrors on either side: slow moving ones downstream and fast ones upstream. Such a ping pong mechanism is a very powerful accelerator. Consider a process which increases the energy of a particle by a fraction ϵ, such that the energy after the encounter is $E_1 = E_0(1 + \epsilon)$, compared to the original energy E_0. After n such encounters, the particle energy will thus be $E_n = E_0(1+\epsilon)^n$. Conversely, the number of encounters it takes to reach a given E_n is:

$$n = \frac{\log E_n/E_0}{1 + \epsilon}$$

Suppose that the probability to escape from further encounters after each one is p. Than the number of particles which reach at least an energy of E_n out of a total of N_0 particles is:

$$
\begin{aligned}
N(E > E_n) &= N_0 \sum_{m=n}^{\infty} (1 - p)^m \\
&= N_0 \frac{(1 - p)^n}{p} \\
&\propto \frac{1}{p} \left(\frac{E_n}{E_0} \right)^{-\gamma}
\end{aligned}
$$

A power law spectrum thus naturally arises from a sequence of stochastic acceleration processes. The exponent of the power law is $\gamma = \log \frac{1}{(1-p)} / \log (1 + \epsilon) \simeq p/\epsilon$. However, this prediction is based on the assumptions that both the relative energy gain per collision, ϵ, and the escape probability, p, are independent of energy. While the first assumption sounds reasonable, the second is less obviously fulfilled. The escape probability is the ratio of the time interval Δt between two scatters and the time τ it takes to leave the accelerating structure. The average time between two scatters is given by the scattering length, $\Delta t \propto r_g/c$, where r_g is the Larmor radius (see Focus Box 5.1). The escape time, on the contrary, is inversely proportional to the Larmor radius, such that normally, we expect $p \propto r_g^2 \propto E^2$. However, in first order Fermi acceleration this is not the case, and the escape probability is independent of energy, as we will see below.

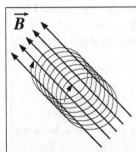

Charged particles spiral around the field lines of a magnetic field as shown in the sketch on the left. In a homogeneous magnetic field of strength B, particles with electric charge q and velocity $v_\perp$ perpendicular to the direction of the field follow a circular path around the magnetic field lines with the gyro-, Larmor or cyclotron radius:

$$r_g = \frac{\gamma m v_\perp}{|q|B}$$

Along the field direction, the particle streams with constant velocity, such that a spiralling motion results. The rotational frequency is $\omega_g = (|q|B)/(\gamma m)$, with the particle mass m and its relativistic factor γ. Under astronomical conditions, the gyroradius is roughly $r_g \simeq 1.7 \times 10^9 \gamma (10^{-6}/B)(m/m_e)$ cm for particles of unit charge, scaling with their mass relative to the electron mass m_e. In a typical interstellar field of $1\,\mu$G, this corresponds to $1.7 \times 10^7 \gamma$ m for an electron or $3.4 \times 10^{10} \gamma$ m for a proton. The gyroradius is thus small enough on an astronomical scale to bind most particles to the field direction, provided that the field extends over dimensions much larger than r_g.

When the field strength increases, r_g decreases accordingly. If the energy of the particle is constant, its transverse velocity increases and the streaming velocity $v_\parallel$ slows down. To lowest order, the magnetic moment of the particle $\mu = m v_\perp^2/(2B)$ stays constant. The total energy of the particle is:

$$\begin{aligned} E &= qU + \frac{1}{2}mv_\parallel^2 + \frac{1}{2}mv_\perp^2 \\ &= qU + \frac{1}{2}mv_\parallel^2 + \mu B \end{aligned}$$

where U is an electric potential, which may be present in addition. The streaming velocity as a function of energy is thus:

$$v_\parallel = \pm\sqrt{\frac{2}{m}(E - \mu B - eU)}$$

In the absence of an electric potential, the streaming thus stops and reverses direction at bounce points, where $E = \mu B$.

Consider a region of a roughly constant field with strength B_{min} and a converging region approaching B_{max}, as sketched on the right. Particles are reflected at a mirror point under the so-called trapping condition:

$$\left|\frac{v_\parallel}{v_\perp}\right| < \sqrt{\frac{B_{max}}{B_{min}} - 1}$$

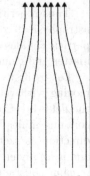

Note that the energy of the particle in the rest system of a magnetic field constant in time does not change, particles are thus reflected elastically. The same reversal of particle direction occurs when the field lines curve with a bend radius larger than the gyroradius.

When field lines converge at two ends, like in a dipolar field, they form a magnetic bottle which can store charged particles, as shown in Section 8.9.1 for the example of the Van Allen belts in the terrestrial magnetic field.

Focus Box 5.1: Magnetic mirrors

One often reads that the magnetic flux is "frozen" inside a collision-free plasma. This means that the magnetic field lines move with the plasma itself. During plasma motion, the magnetic flux Φ, the number of field lines crossing a perpendicular surface, stays constant:

$$\Phi = \int \vec{B}\, d\vec{S} = \text{const}$$

Here $\vec{B}$ is the magnetic field vector and $\vec{S}$ is a vector perpendicular to a surface of size S, as shown in the sketch below. Strictly speaking, the magnetic flux only stays constant for a plasma with infinite conductivity.

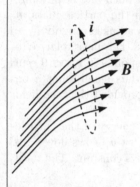

The electromotive force $\mathcal{E}$ along a (virtual) closed conductor loop delimiting the surface is $\mathcal{E} = -d\Phi/dt$. Since charges fill the plasma volume, such loops indeed exist everywhere. The magnetic flux is the sum of the externally supplied one, Φ_{ext} and the one generated by the conducting loop itself, $\Phi_{int} = Li$, where L is the loop's inductance and i the induced current:

$$\Phi = \Phi_{int} + \Phi_{ext}$$
$$\mathcal{E} = -\frac{d(\Phi_{int} + \Phi_{ext})}{d}$$
$$Ri = -L\frac{di}{dt} - \frac{d\Phi_{ext}}{dt}$$

As the resistance of the loop approaches zero, $R \to 0$, any change of the external magnetic flux will be fully compensated by a corresponding change in the flux generated by the loop itself:

$$\frac{d\Phi}{dt} = \frac{d}{dt}\left(\Phi_{ext} + \Phi_{int}\right) = 0$$

Thus when a perfectly conducting plasma moves, the magnetic flux and thus the field lines move with it. Perfect conduction may remind you of solid state superconductors, but there is an important difference: superconductors expel magnetic fields completely, due to the Meissner effect [818].

Focus Box 5.2: Magnetic flux freezing

Thus repetitive small accelerations lead to a power law distribution of energies. In the astrophysical context, such accelerations arise in two likely conditions: when a charged particle collides with a moving plasma cloud, and when a shock wave rips through interstellar matter and clouds.

Let us first consider what happens when a charged particle encounters a plasma cloud moving with velocity β_{cl} and containing turbulent magnetic fields. Turbulent means that the field directions are more or less randomly distributed inside the cloud. The cloud thus contains multiple magnetic mirrors reflecting the incoming particle and randomising its direction as sketched in Figure 5.1. When the particle is encountering the cloud head-on and is finally emitted backwards, it gains energy. In the rest system of the cloud, its energy remains constant, but the motion of the high mass cloud accelerates the particle like a table tennis bat accelerates the light ball. The energy gain in the laboratory system is proportional to β_{cl}^2, as quantified in Focus Box 5.3.

Things change substantially when a shock wave rips through the interstellar matter and plasma. In general, a shock wave is the disturbance of a medium by an object which moves

We discuss the energy gain of a particle colliding with a plasma cloud following the approach of Gaisser and Engel [660, p.239 ff]. Consider a particle moving in a "laboratory" system arbitrarily fixed to some close-by star or galaxy. Let it have four-momentum $p_0 = (E_0, \vec{p}_0)$ and already be relativistic such that $E_0 \simeq |\vec{p}_0|$. Let it encounter a cloud of infinite mass which moves with a velocity $\vec{v}_{cl}$ under an angle θ_0, as sketched in Figure 5.1. In the rest system of the cloud, the energy of the particle is:

$$E_0^* = \gamma_{cl} \left(E_0 - \beta_{cl} \cos \theta_0 p_0 \right) = \gamma_{cl} E_0 \left(1 - \beta_{cl} \cos \theta_0 \right)$$

Since there are only magnetic interactions in the cloud, the energy of the existing particle will be unchanged, $E_1^* = E_0^*$, while its outgoing angle will be θ_1^*, all in the cloud's rest system. The energy in the "laboratory" will thus be:

$$E_1 = \gamma_{cl} E_0^* \left(1 + \beta_{cl} \cos \theta_1^* \right)$$

Thus the energy gain in a single backscatter is:

$$\epsilon = \frac{E_1 - E_0}{E_0} = \frac{1 - \beta_{cl} \cos \theta_0 + \beta_{cl} \cos \theta_1^* - \beta_{cl}^2 \cos \theta_0 \cos \theta_1^*}{1 - \beta_{cl}^2} - 1$$

Note that for relativistic clouds, the maximum energy of the reflected particle is as high as $E_1 = 2\gamma_{cl}^2 E_0$. However, the energy change strongly depends on the angles of incidence and exit. The final direction in the cloud rest system is fully random. Thus, even in a head-on collision, the particle can leave in its original direction, with unchanged energy. It can also loose energy when it overtakes the cloud. However, the flux is larger in the head-on case, such that on average there will be an energy gain.

Let us consider this average case. Since there are many scatters with the turbulent magnetic fields inside the cloud, the final angle is distributed randomly over $-1 \leq \cos \theta_1^* \leq 1$ and $\langle \cos \theta_1^* \rangle = 0$. Thus the average energy gain will be:

$$\langle \epsilon \rangle = \frac{1 - \beta_{cl} \cos \theta_0}{1 - \beta_{cl}^2} - 1$$

For the incident angle, we have to consider the flux of incoming projectiles, cosmic rays with the speed of light, relative to the moving target, the plasma cloud. The number of collisions n is proportional to the relative velocity of cloud and particle, like for two incident particles (see Focus Box 4.1):

$$\frac{dn}{d\cos\theta_0} = \frac{1 - \beta_{cl} \cos \theta_0}{2}$$

The incident angle extends over the range $-1 \leq \cos\theta_0 \leq 1$, since the cosmic ray can impact the cloud also from the sides or from behind. The average angle is thus $\langle \cos\theta_0 \rangle = -\beta_{cl}/3$ and we expect an average energy gain of:

$$\langle \epsilon \rangle = \frac{1 + \frac{1}{3}\beta_{cl}^2}{1 - \beta_{cl}^2} - 1 \simeq \frac{4}{3}\beta_{cl}^2$$

The approximation is valid for non-relativistic clouds, the energy gain is proportional to the square of the cloud velocity. Therefore the mechanism is called second order Fermi acceleration. Since typically $\beta_{cl} \simeq 10^{-4}$, it is in general not a very efficient accelerator. However, exceptions exist for highly relativistic clouds (see Section 5.4).

Focus Box 5.3: Second order Fermi acceleration

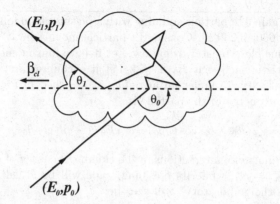

Figure 5.1 Sketch showing the acceleration of charged particles by multiple magnetic mirrors contained in a moving plasma cloud. Note that particles entering the cloud from behind, i.e. in the direction of the cloud velocity, will be decelerated.

faster than the speed of sound, i.e. the speed of ordinary density waves. In our context, we are interested in shock waves caused by supernova explosions, which eject many solar masses worth of matter, with velocities which reach several percent of the speed of light. The shock front thus moves orders of magnitude faster than the usual accelerating plasma clouds and leaves highly perturbed plasma behind. How this contributes to particle acceleration is sketched in Figure 5.2. The shock front, which is practically planar at sufficient distance from the supernova, moves with high velocity, of the order of $\beta_s \simeq 0.1$. It leaves behind multiple shocked plasma clouds. One of them moves away from the shock front with velocity β_{cl} relative to the shock front. However, $\beta_{cl} \ll \beta_s$, it is thus dragged forward by the shock wave. How an encounter with such a shocked cloud accelerates particles is explained in Focus Box 5.4. It turns out that the energy gain is linear in β_{cl}, the cloud velocity, which is now just a little smaller than the large shock speed. Thus, this accelerator is much more efficient than the encounter without shock. It can indeed take particles to the high energies observed in comic rays.

As further added value, the escape probability is independent of the particle energy and only depends on the shock velocity. Due to the ping-pong mechanism, p is indeed exponential in the shock velocity, $p \propto e^{-\beta_s}$ [760].

In any case, multiple passages through the shock front are still required to reach very high energies. But interstellar matter and plasma accommodates such a ping-pong motion back and forth through the shock front, as sketched in Figure 5.2. The acceleration then resembles a particle bouncing back and forth between two converging walls.

Much more insight into the complex plasma physics of shock acceleration is gained through detailed simulations. They generally confirm and substantiate the qualitative picture sketched above; for a review see e.g. [546]. Simulations show that magnetic fields are substantially perturbed and enhanced through the action of the ion current on the surrounding plasma [341], especially for high Mach numbers characterising the shock.[1] Highest energies are obtained for shocks proceeding parallel to the background magnetic field around which ions and electrons gyrate, and for ions reflected backwards on their first encounter, i.e. the situation shown in Figure 5.2.

---

[1]The Mach number is defined as the ratio of the shock speed and the speed of sound in the medium.

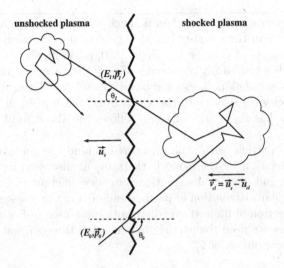

Figure 5.2 Sketch showing the acceleration of charged particles by plasma clouds perturbed by a passing shock front, symbolised by the thick zigzag line.

5.2 SUPERNOVAE

Supernovae[2] are classified by their observational properties since the time of Baade and Zwicky [76, 77, 113]. Rather than this taxonomy, what is of interest for us here is the physical mechanism leading to the supersonic shock waves, which efficiently accelerate cosmic rays, as we have seen above. There are two basic types of ignition mechanisms – and thus progenitors – which cause these powerful explosions and eject vast amounts of nuclei: thermal runaway in binary star systems, responsible for type Ia supernovae; and core collapse of massive stars at their end of life, responsible for all other types.

Let us start with the first mechanism, which causes type Ia supernovae. A binary system consists of two normal (i.e. main sequence) stars rotating around a common center-of-mass. Such systems are quite common. At the end of life of the more massive star, it will become a red giant, expanding its radius enormously. Part of its outer envelope will be attracted to its companion, which in turn becomes more massive, until both rotate inside a common gas envelope. The more massive companion collapses to a white dwarf and the envelope is ejected outwards. The two stars are now much closer to each other. As the less massive companion ages and expands, the white dwarf accretes enough matter to re-ignite fusion in a thermal run-away process when approaching the Chandrasekhar limit defined in Focus Box 4.3. This can cause its complete disruption, while the companion is ejected away. Alternatively, only part of the mass can be ejected and the remainder can form a neutron star or a black hole, depending on mass. By angular momentum conservation, both will rotate quickly since the system became more and more compact during the process. Fast rotating pulsars are contenders for the production of energetic electron-positron pairs as explained in Focus Box 4.7.

The shock wave created by a type Ia supernbova explosion is powered by of the order of a few 10^{44} J and reaches a few percent of the speed of light, much faster than the speed of sound in the surrounding plasma (see Focus Box 5.5). Due to the more or less fixed mass limits at which the event is initiated, type Ia supernovae follow a characteristic light curve,

[2]A concise account of the history of the supernova concept in relation to neutron stars and cosmic rays is given by Adam S. Brown [545].

First order Fermi acceleration occurs when a shock wave rips through interstellar matter with a velocity larger than the speed of sound, e.g. after the explosion of a supernova. We again follow the arguments of Gaisser and Engel [660, p.239 ff], sketched in Figure 5.2.

The shock front, which is practically planar at sufficient distance from the supernova, moves with velocity $\vec{u}_s$ and leaves behind a shocked plasma cloud. The latter moves away from the shock front with velocity $\vec{u}_{cl}$ relative to the shock front, but $|u_{cl}| \ll |u_s|$. In the laboratory frame, the cloud thus has $\vec{v}_{cl} = \vec{u}_s - \vec{u}_{cl}$ and follows the shock front with slightly lower velocity.

The encounter of a particle with a plasma cloud behind the shock front is sketched in Figure 5.2. The acceleration by the cloud is the same as discussed in Focus Box 5.3, but the relevant velocities and angular distributions are very different.

First of all, the angular distribution of particle emission in the rest system of the shocked cloud is now the projection of its isotropic distribution (as used in Focus Box 5.3) onto the plane shock front, since we need the particle to exit again through that font. We thus have to average over the distribution of θ_1^*:

$$\frac{dn}{d\cos\theta_1^*} = 2\cos\theta_1^*$$

in the restricted range $0 \leq \cos\theta_1^* \leq 1$ such that $\langle\cos\theta_1^*\rangle = 2/3$. Substituting this into the general expression derived in Focus Box 5.3, we thus have a relative energy gain of:

$$\langle\epsilon\rangle = \langle\frac{E_1 - E_0}{E_0}\rangle = \frac{1 - \beta_{cl}\cos\theta_0 + \frac{2}{3}\beta_{cl} - \frac{2}{3}\beta_{cl}^2\cos\theta_0}{1 - \beta_{cl}^2} - 1$$

as a function of the angle of incidence θ_0 in the laboratory system. This encounter is the one of a particle moving at the speed of light with random orientation and the plane shock front. The angular distribution thus follows from the projection of an isotropic flux onto a plane, with the range $-1 \leq \cos\theta_0 \leq 0$. We thus have $\langle\cos\theta_0\rangle = -2/3$, regardless of the shock wave velocity. The average energy gain is then:

$$\langle\epsilon\rangle = \frac{1 + \frac{4}{3}\beta_{cl} + \frac{4}{9}\beta_{cl}^2}{1 - \beta_{cl}^2} - 1 \simeq \frac{4}{3}\beta_{cl}$$

with the approximation valid for non-relativistic shock waves. The mechanism is thus called first order Fermi acceleration, since the energy gain is proportional to velocity.

Moreover, the cloud velocity is now $\beta_{cl} = (u_s - u_{cl})/c$ and thus contains the velocity of the shock front, which is of the oder of a few percent of the speed of light, much larger than the cloud velocity in unshocked interstellar matter. This is thus an efficient accelerator to propel cosmic rays to high energies.

Focus Box 5.4: First order Fermi acceleration

such that they can serve as standard candles used in measuring the speed at which the Universe expands [412, p. 45–57].

A second proposed mechanism involves the merger of two white dwarfs, from a binary system which has lost rotational energy over time. The combined mass then exceeds the Chandrasekhar limit and thermal runaway is possible. The light curve of such a supernova will not resemble the standard candle described above, but will vary more widely [475].

Single stars can undergo core collapse when their nuclear fuel runs out, as already discussed in Chapter 4. Such an evolutionary stage is responsible for all supernova types except the type Ia discussed above. There are several pathways to core collapse. When a

Plasma evolving in vacuum behaves much like a liquid. Though it has low density, it resists against compression more than a neutral gas. It also carries its own currents and thus transports electric and magnetic fields. Treating plasma in an approximation using collective variables like density and pressure is called magneto-hydrodynamics (MHD).

Much like sound in gases and liquids, density waves can travel through a plasma. In MHD they are called Alfvén waves [117]. The inertia is created by the mass of the plasma ions; the restoring force comes from the "tension" of the magnetic field lines, i.e. their resistance to being bent. The wave appears in the interplay between ion motion and perturbation of the magnetic field, both transverse to the wave propagation. The wave propagates along the general direction of the magnetic field lines and suffers no dispersion.

In the non-relativistic case relevant for us, propagation (group) velocity and phase velocity of the wave are the same. The propagation speed is called the Alfvén speed:

$$v_A = \frac{B}{\sqrt{\mu_0 \rho}}$$

where B is the magnetic field strength, μ_0 the magnetic permeability of free space and ρ the total mass density. In a plasma with several components, the density it is given by $\rho = \sum_s n_s M_s$, with the number density n_s and the mass M_s of each species. At high magnetic field and low density, the Alfvén speed approaches the speed of light and the wave becomes an ordinary, field-dominated electromagnetic wave.

Focus Box 5.5: Alfvén waves

massive star – above about 10 solar masses – develops an iron core with a mass exceeding the Chandrasekhar limit, its core will no longer withstand gravitational collapse by electron degeneracy pressure (see Section 4.2). It will violently eject its outer layers and collapse to a neutron star or a black hole depending on its mass and elemental composition. Other mechanisms leading to a failure to maintain electron degeneracy include the following:

- Electron capture in an oxygen-neon-magnesium core.

- Electron-positron pair production in a large active core, taking away thermal pressure and causing collapse towards thermal runaway fusion.

- A large flux of energetic gamma rays in very massive stars, disintegrating fused nuclei and causing a complete collapse with or without supernova explosion.

Except for the first of these alternatives, which occurs for stars of less than about 10 solar masses, these processes are only relevant for very heavy stars. If the star is not all that massive, neutron degeneracy may halt the collapse at a radius of some 30 km, with a density comparable to that of an atomic nucleus. If the star is heavier than about 15 solar masses, neutron degeneracy pressure is insufficient and a black hole results. Details about the end-of-life of massive stars are found e.g. in reference [331]. Possible outcomes include Gamma Ray Bursts (GRB) as well as supernovae.

Core collapse supernovae exceed the total energy of type Ia supernovae by two orders of magnitude, reaching 10^{46} J, while the ejected mass is comparable. The remaining compact object may receive a large push away from the epicentre of the system. As an example, Figure 5.3 shows the inner region of the Crab nebula, with its characteristic hat-shaped plasma hiding a pulsar moving with several hundred km/s relative to the nebula. It rotates with about 30 Hz sending out photons of all wavelengths which are received at Earth as a pulsating signal.

Figure 5.3 The central region of the Crab nebula, around the remnant pulsar. A jet emerges towards the bottom from the hat-shaped surrounding plasma. The nebula was produced by a type II supernova explosion in A.D. 1054. (Credit: NASA/CXC/ASU/J. Hester et al., HST/ASU/J. Hester et al. [403])

Stellar bubbles surrounding high mass, very bright stars provide a good environment for the injection of ions into the shock front, when the central star explodes as a supernova [681]. There may thus be two types of SNR: in interstellar medium and in wind-blown bubbles.

5.3 WINDS

Stars lose part of their mass constantly throughout all phases of their evolution, in the form of photons, electrons, neutrinos, protons and heavier elements (see Section 4.2). Basically all types of stars emit X-rays signalling the existence of hot coronae and stellar winds. The coronae are heated by radiation from the burning part of the star until portions are no longer gravitationally bound. They move away from the star at speeds between 20 and 2000 km/s. A good example for this more or less quiescent type of mass loss is the Sun. The solar wind blows at modest speeds around 200 km/s from quiet regions, but up to 700 km/s from more active ones. The mass loss of light stars like the sun is tiny, about $10^{-13} M_\odot/a$, such that they loose only a small percentage of their total mass throughout their life on the main sequence.

The mass loss is more dramatic for heavy stars. Over their short lifetimes they lose a substantial fraction of their mass, blown off at speeds up to 2000 km/s. Beyond the helium burning phase, heavy elements are produced and subsequently lost by radiation pressure, thus enriching the elemental composition of interstellar matter in heavy elements. As an example, Figure 5.4 shows a Hubble telescope image of Eta Carinae and its twin outflow. Hard X-rays are observed from this extraordinary structure [459], powered by a supermassive star in a central binary system.

It has been argued that the total kinetic energy provided by such winds in the galaxy is up to a quarter of the one produced by supernovae explosions [673]. Winds may thus contribute significantly to the acceleration of cosmic rays. How they may do that is best explained taking our own solar system as an example. Figure 5.5 shows an artist's impression of the heliosphere and its components. The heliosphere is the bubble constantly inflated by the solar wind inside the interstellar medium. Solar wind travels at modest velocities under normal circumstance, but still with supersonic speed. The Earth itself is protected from the corresponding particle flow by its own magnetic field. The physics of the solar wind and the influence of the various magnetic fields on solar and galactic cosmic rays are discussed in Chapter 8.

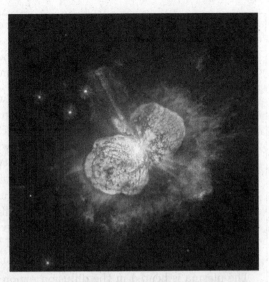

Figure 5.4 Hubble Space Telescope image of the giant star Eta Carinae. It has undergone violent outbursts, including an episode in the 1840s during which ejected material formed the bipolar bubbles of outflow. The bright UV light comes from magnesium rich regions. The intermediate regions are filled with warm gas, whereas the outer regions are nitrogen enriched and accelerated by the outflow's shock front. (Credits: NASA, ESA, N. Smith (University of Arizona) and J. Morse (BoldlyGo Institute))

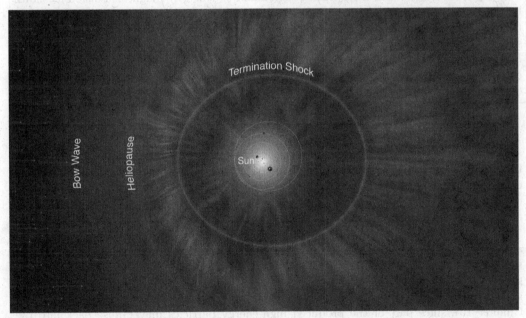

Figure 5.5 Artists impression of the heliosphere with the solar system surrounded by the solar wind and its termination shock. Downstream from the shock front, the solar wind extends until it is stopped at the heliopause by the interstellar medium. The motion of the solar system in the galaxy drives a bow wave in the interstellar medium. (Credit: NASA/IBEX/Adler Planetarium)

What interests us here are the potential acceleration mechanisms associated to stellar plasma. The plasma flow is slowed down by interaction with interstellar matter. At the so-called termination shock front it reaches subsonic velocities. It continues outward until it is stopped (relative to the Sun) at the so-called heliopause. If the sun moves with supersonic speed relative to the interstellar medium, its motion causes a bow shock similar to a fast boat in water. In the more likely case that the Sun's velocity is subsonic, a bow wave is caused instead. As described above in Section 5.1, multiple passages through shock fronts can

Magnetic reconnection is a process which converts magnetic energy into kinetic energy of plasma particles, often in an explosive way. It is associated to astrophysical processes on vastly varying scales, from planetary to stellar and galactic phenomena. It occurs, when opposite magnetic fields frozen into sheets of plasma move towards each other:

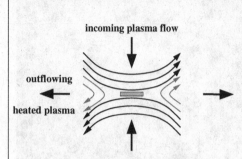

Two plasma sheets carrying oppositely directed magnetic fields convect towards each other as sketched on the left. The magnetic field lines are shown in black. At the boundary layer between the two sheets, a diffusion region forms (grey rectangle). It carries a current density $\vec{J}$ necessary to sustain the change in magnetic field direction, according to Maxwells law $\vec{\nabla} \times \vec{B} = \mu_0 \vec{J}$. The resistivity of the current layer will allow magnetic field lines from opposite sheets to bend over and recombine as shown by the arrows.

The plasma is heated in the diffusion region and expelled by magnetic tension as shown by the horizontal arrows. Magnetic tension occurs for strongly bent field lines and amounts to $(\vec{B} \cdot \vec{\nabla})\vec{B}/\mu_0$. It leads to a rapid outflow of plasma with a velocity close to the Alfvén speed (see Focus Box 5.5). Because of the conservation of matter in the process, the outflow compensates and sustains the inflow of plasma into the diffusion region. Details of the processes happening inside the diffusion region are, however, less clear than the basic principles outlined above [752].

Focus Box 5.6: Magnetic reconnection

accelerate particles [216]. However, since the energy gain per passage is at best proportional to the shock velocity $\beta_{sh} \simeq 10^{-3}$, as shown in Focus Box 5.4, high energies cannot be reached during quiescent phases. But during active phases of the Sun's corona, the magnetic fields of Earth and the solar wind may reconnect as described in Focus Box 5.6, such that high energies and fluxes may be reached for a short period.

Stars approaching their end of life can produce more powerful winds. Magnetic bubbles and their termination shocks have been considered as cosmic rays accelerators [681]. The same is true for galactic winds, resulting from the accumulation of outflow from a complete galaxy [597]. These phenomena may be populating the energy region of cosmic rays between galactic and extragalactic sources, i.e. the region between the "knee" and the "ankle" (see Chapter 10), since part of the produced high-energy particles may diffuse back into the galaxy.

5.4 JETS

In high-energy astrophysics, jets of plasma are a ubiquitous phenomenon. Whenever outflow from a compact object arises in the presence of a strong magnetic field, the outflow gets organised in a linear, transversely limited structure, a jet. A first example we have already noted in the context of pulsars in Focus Box 4.7. There, the magnetic field of a white dwarf remnant from a supernova explosion concentrates and accelerates the outflowing plasma. Indeed, jet-like structures are quite often associated to supernovae, including those powered by accretion in type Ia. We already noticed there how a toroidal component of the magnetic field arises through the rotation of a polar magnetic field.

Jets also arise in the process of star formation [830]. They appear to be instrumental in removing angular momentum from the protoplanetary disk surrounding young stars, such that accretion from the disk may proceed.

Accretion onto black holes is another example. The magnetic field generated around a rotating black hole leads to the formation of jets around the rotation axis, fed by the accretion disk. The process can be approximated in an analytical treatment [780] which exposes the main features: the launching of jets through the accretion process, the acceleration and collimation of jets by the plasma motion. Figure 5.6 shows the magnetic field lines and plasma flow predicted by this magneto-hydrodynamic treatment. The dimensions scale with the gravitational radius of the object, $r_g = GM/c^2$, and so do dimensions in the figure. The magnetic field forms stream surfaces, where the magnetic stream function – analogous to the stream function in a liquid flow – is constant. The stream surface is shown as a transparent surface with an embedded magnetic stream line. In zero force approximation, this surface is also where the plasma and its currents flow. The growing velocity of the plasma flow is shown through the hue of the plasma flow line, giving an example of the centrifugal acceleration already noted in Focus Box 4.7. The magnetic field is dominantly poloidal up to the Alfvén critical surface – where the plasma velocity reaches the Alfvén speed – and dominantly toroidal beyond.

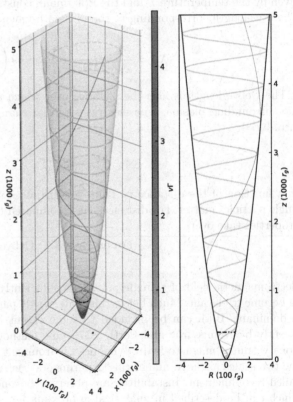

Figure 5.6 Configuration of magnetic field lines and velocity on a magnetic stream surface [780], indicated by the transparent surface on the left panel, emerging from a spinning black hole. It contains the spiralling magnetic field line, the grey scale of which indicates the fourvelocity γv (central scale). The right panel is a 2D projection seen edge-on. The dimensions are scaled by the gravitational radius of the spinning central object. The horizontal dashed line indicates the Afvén critical surface, where the flow velocity is equal to the Alfvén speed.

Since the accretion of the compact object is not necessarily a continuous process, one can also expect time structure in the jet formation. This is signalled e.g. by X-ray flares due to the synchrotron radiation mainly from the electron component of the plasma. Since the amount of matter supplied by accretion may also vary, density variations of the plasma can result. We will come back to this space-time structuring of the jet in the form of plasmoids with closed field lines a bit further on.

The bulk motion of the plasma is, however, not the end of the story. When one looks at what the bulk does to single particles, one needs to change scale. The relevant dimension is the Debye length, defined in Focus Box 5.7. It is the distance, at which the

The length over which a plasma can sustain a charge imbalance by its own dynamics is called the Debye length λ_D. Such an imbalance can e.g. be created when the centre-of-mass of the negative charges is displaced with respect to that of the positive charges by a distance δx. This creates a collective harmonic oscillation of the negative against the positive charges (like on a parallel plate capacitor) with the so-called plasma frequency:

$$\omega_p^2 = \frac{e^2 n_e}{\epsilon_o m_e}$$

where e and m_e are mass and charge of the electrons, n_e their number density and ϵ_0 the vacuum permittivity. To convert this frequency into a wave length, we need the velocity of the wave. In thermal equilibrium, the average velocity of electrons in the x direction is:

$$\frac{1}{2} m_e \overline{v_x^2} = \frac{1}{2} kT$$

given by the temperature T and the Boltzmann constant $k = 8.617333262145 \times 10^{-5}$ eV/K. The wavelength corresponding to the plasma frequency is the Debye length:

$$\lambda_D = \frac{v_x}{\omega_p} = \left(\frac{kT\epsilon_0}{n_e e^2} \right)^{\frac{1}{2}} \simeq 69 \left(\frac{T}{n_e} \right)^{\frac{1}{2}} \text{ m}$$

The Debye length is also the distance over which a single electric charge is shielded by the surrounding plasma. The electric potential changes by δV when a single charge q is introduced:

$$\delta V = \frac{q}{4\pi\epsilon_0} e^{-\frac{\sqrt{2}r}{\lambda_D}}$$

as a function of the distance r. Thus, for distances much smaller than the Debye length, single particles matter. For distances large compared to the Debye length, collective plasma properties take over.

Focus Box 5.7: Debye length

electromagnetic field of a single particle is sufficiently shielded by the surrounding plasma to become irrelevant. Much below this scale, single particles matter, much beyond, the bulk is dominant. Both can be included to some extent in numerical simulation, as reviewed recently by Komissarov and Porth [817]. One method is to embed particle-in-cell simulations within a magneto-hydrodynamics environment and consider the couplings between the two scales [628]. In numerical "experiments", jets show indeed an auto-focussing effect called reconfinement. Instabilities are observed beyond the smooth steady state behaviour which can be described in analytical approximation. Simulation results can thus confront actual observations.

As far as our subject is concerned, jets contain the two accelerating structures we have commented on in Section 5.1 above: relativistic magnetic clouds and shock fronts in which collisionless acceleration can occur [760]. Active galactic nuclei (AGN), also called quasars, are extragalactic examples of supermassive black holes accreting vast amounts of matter and spitting out jets reaching far beyond the host galaxy, with substantial fractions of the speed of light. Inside our galaxy, neutron stars and so-called microquasars – accreting black holes of lesser mass than AGN – beam jets of relativistic velocities. All of these are powerful enough to accelerate particles to high and extreme energies.

Some theorists go so far as to propose jet-like phenomena as the only relevant accelerating structures for cosmic rays. In the so-called "cannonball" model [397], relativistic plasmoids emitted in supernovae, the cannonballs, are held responsible for acceleration by a single magnetic reflection (see Focus Box 5.1). In order to match the energy balance, it is conjectured that most core collapse as well as accretion-driven supernovae emit such highly relativistic closed plasma structures. If the bulk relativistic factor of the cannonball is γ_{cb}, a single elastic reflection can convey to a cosmic ray of mass m a maximum energy of $2\gamma_{cb}^2 mc^2$ (see Focus Box 5.3). This acceleration is the relativistic version of the energy transferred by a racket to a tennis ball. It is thus argued, that observed relativistic factors of the order of $\gamma_{cb} \simeq 10^3$ allow to reach a "knee" energy of a few $(10^6 \times A)\,\text{GeV}$. Note that in contrast to Fermi acceleration, a single scatter suffices to reach high energies. Also, the highest reachable energy of a nucleus is proportional to its mass number A and not its nuclear charge Z as in Fermi acceleration. If sensitive charge and mass measurements for cosmic rays ever reach knee energies, this prediction may be verified. To reach even higher energies, however, re-acceleration according to first order Fermi processes is still required, such that the absolute end of the cosmic ray spectrum is still Z-dependent.

In addition to cosmic rays, cannonballs are conjectured to be responsible for many astrophysical phenomena, like gamma background radiation, gamma-ray bursts and X-ray flashes [381]. A recent review of the status of the cannonball model and a complete list of references may be found in [723].

FURTHER READING

Claudio Chiuderi and Marco Velli, *Basics of Plasma Astrophysics*, Springer (2015)

Malcolm S. Longair, *High-Energy Astrophysics*, 3rd Edition, Cambridge Univ. Press (2011)

Todor Stanev, *High Energy Cosmic Rays*, 3rd Edition, Springer (2021)

6 Particle Transport

In this chapter, we describe what happens to particles as they travel from their acceleration site to Earth. Basic properties of galaxies relevant for particle transport are described with special emphasis on the Milky Way. We then discuss the basic principles of diffusive transport. Interactions with matter and fields along the particle path are discussed. Numerical models of transport are also touched.

Figure 6.1 Image of the Milky Way from the GigaGalaxy Zoom project of the European Southern Observatory, shown edge-on from our perspective on Earth. The projection used shows the general components of our spiral galaxy, including its disc populated by bright, young stars, as well as the central bulge and its satellite galaxies. The disk is about 0.2 to 0.3 kpc thick, its radius extends to about 20 kpc. The bulge around the galactic centre has a few kpc radius and is mostly populated by old stars. (Credit: ESO/S. Brunier)

6.1 GALACTIC PROPERTIES

The physics of matter and magnetic fields in the Milky Way and observational techniques to assess these are discussed in detail e.g. in [462, Cha. 12]. Here we provide a short overview of properties relevant for particle transport. The arms of a spiral galaxy like the Milky Way have a shape close to logarithmic spirals, with distance $r = ae^{\phi\beta}$ to the galactic centre, where a and β are constants and ϕ is the azimuthal angle in the galactic plane. The constant in the exponent, $\beta = 1/\tan p$, is related to the pitch angle p, a measure of how tightly the spiral arms of a galaxy are wound.[1] The interstellar matter of the spiral arms consists of $\sim 90\%$ hydrogen and $\sim 9\%$ helium with the rest heavier elements [316]. Hydrogen comes in the form of atomic hydrogen ($\simeq 85\%$), molecular hydrogen (H_2) and a small faction of ionised hydrogen (H^+). The total density is roughly 1 nucleon/cm^3. Between the arms, the

---

[1]The pitch angle is defined as the angle between the tangent to a spiral arm and a circle about the galactic centre at the same radius. The pitch angle in the Milky Way is about $-11°$.

DOI: 10.1201/9781003181385-6

109

When relativistic charged particles are accelerated, they radiate light. When the acceleration reduces the magnitude of their velocity, this process is called bremsstrahlung. When the acceleration is orthogonal to the velocity vector, as is always the case for magnetic forces, it is called synchrotron radiation. The nomenclature is due to the fact, that the latter effect was first discovered at this type of accelerator [119].

The full treatment is rather complex and can be found e.g. in [295, Cha. 14] and [479, Sec. 171]. The total power P emitted by a particle of mass m, charge e, momentum p, velocity β and relativistic factor γ in circular motion is found to be:

$$P = \frac{2}{3}\frac{e^2}{m^2}\gamma^2\left(\frac{d\vec{p}}{dt}\right)^2 = \frac{2}{3}e^2\frac{\beta^4\gamma^4}{\rho^2}$$

The main features are the following:

- The total power is proportional to $\gamma^4 = (E/m)^4$, such that synchrotron emission mainly concerns high-energy electrons. It is inversely proportional to the square of the bending radius ρ^2.

- The radiation is strongly peaked in the forward direction of a highly relativistic emitting particle. A narrow beam of opening angle $\propto 1/\gamma$ results, which circulates like the beacon of a lighthouse with frequency c/ρ.

- The emitted light is highly polarised parallel to the orbital plane, with a power ratio $P_\parallel/P_\perp \simeq 7$.

- The spectrum of synchrotron light is characterised by the critical frequency $\omega_c = (3/2)(E/m)^3/\rho$. For angular frequencies ω much below ω_c, the intensity is proportional to $\omega^{2/3}$. It has a maximum for $\omega \simeq \omega_c$ and cuts off exponentially much above, $\propto e^{-\omega/\omega_c}$.

Synchrotron emission from electrons with a power law spectrum is treated in Focus Box 7.6. For astrophysical sources, frequencies start in the radio range from $1\,\mathrm{GHz}$ to $1000\,\mathrm{GHz}$, observed with ground-based antennae. Synchrotron spectra extend upwards in frequency over 10 orders of magnitude, observed by space telescopes, covered in Section 7.1.

Focus Box 6.1: Synchrotron radiation

density of matter is reduced by more than half. The total mass of the Milky Way interstellar matter is estimated to be of the order of $6 \times 10^9 M_\odot$, its total mass is about $10^{12} M_\odot$.

In addition to interstellar matter, the interstellar medium (ISM) in which particles propagate includes the local magnetic field. Inside the galactic disk the latter follows the spiral arms. It is measured by two main observation techniques, intensity and polarisation of synchrotron radiation from electrons (see Focus Box 6.1), and polarisation of photons emitted by pulsars. The first method is based on the hypothesis of equipartition of energy between cosmic rays and magnetic fields, as sketched in Focus Box 6.2. It measures an average strength and direction of the magnetic field in the field of view of the measurement. The second method measures strength and direction of magnetic fields along the line of sight towards a point source, through the so-called Faraday rotation of linear polarisation briefly discussed in Focus Box 6.3.

Of course neither matter nor magnetic field simply disappear at the edge of the luminous galaxy. Disk and bulge are surrounded by two roughly spherical halos, one made of normal matter, the other consisting of dark matter. The first consists of stars, sometimes bound in globular clusters. It represents about 1% of the galaxy's mass. The mass of the dark matter

The total and regular magnetic field in the plane of the sky can be determined from synchrotron radiation, if one assumes a relation between the energy of radiation from cosmic ray electrons and that of magnetic fields. The choice of such an equipartition hypothesis is not unique. One can postulate equipartition between the energy density of cosmic rays ϵ_{CR}, and that of the magnetic field ϵ_B, i.e. $\epsilon_{CR} = \epsilon_B$. Or one can assume that the sum of their energy densities reaches a minimum. Both assumptions give very similar results, since there is tight coupling between cosmic ray motion and magnetic fields, such that they ought to exchange energy until an equilibrium is reached. Deviations from equilibrium are however expected, whenever composition or energy change either locally or in time, e.g. when and where a supernova remnant injects fresh particles.

The cosmic ray energy density is given by the sum of all nuclei $\mathcal{N}$, which do not radiate (see Focus Box 6.1), and electrons $e^{\pm}$, which are responsible for the observed radiation. Let the ratio of their energy densities be $\mathcal{K} = \epsilon_{\mathcal{N}}/\epsilon_e$. The minimum total energy hypothesis then gives the magnetic field strength B as a function of the synchrotron luminosity L_ν:

$$B = \left[6\pi G(\mathcal{K}+1)\frac{L_\nu}{V} \right]^{2/7}$$

G is a function of the frequency limits [359], between which the synchrotron luminosity is integrated, as well as the spectral index of the radiation. V is the volume of the source, such that L_ν/V is the synchrotron emissivity, averaged over the observed volume. The equipartition assumption gives a value which is 8% higher. Variations of this formula with lower sensitivity to unknown factors are considered e.g. in [313].

Since the synchrotron spectrum is integrated over the source volume as well as the frequency spectrum, such equipartition arguments indicate the total magnetic field. Observations of polarisation (see Focus Box 6.3), on the other hand, measure the regular field orthogonal to the line of sight [313].

Focus Box 6.2: Equipartition

halo, on the other hand, exceeds the visible mass by about a factor of five and extends radially far beyond the normal matter halo, to several hundred kpc radius [738].

A successful theory of how galactic magnetic fields are generated and maintained by rotating plasma is the dynamo model. The basic mechanism is sketched in Figure 6.2. A disk of conductive matter is rotating around an axis normal to the disk, inside a primordial magnetic field lying in the disk plane. The rotation "drags" the plasma along, such that the angular velocity decreases as a function of radius. The resulting differential rotation deforms the originally parallel field lines, since they are frozen in the plasma, in the way sketched on the right in Figure 6.2. In addition to its regular motion, the plasma is subject to turbulence, caused by the kinetic energy of winds and shocks. In principle, turbulence should weaken the regular field and make it decay rather quickly. However, given the random orientation of turbulent flow, some will bulge magnetic field lines out of the plane. A coupling between poloidal and toroidal field components results, with positive feed back so as to maintain the regular field. Details of the dynamo mechanism, observational methods as well as alternative mechanisms are very pedagogically covered by Widrow [330]. How the model confronts observations is discussed in a recent review by Beck and others [747].

In polar coordinates, the magnetic field strength of the Milky Way is roughly [283]:

$$B(r,\phi) = B_0(r) \cos\left(\phi - \beta \log \frac{r}{r_0} \right)$$

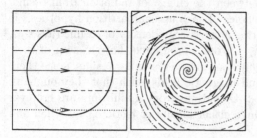

Figure 6.2 Sketch of the dynamo mechanism [330] applied to the generation of galactic magnetic fields. A disk of conducting material is imbedded into an originally homogeneous magnetic field (left graph). Differential rotation deforms the external field as shown in the right graph. The different line styles allow to follow the deformation. The result is a bi-symmetric magnetic field configuration.

with $B_0(r_\odot)$ of a few μG, falling off with inverse distance to the galactic centre, and $r_0 \simeq 10.55$ kpc. Between spiral arms it has a reversed direction, as shown in Figure 6.3. Its symmetry appears to be a bi-symmetric one, as sketched in Figure 6.4. Transverse to the plane it falls off exponentially. Outside the galactic disk, the magnetic field is harder to assess but appears to be poloidal plus toroidal, the latter reversing direction when crossing the galactic plane [328], as schematically shown on the bottom right sketch of Figure 6.4.

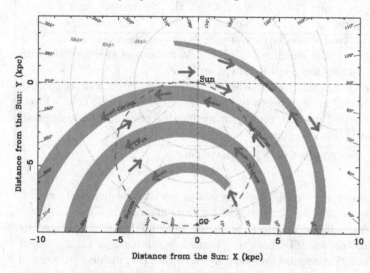

Distance from the Sun: X (kpc)

Figure 6.3 Schematic of the magnetic field directions (arrows) in and between the spiral arms of the Milky Way [377] (grey bands). The position of the Sun and the galactic centre (GC) are indicated. The dashed circle marks the locations of tangential points for equiangular spirals of pitch angle $-11°$.

Measurements are in qualitative agreement with the dynamo theory, although details remain hard to confirm [747]. To give an example [313], the global strength appears to be of the order of $(6 \pm 2)\,\mu$G, with even $(10 \pm 3)\,\mu$G at 3 kpc galactic radius, in excess of what was quoted above. It thus appears that, in addition to the local regular field, important random components are present. However, biases inherent to the observational methods may also exist and explain the discrepancy between measurements based on synchrotron light and pulsar radiation [747].

6.2 DIFFUSION

In our previous discussion, we have identified localised sources of cosmic ray acceleration, like supernova remnants and winds at various scales. So how come the cosmic rays observed e.g. near Earth do not remember their original direction, but arrive at us with a basically isotropic angular distribution? The answer is obviously to be sought in their interaction with the magnetised plasma filling interstellar space. We had found in Chapter 5 that cosmic rays gyrate around the field lines of the ambient magnetic field (see Focus Box 5.1), while

The density of electrons in the interstellar plasma and the typical magnetic field strength are such, that both the plasma frequency ω_p (see Focus Box 5.7), and the cyclotron frequency ω_g (see Focus Box 5.1) are much less than the typical radio frequencies of synchrotron radiation. This means, that synchrotron light, which is strongly polarised, interacts appreciably with plasma electrons.

Imagine a magnetic field pointing along the line of sight between a synchrotron light source and an observer. Electrons gyrate in a plane orthogonal to the line of sight. We can decompose the linearly polarised synchrotron light, travelling along the line of sight, into two circularly polarised waves of opposite helicity. Thus one of them will have an electric field circulating in the direction of electron gyration, the other one opposite to it. It is thus plausible that the velocities of propagation of the two waves in the medium will be different. The plasma is a birefringent medium and the refractive indices of the two partial waves are [462, Sect. 12.3.4]:

$$n_{\pm}^2 = 1 - \frac{(\omega_p/\omega)^2}{1 \pm (\omega_g/\omega)}$$

where ω is the angular frequency of the synchrotron light. For small gyration and plasma frequencies, the difference in refractive index is thus $\Delta n = |n_+ - n_-| = (\omega_p^2 \omega_g / \omega^3)$. After propagating through the medium for a distance l, the angle of linear polarisation will thus have changed by $\Delta\theta$:

$$\Delta\theta = \frac{\pi}{\omega} \int_0^l (\omega_p/\omega)^2 \, dl = (8.12 \times 10^3)\lambda^2 \int_0^l n_e B_{||} \, dl$$

where λ is the synchrotron light wave length in meters, n_e is the electron density in particles per m^3, $B_{||}$ is the component of the magnetic field along the line of sight in Tesla and l is in parsec. In astrophysics jargon, the quantity RM $= \Delta\theta/\lambda^2$ is called the *rotation measure*, directly proportional to the effective magnetic field.

To convert the rotation measure into a magnetic field estimate, the electron density has to be known. It can be independently measured by observing the (column-) density-dependent dispersion of the synchrotron light. The dispersion can be readily measured by observing the arrival time of the periodic pulsar signal as a function of frequency. The arrival time t_1 of an earlier signal at frequency $\nu_1 = \omega_1/(2\pi)$ advances a later signal t_2 at frequency ν_2, such that a dispersion measure DM can be defined as:

$$\mathrm{DM} = \frac{t_2 - t_1}{k_{\mathrm{DM}}\left(\frac{1}{\nu_2^2} - \frac{1}{\nu_1^2}\right)} \propto \int_0^l n_e \, dl$$

with time measured in milliseconds and the proportionality constant $k_{\mathrm{DM}} \simeq (4.149 \times 10^{-6})\,\mathrm{GHz}^2\,\mathrm{pc}^{-1}\,\mathrm{m}^3\,\mathrm{ms}$. The component of the magnetic field in the line of sight is then evaluated from the ratio of rotation and dispersion measures.

Focus Box 6.3: Faraday rotation

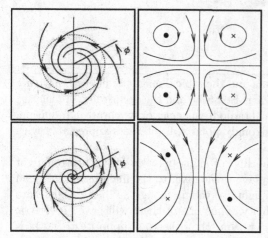

Figure 6.4 Left: Schematics of axisymmetric (top) and bi-symmetric (bottom) magnetic fields in a galactic disk [330], represented in top view onto the galactic plane. Right: Poloidal (lines) and toroidal fields (• and ×) in the galactic halo, with even (top) and odd (bottom) symmetry with respect to the galactic plane [330], represented by the horizontal line.

they stream along the same field line at constant velocity. When the field is regular in nature, their pitch angle, i.e. the angle between their velocity vector and the magnetic field direction, remains constant. However, like all liquid-like matter, plasma motion is subject to excitations in the form of waves. Examples are shown in Figure 6.5. Magnetosonic waves concern the density of plasma, in the direction of the ambient field, transverse to it or both. Alfvén waves periodically wiggle the magnetic field lines transverse to the regular field by separating electrons and ions as explained in Focus Box 5.5.

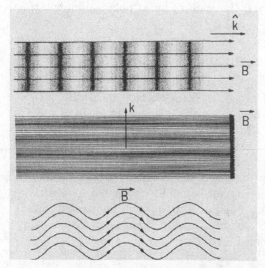

Figure 6.5 Schematic representation of plasma wave types. Magnetosonic waves evolve parallel (top) or perpendicular (middle) to the regular magnetic field direction. Alfvén waves (bottom) evolve parallel or oblique to the regular magnetic field direction.

Let us concentrate on the latter type. An Alfvén wave introduces a component of the magnetic field which is transverse to the streaming direction. If the frequency of the wave is close enough to the Larmor frequency of a cosmic ray, the streaming velocity is diminished by the magnetic force as sketched in Figure 6.6. To conserve energy, the transverse component of the velocity must increase and so does the Larmor radius. The particle pitch angle is thus changed, a scatter has occurred. Repeated scatters of this type thus randomise the particle direction and make it forget where it comes from.

We must not forget that not only the plasma acts on cosmic rays, but the streaming cosmic rays themselves constitute a current acting back on the plasma. This way, instabilities occur which cause turbulence and amplify the scattering process. It is thus plausible that the transport of cosmic rays through interstellar matter and fields is best described as a

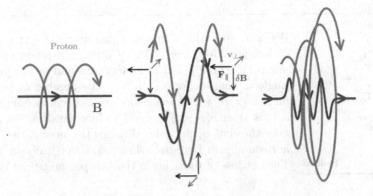

Figure 6.6 Left: Path of a proton, gyrating around and streaming along a regular magnetic field line. Middle: Interaction with a single cycle of an Alfvén wave, which excites a transverse component δB of the magnetic field. If Larmor frequency and frequency of the wave are sufficiently close, the magnetic force $F_{\parallel}$ will act against the streaming direction of the proton. The streaming velocity will decrease, increasing transverse velocity $v_{\perp}$ and Larmor radius to conserve energy. Right: Successive scatters will thus modify the particle pitch angle. (Credit: S. Jacob and C. Pfrommer, 2021, reprinted by permission)

random walk, a diffusive process. The basic physics of diffusion in space and in momentum is sketched in Focus Box 6.4. The resulting differential transport equation is generically called diffusion equation or transport equation. Since it is a linear differential equation, it can in principle be solved analytically. However, analytical solutions only exist in special cases, which are not applicable to the problem of cosmic ray transport. Moreover, most of the intervening parameters are neither independent of position and momentum, nor are they well known in theory or experimentally. Instead, they must be adjusted using cosmic ray properties measured near Earth, e.g. by the spectra of nuclei which are known to be mostly produced as secondaries during particle transport. In Chapter 9, we will explain how this is done. In addition, as mentioned earlier, the low-energy part of cosmic ray spectra observed in the Solar system is modulated by the Sun's magnetic field. We will explain the mechanism and how its effects are corrected in Chapter 8.

·In commonly adopted diffusion models of cosmic rays in the galaxy, cosmic rays are assumed to diffuse in a large galactic halo of scale height[2] H enclosing a thin galactic disc of scale height $h \ll H$ and radius R_d which hosts cosmic ray sources and the solar system. Cosmic rays propagate inside a cylindrical box of height $2H$ and radius R_d. The diffusion transport equation is then solved imposing that cosmic rays at the boundary of the galactic halo escape freely, that is $\rho = 0$ [638]. We will show in Chapter 9 how cosmic ray spectra observed near Earth relate to the spectra injected at the source and to the diffusion coefficient, the galactic halo size parameters and the cosmic ray residence time in the galaxy.

There is an extremely simplified diffusion model, the leaky box model, which can be solved analytically. This model assumes that both cosmic ray density and the density of interstellar matter are uniform in the propagation volume, which is obviously not the case. It replaces the diffusion term in the transport equation with a phenomenological mean path length λ_{esc} through the ISM. Cosmic rays propagate freely inside a cylindrical box. When

---

[2]A scale height is the distance over which a quantity decreases by a factor of e. Here, this quantity is the density of interstellar matter.

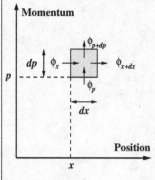

We introduce the diffusion equation following the approach of Longair [462, Sec. 7.5.2]. Consider a phase space density $\rho(x,p,t)$, where p is the momentum and x the position in a single dimension for simplicity. The number of particles in a box at (x,p) with dimensions dx and dp, as sketched on the left, is then $\rho(x,p,t)\,dx\,dp$. The flux densities Φ_x and Φ_p denote the flow of particles through the boundaries of the box, horizontally in the space dimension, vertically in momentum. The net loss of particles in the box per unit time is then:

$$\frac{d\rho}{dt} = Q(x,p,t) - \frac{\partial \Phi_x}{\partial x} - \frac{\partial \Phi_p}{\partial p} = Q(x,p,t) + \frac{\partial}{\partial x}\left(D_{xx}\frac{\partial \rho}{\partial x}\right) - \frac{\partial}{\partial p}(\dot{p}\rho)$$

where $Q(x,p,t)$ is the injection rate of particles inside the box. A steady state is reached when the left side of the equation is zero. The spatial flux density is proportional to the gradient of the particle density, $\Phi_x = -D_{xx}\partial\rho/\partial x$. The proportionality factor is called the diffusion coefficient, with dimension length^2/time. In general, it depends on the location x and the particle velocity and rigidity, which determines the Larmor radius. In particular, it scales with the particle rigidity as $D_{xx} \propto R^\delta$, where δ is related to the spectral index of the energy density of interstellar turbulence [393]. Similarly, the momentum flux density is proportional to the rate of momentum loss or gain, $\dot{p} = \partial p/\partial t$, $\Phi_p = \dot{p}\rho$.

In the general case, more effects enter the diffusion equation. For a cosmic ray species i it can be written as [393]:

$$\frac{d\rho_i}{dt} = \underbrace{Q_i(\vec{r},p)}_{\text{source}} + \underbrace{\nabla\left(D_{xx}\nabla\rho_i - v_p\rho_i\right)}_{\text{diffusion-convection}} + \underbrace{\frac{\partial}{\partial p}\left[\dot{p}\rho_i - \frac{p}{3}(\nabla v_p)\rho_i\right]}_{\text{momentum loss}} + \underbrace{\frac{\partial}{\partial p}\left[p^2 D_{pp}\frac{\partial}{\partial p}\frac{\rho_i}{p^2}\right]}_{\text{reacceleration}}$$

$$\underbrace{- \frac{\rho_i}{\tau_i} + \sum_j \frac{b_{ji}\rho_j}{\tau_j}}_{\text{decay}} \underbrace{- \rho_i v_i \rho_g \sigma_i + \sum_j \rho_j v_j \rho_g \sigma_{ji}}_{\text{spallation}}$$

The motion of the plasma itself is taken into account by a convection term $-v_p\nabla\rho$, where v_p is the plasma velocity. If diffusion coefficient or momentum loss rate are not isotropic, they need to be represented by tensors, as e.g. shown in Focus Box 8.2. Momentum loss can come about by ionisation, bremsstrahlung or synchrotron radiation (first term), or by adiabatic interactions with the plasma (second term). Reacceleration processes like those discussed in Chapter 5 are described by a momentum diffusion term, with a diffusion coefficient $D_{pp} \propto p^2/D_{xx}$ where the proportionality factor depends on the specific reacceleration model.

For an unstable species i, the density ρ_i is diminished by decay, with a loss rate $-\rho_i/\tau_i$, where τ_i is the particle lifetime. The decays of heavier mother particles j cause a gain rate of $+\sum_j \rho_j b_{ji}/\tau_j$, where b_{ji} is the branching fraction of the decay $j \to i$. Analogous gain and loss terms come from inelastic scattering processes off interstellar matter, like spallation. The cross section σ_i applies to loss reactions of species i, σ_{ji} to processes feeding $j \to i$. In either case, the product ρv is the incoming flux of projectiles, ρ_g is the target density. All of the above processes cause a distortion of the momentum spectrum. Analytical solutions of the diffusion equation exist for very simple cases. In general, it is better solved numerically.

Focus Box 6.4: Cosmic ray transport equation

they encounter the boundary of the box, they are either scattered back, or they leak from the box with a probability which is the inverse of a constant, the escape time:

$$\tau_{esc}(E) = \frac{\lambda_{esc}}{\langle M \rangle n v}$$

where $\langle M \rangle$ and n are the average atomic mass and number density of interstellar matter and v is the cosmic ray particle velocity. The escape time depends on energy E (or rigidity R), and so does a constant source rate $Q(E)$ corresponding to the rate at which particles are injected inside the box. When one neglects all interactions and decays, the cosmic ray particle density $\rho(E)$ evolves as

$$\frac{\partial \rho}{\partial t} = -\frac{\rho}{\tau_{esc}} + Q$$

In steady state the first loss term is equal to the second source term. This simplistic model can be used to get rough estimates of basic properties like the total power and residence time of cosmic rays in the galaxy which we mentioned in Chapter 5.

Numerical propagation models go much beyond the primitive approach of the leaky box model in following cosmic rays from their acceleration site across the Milky Way. Well known codes include the following:

- GALPROP [289, 393] is a market leader in the field, which is regularly updated. The web site https://galprop.stanford.edu contains all necessary information, a complete bibliography and access to code and input data.

- DRAGON [398] is a contender, which shares data, some software and cross section tables with GALPROP. The released version and information about more recent beta-test versions are stored on the server https://github.com/cosmicrays/DRAGON.

- USINE [739] is a library of semi-analytical propagation models which range from simple leaky box to 1- and 2-dimensional diffusion models. Access to the code and documentation is provided through https://dmaurin.gitlab.io/USINE/.

- PICARD [527] explicitly computes steady state solutions to the diffusion equation, using its own numerical solver. The code is relatively new and constantly updated. Information is available at the author's home page:
  https://www.uibk.ac.at/astro/research_groups/ralf-kissmann/.

- CRPropa 3 [584] is a code designed for efficient development of astrophysical predictions for ultra-high-energy particles. See https://crpropa.desy.de.

These propagation codes allow for – and require – separate models of particle production and acceleration. Back-reaction of the cosmic rays on the transporting plasma is not considered. In contrast to this, Pfrommer and collaborators [798–800] are pursuing a more integrated approach. Starting from the regions where young stars are formed in a galaxy, they predict the density of supernovae. Treating the emitted cosmic rays as a relativistic fluid, they perform a joint three-dimensional magneto-hydrodynamics simulation of cosmic ray flux and plasma. This then leads to predictions of the steady state interstellar cosmic ray spectra of Figure 3.4, as well as their non-thermal photon emission. For the moment, the simulation is limited to protons and electrons. Extending it to all cosmic ray species, production and interaction mechanisms would qualify it as a complete and self-consistent model of galactic cosmic rays in the supernova paradigm.

6.3 INTERACTIONS

Spatial and directional diffusion is not the only important process affecting the transport of cosmic rays from their production and acceleration site to the outskirts of the solar system. Direct interactions with interstellar matter add to the collisionless scattering processes discussed so far and alter not only the direction but also energy and identity of cosmic ray particles. We fly over the most important types of collisions in this section, without insisting too much on details which are easily found in any introductory text on particle physics. The propagation codes quoted above contain descriptions of these processes, based on up-to-date theory and experimental data. However, as we will see in Focus Box 9.3, uncertainties in nuclear cross sections compromise the accuracy of predictions.

All charged particles lose energy when penetrating matter by exciting or ionising atoms they pass by, through multiple Coulomb interactions. Although interstellar matter is sparse, energy loss by this electromagnetic process is not negligible because of the long road from origin to observation. We already encountered the energy loss per unit path length $-dE/dx$ in Focus Box 1.2, when we discussed radiation damage by cosmic nuclei. During transport through the galaxy, the same process causes a steepening descent of the energy distribution of the cosmic ray flux.

Electrons and positron lose energy most easily because of their small mass. In ambient magnetic fields, they lose energy by synchrotron radiation, as detailed in Focus Box 6.1. They of course also lose energy by multiple Coulomb interactions like all other charged particles. In addition, they radiate photons in the electromagnetic field of ambient charged particles, nuclei in particular. This process is called bremsstrahlung, it dominates the ionisation energy loss of electrons and positrons at energies above a few MeV as shown in Focus Box 2.7. Their small mass also enhances the angles by which electrons and positrons are diffused through multiple Coulomb interactions.

Positrons are subject to annihilation with ambient electrons. Both electrons and positrons interact with single ambient electrons by elastic scattering, called Møller scattering ($e^-e^- \rightarrow e^-e^-$), and Bhabha scattering ($e^+e^- \rightarrow e^+e^-$), respectively. Focus Box 2.7 summarises the fractional energy loss, $-(1/E)dE/dx$, suffered by electrons and positrons per unit length and per unit grammage, differentiating between the contributions mentioned above [765]. Beyond about 10 MeV, bremsstrahlung losses dominate.

In summary, because of their fast energy loss, energetic electrons and positrons observed near Earth generally come from closer sources than e.g. protons. As rule of thumb, a TeV electron or positron cannot come from a source farther away than a few 100 pc.

A particularly interesting process for electrons is Compton scattering off ambient photons, $e^-\gamma \rightarrow e^-\gamma$. In the laboratory, this process normally involves an atomic electron, nearly at rest, as a target for a high-energy photon projectile; the photon transfers energy to the electron. In an astrophysical context, the roles are inverted, as explained in Focus Box 6.5. The cosmic microwave background populates all of space with very low-energy photons, with a mean energy of 6.626×10^{-4} eV. In addition, there are abundant thermal, radio and optical frequency photons, all with energies much less than the electron mass. These photons take the role of the target, high-energy electrons the role of the projectile. The energy transfer thus goes from the electron to the photon, which is why the process is often called "inverse" Compton scattering. This way electrons lose additional energy and photons are upgraded to high energies, in the X-ray and γ-ray range. Such photons thus point back to regions where high-energy electrons are abundant and are an excellent astronomical tool to probe regions of cosmic ray production and acceleration, as detailed in Chapter 7.

Protons and heavier nuclei lose energy by multiple Coulomb scattering like electrons and positrons do, as we have seen in Focus Box 1.2. However, due to their higher mass,

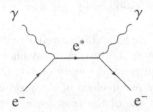

The Feynman graph on the left shows the first order process for Compton scattering, $e^-\gamma \to e^-\gamma$. It proceeds via a virtual excited state e^* of the electron. A full treatment in the astrophysical context is found e.g. in [178, 356]. Our discussion follows [462, Sec. 9.3].

In "normal" Compton scattering (see Focus Box 2.7), the photon transfers energy to the electron, which is practically at rest, e.g. bound in an atom. What we discuss here is "inverse" Compton scattering, where the opposite is true. The energy of photons populating interstellar space is often much smaller than the electron mass, $E_\gamma \ll m_e$, let alone the electron energy γm_e. Thus the cross section for Thomson scattering, σ_T, applies:

$$\sigma_T = \frac{8\pi}{3}\left(\frac{\alpha}{m_e}\right)^2 \simeq 6.66 \times 10^{-29}\,\mathrm{m}^2$$

where α is the fine structure constant. This is a gigantic cross section[3] of more than half a barn, so that the process probability is appreciable even in a sparsely populated environment. The energy loss rate for the electron is then:

$$\frac{dE_e}{dt} = \frac{4}{3}\sigma_T \beta^2 \gamma^2 U_\gamma$$

where β and γ are the electron velocity and relativistic factor. U_γ is the energy density of photons. For the cosmic microwave background, $U_\gamma \simeq 2.6 \times 10^5\,\mathrm{eV/m}^3$.

For low-energy photon "targets", the maximum energy of the outgoing photon is a factor of $\sim 4\gamma^2$ higher than the original energy. For typical electron relativistic factor of order 1000, radio wave photons are thus upscattered to UV energies, far infrared photons to X-rays and optical frequency photons to gamma rays. The average photon energy is then a third of this value by equipartition.

For Compton scattering off photons with energies comparable to the electron energy, one must use the Klein-Nishina cross section:

$$\sigma_{KN} = \frac{\pi^2 r_e^2}{E_\gamma}\left(\ln 2E_\gamma + \frac{1}{2}\right)$$

which decreases as the inverse of the photon energy at high energies. The mean energy of the outgoing photon is then about half the mean electron energy.

Focus Box 6.5: Compton scattering and its "inverse"

they do not suffer energy loss by other radiative processes like synchrotron radiation or bremsstrahlung, except at extreme energies. For protons and nuclei, the relevant processes are rather inelastic scattering off other ambient nuclei [528]. Inelastic nuclear interactions produce low mass hadrons like pions. Examples of simple proton reactions are $pp \to pn\pi^+$ or $\to pp\pi^0$. At modest energy transfer, such reactions proceed resonantly through virtual excited states of the proton and have rather high cross sections,[3] reaching several 10 mb. The high-energy photons from neutral pion decay, $\pi^0 \to \gamma\gamma$, signal sites of hadron production and acceleration. So do the neutrinos from the decay chain of charged pions, $\pi^+ \to \mu^+ \nu_\mu$

---

[3]The unit barn, $1\,\mathrm{b} = 10^{-29}\,\mathrm{m}^2$, is the unit for cross sections used in particle physics. It is a very large unit indeed. Cross sections in nuclear reactions are typically of the order of mb or μb. Cross sections for weak interactions are more of the order of nb or even pb.

followed by $\mu^+ \rightarrow e^+ \nu_e \bar{\nu}_\mu$. We will come back to pointing messengers of this sort in the next Chapter 7.

A subclass of inelastic reactions between heavy cosmic ray nuclei and interstellar matter are the so-called spallation reactions [547]. In these, the reaction splits the cosmic ray nucleus into two or more lighter, tightly bound nuclei. The spallation products more or less keep the original velocity of the projectile. Thus the kinetic energy per nucleon of the products is roughly the same as that of the parent projectile. This is one of the reasons, why the variable energy/nucleon was used in Chapter 3 to present cosmic ray spectra.

Interactions of protons and nuclei with ambient photons of course also exist. Among these, photoproduction of low-mass baryonic resonances has a particularly high cross section. An example is the reaction $p\gamma \rightarrow \Delta^+ \rightarrow p\pi^0$ or $\rightarrow n\pi^+$, which has a cross section of the order of 0.6 mb. The mass of the delta resonance Δ^+ is about 1.2 GeV and its width is roughly a tenth of its mass. To excite it by interacting with a photon from the cosmic microwave background requires a proton with an energy of at least 5×10^{19} eV. This energy is called the Greisen-Zatsepin-Kuzmin (GZK) limit [167, 171]. Higher energy protons will interact with the cosmic microwave background until their energy falls below the threshold. This limits the propagation distance of protons with an energy in excess of the GZK limit to about 50 Mpc. Since this is beyond the limits of our galaxy, we will come back to such a potential end of the cosmic ray spectrum in Chapter 10.

FURTHER READING

Claudio Chiuderi and Marco Velli, *Basic Plasma Astrophysics*, Springer (2015)

Daniele Gaggero, *Cosmic Ray Diffusion in the Galaxy and Diffuse Gamma Emission*, Springer (2012)

Elena Amato and Pasquale Blasi, *Cosmic Ray transport in the Galaxy: a Review*, Adv. Space. Res. (2018) 2731–2749

George B. Rybicki and Alan P. Lightman, *Radiative Processes in Astrophysics*, Wiley-VCH (2004)

Katia M. Ferrière, *The interstellar environment of our galaxy*, Rev. Mod. Phys. 73 (2001) 1031–1066

Malcolm S. Longair, *High Energy Astrophysics*, 3rd Edition, Cambridge Univ. Press (2011)

7 Pointing Messengers

In this chapter, we describe how photons and neutrinos from astrophysical sources are observed. We discuss what can be learned from these messengers pointing back to cosmic ray sources and acceleration sites. We concentrate on high-energy photons indicating violent processes. Neutrinos signal hadronic processes.

7.1 PHOTONS

Photons emitted by non-thermal processes witness release, acceleration and transport of cosmic rays. Figure 7.1 shows the opacity (in %) of the Earth atmosphere as a function of wave length. The only windows, where the atmosphere is more or less transparent, are in the visible, infrared and radio range. Detection techniques thus depend on the photon energy:

- Radio waves, with wavelengths between 10 cm and 10 m (energies 10^{-5} to 10^{-7} eV), can be measured with large ground-based parabolic antennae, called radio telescopes.

- X-rays, typically 100 eV to 100 keV, are covered by space-based instruments, since the atmosphere is opaque to X-rays.

- "HE or GeV photons", typically 30 MeV to 100 GeV, are covered by space-based instruments for the same reason.

- "VHE or TeV photons", typically 100 GeV to very rare PeV energies, are covered by ground-based imaging telescopes which use the atmosphere itself as detection medium, since low fluxes require large areas and long exposure.

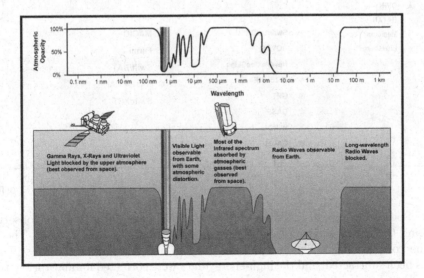

Figure 7.1 Opacity of the Earth atmosphere as a function of the photon wavelength. Windows where the atmosphere transmits light exist in the visible and near-infrared ranges, as well as for short wavelength radio waves. (Credit: Wikipedia Commons)

DOI: 10.1201/9781003181385-7

The observables also depend on photon frequency. One can distinguish two basic methodologies:

- With wave-like detection (radio waves, visible and infrared light and soft X-rays) one measures the spectral flux density $F(\nu)$ (units: energy per unit surface, time and frequency) or energy flux density $J(\nu)$ (units: energy per unit surface and time).

- With single photon counting (hard X-rays, HE and VHE gamma rays) one measures the spectral flux density as $E_\gamma \cdot dN_\gamma/dE_\gamma$ and energy flux density as $E_\gamma^2 \cdot dN_\gamma/dE_\gamma$, with N_γ the number of photons per unit area and time.

Of course, frequency and energy are related by Planck's constant, $E_\gamma = h\nu$. The relation between spectral flux density and energy flux density is:

$$J(\nu) = \int_{\nu_1}^{\nu_2} F(\nu)\, d\nu$$

which simplifies to $J(\nu) = \nu F(\nu)$ in case of a power law spectrum. When flux density J is plotted against frequency or energy, this is called the spectral energy distribution. It typically has a "double hump" shape, as shown for the example of the Markarian 421 active galactic nucleus in Figure 7.2.

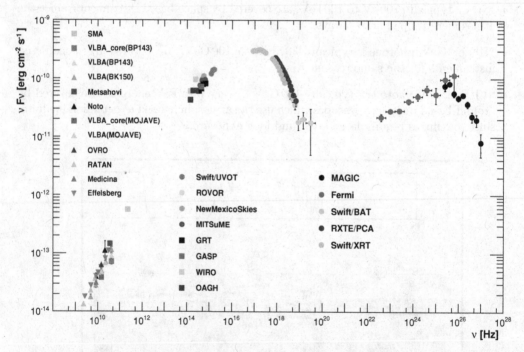

Figure 7.2 Spectral energy distribution of Mrk 421 [450] averaged over all the observations taken during the multifrequency campaign from January 19 to June 1, 2009. The legend reports the correspondence between the instruments and the measured fluxes. The host galaxy has been subtracted, and the high-energy data were corrected for absorption. (Credit: American Astronomical Society)

X-ray telescopes cover wavelengths from about 10 pm to 10 nm, thus from energies of about 100 eV to 100 keV. This range contains the typical excitation levels of atoms, 0.1 keV to 10 keV. X-rays are thus crucial to measure the elemental composition of cosmic matter.

Soft X-rays can be focused onto the telescope detection plane by total reflection at grazing incidence in a Fresnel lens, as shown in Figure 7.3. Missions must combine imaging, e.g. by a CCD camera, with spectrometry. Important X-ray missions include Chandra[1] (transmission grating spectrometer), XMM-Newton[2] (reflection grating spectrometer) and Hitomi (microcalorimeter, pioneering mission). Focus Box 7.1 describes the principles of grating spectrometers. Focus Box 7.2 introduces microcalorimeters.

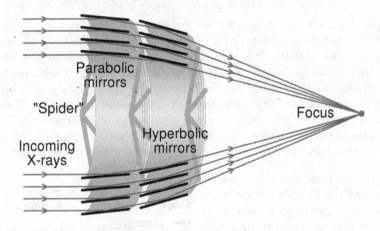

Figure 7.3 Wolter-type X-ray telescope lens with multiple parabolic and hyperbolic mirrors, held in place by a "spider" structure. (Credit: Wikipedia Commons)

Hard X-rays and gamma rays cannot be focalised and thus require large areas of photosensors. The transition region between X-rays and γ rays is covered by missions which are especially interested in transient phenomena like gamma ray bursts. Examples are the European INTEGRAL satellite [789] and NASA's Neil Gehrels Swift Observatory. INTEGRAL combines the imager IBIS, using a coded-aperture mask to locate sources with an angular resolution of about 12 arc minutes and a large angular coverage, with the semiconductor spectrometer SPI [806] with typically permille level energy resolution. The use of semiconductors for spectroscopy is covered in Focus Box 7.3. Both instruments are pointing into the same direction and cover an energy range of some 10 keV to about 10 MeV. Similarly, Swift combines the X-ray telescope XRT [363], using a reflection grating of the Wolter type similar to the one used on XMM-Newton, with the large coded-aperture mask burst array telescope BAT. The latter covers a large fraction of the sky and allows the other instruments to be swiftly pointed to a transient source; hence the name of the mission. Due to its technology, XRT has a lower energy range than SPI, 0.2 to 10 keV, but a better energy resolution.

Gamma ray energies start at about 100 keV and have no upper limit. The region up to a few hundred GeV has a high enough rate to be observed through the limited acceptance of space telescopes. They typically track the electron-positron pair from photon conversion to determine the photon direction. The photon energy is measured by total absorption calorimetry. Examples are the pioneering EGRET instrument on the Compton Gamma Ray Observatory. More modern implementations of the same principle are carried by the AGILE and Fermi satellites, described in Focus Box 7.4. The large acceptance of the Fermi Large Area Telescope (LAT) in particular has enriched the field of gamma ray astronomy in a spectacular way. It is shown in Figure 7.4.

---

[1] See https://chandra.harvard.edu/about/science_instruments.html
[2] See https://www.cosmos.esa.int/web/xmm-newton/about-xmm-newton

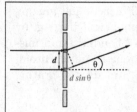

In a grating spectrometer, angular diffusion is used to separate light with different wavelengths λ. A transmission grating consists of a regular pattern of absorbing and transmitting strips. The transmitting slits have a tiny width and pitch d of a few hundred nm. For X-rays, the strip shaped absorber is a material with high atomic charge and high density, like gold. It has to be as thick as possible, typically several hundred nm.

The condition for constructive interference between neighbouring slits is $d\sin\theta = m\lambda$, where the integer $\pm m$ is called the order of the resulting maximum. The angular offset θ of intensity maxima measures the wavelength. A bright maximum is always present at $\theta = 0$ regardless of wavelength. When many slits are arranged with constant pitch, faint secondary maxima of intensity appear between the primary ones. The primary ones become brighter, since more slits contribute. A substantial fraction of X-ray photons will penetrate through the absorber. However, the absorber thickness can be adjusted such that penetrating waves have a phase shift of $\simeq \pi$ with respect to the transmitted waves. They will thus partially extinguish and the zero-oder beam is diminished in intensity.

The sketch on the right shows the principle components of the Chandra X-ray telescope [362]. Its two high-energy transmission gratings can be flipped into the incoming focalised X-ray beam. They consist of gold strips, 360 nm thick with a pitch of 200 nm and 510 nm thick with 400 nm pitch, respectively. The strips are supported on a thin polyimide foil.

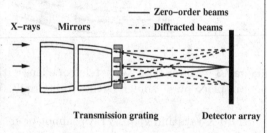

Reflection gratings make reflected beams interfere in the focal plane of a spectrometer, using a grooved reflector. The sketch on the left shows that constructive interference is reached

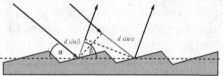

for $d(\sin\alpha + \sin\beta) = m\lambda$, where d is the spacing of the grooves, α and β are the angles of incident and dispersed rays with respect to the grating plane. The grating lets about 50% of the X-rays pass through.

Different orders m may overlap on the detection plane, but they are separated by the inherent energy resolution of the CCD photodetector (see Focus Box 7.3). The sketch on the right shows the optical paths for transmitted and diffracted beams on the XMM-Newton spectrometer [320].

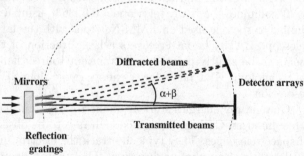

Focus Box 7.1: Grating spectrometers

As an example, Figure 7.5 [751] shows a graphic representation of the rich Fermi-LAT catalog of high-energy gamma ray sources, both inside and outside the galactic plane. Among the known sources, it includes star-forming regions and binary systems, relevant for nucleosynthesis as covered in Chapter 4. More prominently, it includes pulsars and pulsar wind nebulae, galaxies and globular clusters, novae and supernova remnants as well as active

Microcalorimeters are capable of measuring the energy of single absorbed photons by the minute temperature increase in cryogenic materials. The temperature "jump" $\Delta T = E/c_V$ is proportional to the absorbed energy E via the material's heat capacity c_V. Classically, the latter is proportional to the number N of atoms in the sample, $c_V = 3Nk$ with the Boltzmann constant k. For a $1\,\text{mm}^3$ sample of silicon, the heat capacity is $c_V \simeq 10^{16}\,\text{eV/K}$. The temperature increase when absorbing a $1\,\text{keV}$ photon is thus of the order of $10^{-13}\,\text{K}$. It can be measured by extraordinarily sensitive thermometers, using the highly non-linear behaviour of superconductors on the transition edge towards normal conduction. One must choose the transition temperature of the material such that target photon energies drive the material into the transition edge. Such a thermometer is called a transition edge sensor. Its working temperature is typically a few $10\,\text{mK}$.

The X-ray spectrometer XRS on the Japanese Suzaku mission was the first microcalorimeter on an orbiting observatory. It had 30 mercury telluride pixels sensitive in the 0.3 to $12\,\text{keV}$ energy range with excellent resolution of the order of a few eV and high quantum efficiency. Its heritage fed into the SXS spectrometer on the Hitomi mission with a similar energy range and resolution [474]. The spacecraft was lost only weeks after the launch due to a communication failure, but few observations established the excellent performance of the microcalorimeter [588, 618]. A replacement is planned with the future X-Ray Imaging and Spectroscopy Mission (XRISM) aiming at a similar performance.

Focus Box 7.2: Microcalorimeters

The physics of particle detection by semiconductors has been covered in Focus Box 3.5. Here we describe their use as spectrometers for X-rays, especially concerning high purity Ge sensors. The basic facts are the following:

- X-rays ionise atoms and thus create electron-hole pairs in a semiconductor. The result is a current surge. The number of pairs is proportional to the photon energy if the photon is absorbed.

- Germanium is a semiconductor which can be depleted over a thickness ranging up to centimetres. Such a device can thus efficiently absorb photons up to a few MeV energy.

- The resulting pulse height is proportional to the photon energy. The number of pulses per unit time is proportional to the light intensity.

- However, a $1\,\text{MeV}$ photons liberates about 3×10^5 electron-hole pairs. This is much less than the number thermally produced at room temperature, about 2.5×10^{13} at $20°\,\text{C}$ for $1\,\text{cm}^3$ of pure Ge. Any dopant will increase the number of thermally produced pairs.

- Therefore high purity Germanium detectors have to be cooled to liquid nitrogen temperatures, about $80\,\text{K}$, to suppress thermal dark currents.

Focus Box 7.3: Semiconductor X-ray spectrometers

galactic nuclei. All of these are relevant as cosmic ray acceleration sites, as explained in Chapter 5.

In addition to data on high-energy sources, the Fermi-LAT instrument has mapped the sky for more diffuse regions of gamma ray production. An example from the Milky Way are the so-called Fermi bubbles shown in Figure 7.6. They are giant bipolar structures of gamma ray emission above and below the galactic plane [432, 445], forming a counterpart

Beyond an energy of about 100 MeV, photons interact with matter exclusively by converting into electron-positron pairs (see Focus Box 2.7). For high-energy photons, a generic detector thus looks like the sketch on the right. Foils made of high density, high nuclear charge materials serve as photon converters. Tracking detectors for charged particles locate the e^+e^- pairs and measure their direction. The thickness of materials is a compromise between high conversion probability and low multiple Coulomb scattering worsening the angular resolution. A calorimeter at the bottom measures the energy of the pair. The whole set-up is shielded from the high rate background of charged cosmic rays by anti-coincidence counters.

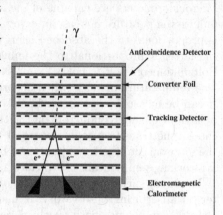

Space γ-ray observation was pioneered by the OSO-3 satellite with a counter telescope in the 1970s [186]. The EGRET instrument on board of the Compton Gamma Ray Observatory was built according to the above scheme with wire-mesh spark chambers as particle tracker [227, 254] and tantalum converters. It delivered the first systematic catalogue of high-energy gamma ray sources [292]. The instrument covered an energy range from 20 MeV to above 10 GeV with an effective area around 1500 cm^2 and a field of view of 0.5 sr. The energy resolution was 20 to 25% over most of its range of sensitivity. Absolute arrival times for photons were recorded with approximately 50 μs accuracy.

The AGILE mission [420] of the Italian Space Agency, launched in 2007, provides silicon semiconductor tracking with its GRID instrument [431] using tungsten converters. It is sensitive in the energy range from 30 MeV to 50 GeV. Its effective area is about 300 cm^2 and its field of view 3 sr. Its angular resolution is 5 to 10 arc minutes depending on energy and source strength. GRID is complemented with a shallow calorimeter MCAL made of cesium iodide scintillator [416]. More details and data products are provided at the AGILE Science Center page https://agile.ssdc.asi.it.

An order of magnitude larger effective area is featured by the Large Area Telescope (LAT) [422] on board of NASA's Fermi satellite, in orbit since 2008 (see Figure 7.4). It covers an energy range from 20 MeV to 300 GeV, with a peak effective area > 8000 cm^2 at energies above 10 GeV and a field of view > 2 sr [795]. It has a modular design with 4×4 identical towers. Each one consists of 18 silicon strip particle detectors measuring the position of traversing charged particles, interleaved with tungsten converter foils. At the bottom of each tower are 12 caesium iodide scintillator bars in hodoscope arrangement measuring shower shape and energy of the electromagnetic shower. The energy resolution is better than 15% for energies above 100 MeV. More information is found on the Fermi home page https://fermi.gsfc.nasa.gov.

Focus Box 7.4: Gamma ray telescopes

to similar structures observed in multiple wavelengths, from radio to X-ray energies. They exhibit a significantly harder spectrum than gamma rays from inverse Compton emission in the galactic disk. Two possible engines driving these structures have been discussed in the literature. The first one is that the supermassive black hole at the centre of the Milky Way, Sr A*, which could have formed jets over the last few million years [486]. Alternatively – or in addition – the bubbles could represent much older star forming regions. Again, both interpretations are relevant for processes behind cosmic rays production and acceleration. For recent reviews, we refer the reader to references [529, 748].

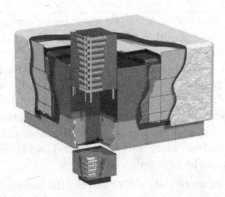
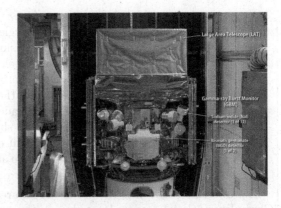

Figure 7.4 Left: View of the Fermi Large Area Telescope. The insulating cover and anti-coincidence counter tiles are partially cut away to expose one of its towers consisting of an electron pair tracker (moved upward) and a calorimeter module (moved downward). Right: The Fermi satellite inside the rocket faring shortly before launch. (Credit: NASA GSFC)

Beyond about 100 GeV photon energy (VHE), events become too rare for satellite observation. Instead, one uses the Earth atmosphere as an imaging calorimeter. The photon direction is estimated by the shower axis, photon energy by the Cherenkov light yield. Important telescopes are MAGIC located on the Canary island La Palma, H.E.S.S. in Namibia and VERITAS in Arizona. The principles of imaging atmospheric Cherenkov telescopes are described in Focus Box 7.5. All have multiple telescope dishes to provide a stereoscopic view of electromagnetic air showers and distinguish them from the much more abundant hadronic showers caused by cosmic rays.

Air shower arrays (see Section 2.5 and Chapter 10) can also be used to measure very high-energy-photons by detecting ionising particles hitting the ground. They are distinguished from hadronic showers, i.e. cosmic rays, through the shower shape and/or the absence of muons. The highest energy photons have been observed by the LHAASO array in Tibet, where energies up to 1.4 PeV have been observed from sources in the Milky Way [811]. This indicates the presence of extremely powerful particle accelerators in our galaxy, but astronomic counterparts have not yet been established.

When directed towards the same target at the same time, the wide variety of photon detectors discussed above can establish the emitted frequency spectrum over an impressive range of energies. As an example, Figure 7.2 [450] already showed the spectral energy distribution from Markarian 421, an active galactic nucleus situated in the Ursa Major constellation. The low frequency hump in the spectrum comes from synchrotron radiation mainly by primary electrons, ranging from radio frequencies to X-rays. The spectral energy distribution of synchrotron light emitted by cosmic electrons, which have a power law spectrum, has itself a power law form. This is argued in Focus Box 7.6.

The high-energy hump can come from photons pushed up in energy by energetic primary electrons through inverse Compton scattering as explained in Focus Box 7.7. In addition, hadronic processes can contribute photons from $\pi^0 \rightarrow \gamma\gamma$ decay or radiated by electrons and muons from meson decay. The joint spectrum can be fitted by a purely leptonic origin of the radiation, i.e. through synchrotron radiation plus Compton scattering from primary electrons alone. It can also be fitted by a mixture of leptonic and hadronic processes [450].

Multifrequency campaigns look into the spectra emitted by relatively close-by objects in great detail. Observations are ideally quasi simultaneous to capture the same physical state of the object. A particularly impressive result is shown in Figure 7.7, obtained by the

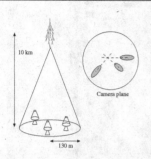

Photons with energies above 50 GeV can be observed by ground based telescopes using the Cherenkov light (see Focux Box 2.11) from atmospheric electromagnetic showers (see Focus Box 2.8). The sketch on the left [765] shows the principle. The emitted cone of light hits the facetted reflective dishes and is focussed onto a camera made of photodetectors (see Focus Box 2.10). Depending on the shower direction, the detected photon intensity forms a regular ellipse. Hadronic showers, which are orders of magnitude more abundant, make a more extended and irregular pattern and can thus be rejected.

The Cherenkov light hits the ground in an area of more than 100 m radius, the so-called Cherenkov pool. A telescope located anywhere inside the pool will receive a part of the light and contribute to imaging the shower. The sensitive area is thus the size of the pool, several $10000\,m^2$. Complete reconstruction of the shower requires a stereoscopic view of at least two telescopes. The light intensity is low; a TeV shower causes of the order of 100 photons per m^2 arriving within a few ns.

The smallest operational array of Cherenkov telescope is MAGIC located on the Spanish island of La Palma [360, 578, 579]. It consists of two telescopes with 17 m diameter dishes and cameras with about 1000 pixels, sensitive to energies between a few 10 GeV and a few 10 TeV. Its lightweight structure allows a reorientation of the telescopes within little more than 20 s. The VERITAS array in the US Arizona desert [350] features four 12 m telescopes and is sensitive in a similar energy range. The largest operational array is the High-Energy Stereoscopic System H.E.S.S. located in the Namibian desert [338, 339, 525].

H.E.S.S. consists of four 12 m diameter telescopes arranged in a square with 120 m sides, and a single 28 m diameter dish located in the middle as shown on the photograph. The smaller dishes are equipped with 960 pixel photomultiplier cameras, the larger one has 2048 pixels. The energy range of the array again covers tens of GeV to tens of TeV, overlapping with the energy reach of satellite telescopes at the lower end and with ground arrays of air shower detectors at the high end.

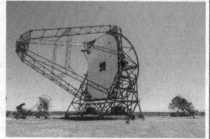

(Credit: Shutterstock)

The performance of all existing arrays will be eclipsed by the Cherenkov Telescope Array CTA currently under construction [647]. It will have two sites, a southern hemisphere array in Paranal, Chile; and a northern one on the Spanish La Palma island. They will consist of hundreds of telescopes in three size categories. The small ones will have dual mirrors of 4.3 m and 1.8 m diameter. They will be sensitive to the highest energies only, 5 TeV to 300 TeV. Only the southern site will deploy 70 such telescopes to observe the rare photons from high-energy sources close to the galactic centre. Medium size telescopes, the work horses of CTA, will have either dual or single mirrors with 11.5 m diameter. The former will feature more than 10000 pixels equipped with silicon photomultipliers. The latter will either have vacuum tube or solid state photomultipliers with close to 2000 pixels. CTA will deploy 25 such telescopes in the southern and 15 in the northern site. Four large telescopes with 23 m dishes will equip each of the northern and southern sites. With their 2000 photomultiplier pixel cameras, they will be sensitive to the most abundant low-energy showers which have the lowest light intensity.

Focus Box 7.5: Imaging atmospheric Cherenkov telescopes

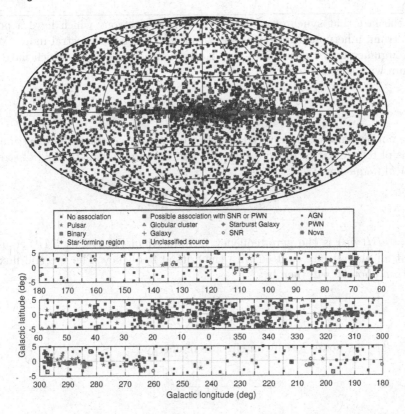

Figure 7.5 Fourth Fermi LAT catalogue of observed sources [751]. Full sky map (top) and blow-up of the Galactic plane split into three longitude bands (bottom) showing sources by source class. (Credit: The American Astronomical Society)

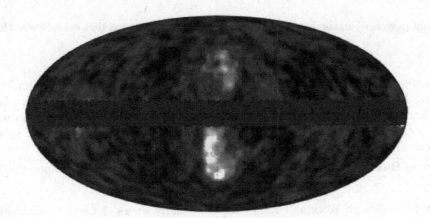

Figure 7.6 Intensity map of the sky in galactic coordinates showing the high intensity regions above and below the galactic plane names "Fermi bubbles". Intensity is shown for photons of more than 6.4 GeV with the galactic plane and identified point sources subtracted. (Credit: Fermi-LAT team, A. Franckowiak, D. Malyshev)

We argue the fact that synchrotron light emitted by electrons which have a power law energy spectrum, inherits an analogous spectrum with a related spectral index. We follow the simple arguments by Spurio [565, Sec. 8.4.1]. Consider electrons which have a power law spectrum with spectral index α:

$$\frac{dN}{dE}\, dE = \kappa E^{-\alpha}\, dE$$

where κ is a normalisation constant. In Focus Box 6.1 we characterised the spectrum emitted by electrons of relativistic factor $\gamma = E/m$, cycling around a magnetic field of strength B, by the critical frequency ω_c:

$$\omega_c = \frac{3}{2}\gamma^3 \frac{1}{\rho} = \frac{3}{2}\gamma^3 \frac{eB}{\gamma m} = \frac{3}{2}\gamma^2 \omega_g$$

where $\omega_g = eB/(\gamma m)$ is the gyrofrequency introduced in Focus Box 5.1. Let us simply assume that all photons are emitted with the critical frequency. The spectral flux density of synchrotron photons will then be:

$$J(\omega)\, d\omega = -\left(\frac{dE}{dt}\right)\left(\frac{dN}{dE}\, dE\right)$$

where $-dE/dt$ is the power lost by an electron of energy E:

$$-\frac{dE}{dt} = \frac{4}{3}\sigma_T \frac{B^2}{8\pi}\gamma^2$$

where σ_T is the Thomson cross section introduced in Focus Box 2.7. The quantity $u_B = B^2/(8\pi)$ is the energy density of the magnetic field. The electron energy can be expressed through the frequency $\omega = \omega_c$:

$$E = \gamma m = \left(\frac{\omega}{\omega_g}\right)^{1/2} m \quad ; \quad dE = \frac{m}{2\omega_g^{1/2}}\, \omega^{-1/2}\, d\omega$$

Counting powers of ω, the frequency dependence of the spectral flux density will thus be:

$$J(\omega) \propto \omega^{-(\alpha+1)/2} \quad ; \quad \nu J(\nu) \propto \nu^{-(\alpha-1)/2}$$

This qualitatively describes the falling part of the synchrotron spectrum at high frequencies seen in Figure 7.2, but not the low frequency behaviour. However, for every emission process there must be an absorption process at the same frequency. For synchrotron radiation, it is called self-absorption. It dominates at low frequencies and takes the spectrum smoothly down to zero at low frequencies like $J(\nu) \propto \nu^{5/2}$.

Focus Box 7.6: Synchrotron light from electrons with a power law spectrum

EHT Multi-Wavelength Working Group [794]. The figure shows the active galaxy Messier 87, including the event horizon of its central black hole seen by the Event Horizon Telescope (EHT), which has the highest spatial resolution in radio frequencies. Images and spectroscopy range from radio frequencies via the visible and X-ray range to gamma ray energies. More than a dozen instruments were involved in the campaign in April 2017, with very different spatial resolution. The morphology shows a jet emerging from its centre, with bright knots clearly visible in radio up to X-ray frequencies. Such jet-like structures play an

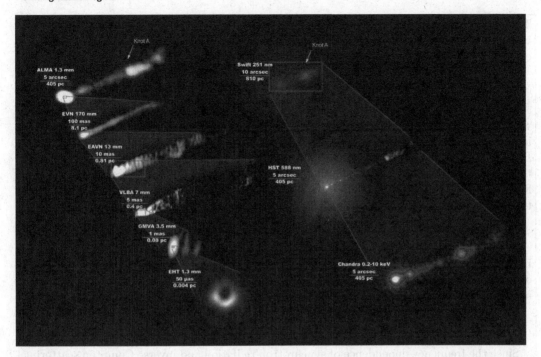

Figure 7.7 Compilation of the quasi-simultaneous M87 jet images at various scales during the 2017 campaign [794]. The instrument, observing wavelength and scale are shown on the top-left side of each image. Location of the Knot A (far beyond the core and HST-1) is indicated in the top figures for visual aid. (Credit: EHT Multi-Wavelength Working Group, American Astronomical Society)

important role in the acceleration of cosmic rays to highest energies, as explained in Section 5.4. M87 was indeed the first astrophysical jet observed in 1918 by Heber Curtis [38, p. 31] and described in a catalogue as a "curious straight ray... apparently connected with the nucleus by a thin line of matter".

7.2 NEUTRINOS

Neutrinos are the most elusive particles of the Standard Model. Possessing no other charge than weak isospin, they only interact with matter through charged and neutral weak interactions. Although abundantly produced in astrophysical context, they are thus extremely hard to detect. However, they do not interact with magnetic fields and rarely with interstellar matter. They thus point back to their origin and traverse dense material without scattering. They are detected by giant detectors underground, under water or under ice, well shielded from cosmic rays and ambient radioactivity. However, the vast majority of detected neutrinos originates from hadronic cascades in the atmosphere. Only at very high energies is there a measurable contribution of astrophysical origin.

Neutrinos signal hadronic processes. They are produced in inelastic $p\mathcal{N}$ and $\gamma\mathcal{N}$ interactions, mostly through light mesons like $\pi^\pm$ and $K^\pm$, more rarely through mesons containing heavy quarks c and b. Although the cross sections for hadronic production are two orders of magnitude larger than the photoproduction cross sections, the latter dominate due to the large photon density at astrophysical sources. The mesons decay via weak interactions.

The main properties of Compton scattering, both normal and inverse, have been discussed in Focus Box 6.5. The power loss of an electron in a "bath" of low-energy photons with energy density u_g is:

$$-\frac{dE}{dt} = \frac{4}{3}\sigma_T u_g \gamma^2$$

where σ_T is the Thomson cross section and γ is the relativistic factor of the electron. This is very similar to the energy loss of electrons through synchrotron radiation quoted in Focus Box 7.6. The similarity is no surprise: the difference between the two processes is essentially the wavelength of the incoming photons. We are thus entitled to directly transfer the results for the spectrum of photons found for synchrotron radiation. The frequency dependence of the spectral flux density is thus:

$$\nu J(\nu) \propto \nu^{-(\alpha-1)/2}$$

in situations, where the Thomson cross section is applicable, i.e. when the photon is initially of low energy, $E_\gamma \ll m_e$. For high-energy photons, where the Klein-Nishina cross section applies, one finds

$$\nu J(\nu) \propto \nu^{-\alpha} \ln \nu$$

Since the frequency dependence of synchrotron radiation and inverse Compton scattering is the same, we expect a constant ratio of energy losses by the two mechanisms, independent of frequency.

Synchrotron radiation and Compton scattering are tightly coupled processes. Synchrotron radiation can provide high-energy photon "targets", then upgraded in energy by Compton scattering. This joint chain of processes is often called the synchrotron self-Compton mechanism, to distinguish it from synchrotron radiation due to an externally provided magnetic field.

Focus Box 7.7: "Inverse" Compton photons from electrons with a power law spectrum

Examples are:

$$\pi^+ \to \mu^+ \nu_\mu \quad ; \quad \pi^- \to \mu^- \bar{\nu}_\mu$$
$$\mu^+ \to \bar{\nu}_\mu e^+ \nu_e \quad ; \quad \mu^- \to \nu_\mu e^- \bar{\nu}_e$$

Thus every meson decay chain produces three neutrinos and antineutrinos. Neutrino fluxes and photon fluxes from astrophysical sources are of comparable size and tightly coupled, since hadronic final states also contain copious quantities of $\pi^0 \to \gamma\gamma$.

Neutrino cross sections with matter are among the tiniest known, making neutrinos hard to detect with anything but detectors of huge mass. Figure 7.9 shows representative neutrino cross sections with nucleons and electrons. They are small enough to allow neutrinos to traverse the whole Earth without much attenuation. The so-called charged current processes $\nu_l \mathcal{N} \to l^- X$ – where l stands for the leptons e, μ or τ, $\mathcal{N}$ for neutron or proton and X for a generic hadronic final state – proceed in a non-resonant way by exchange of a $W^\pm$ boson. The neutral current processes $\nu_l \mathcal{N} \to \nu_l X$ proceed via the exchange of the neutral boson Z and are rarer by a factor of $\simeq 2.4$.

Neutrino-nucleon processes dominate over neutrino-electron interactions with notable exceptions. These are the processes $\bar{\nu}_e e^- \to \bar{\nu}_e e^-$ and $\to \bar{\nu}_\mu \mu^-$, which proceed though the s-channel exchange of a W^- boson as shown in Figure 7.8. When the centre-of-mass energy of antineutrino and electron are close enough to the boson mass, the resonance enhances

Figure 7.8 Lowest order Feynman graph for the process $\bar{\nu}_e e^- \rightarrow \bar{\nu}_\mu \mu^-$, which proceeds through s-channel exchange of a W boson. When the centre-of-mass energy of the initial state approaches the mass of the W boson, the process is enhanced by the so-called Glashow resonance.

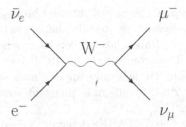

the cross section for this otherwise rare process by many orders of magnitude. This occurs for neutrino energies in the laboratory of $E_\nu \simeq M_W^2/(2m_e) \simeq 6.33 \times 10^6$ GeV, as shown in Figure 7.9, and is known as the Glashow resonance [156].

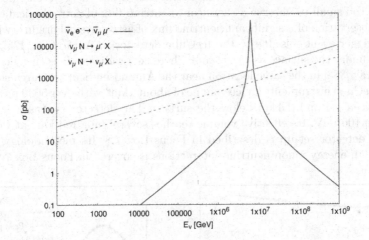

Figure 7.9 Neutrino cross sections for representative processes at high energies [274, 451, 499].

The first occasion to observe neutrinos of astrophysical origin arose with the recent near-by supernova SN 1987A. It was a core collapse supernova, first identified optically at observatories in Chile and New Zeeland [222]. It had a radioactive afterglow and a neutron star remnant later identified in radio waves [611] and X-rays [786, 788]. It was situated in the Large Magellanic Cloud, a dwarf satellite galaxy of the Milky Way at a rather close distance of 51.4 kpc from the Solar system. Since neutrinos are emitted early in the collapse process (see Section 4.2), a neutrino signal was recorded a few hours before the photon signal was observed. Timely neutrino interactions were seen in the water Cherenkov observatories Kamiokande II [219] and Irvine-Michigan-Brookhaven (IMB) [221], as well as the Baksan liquid scintillator detector [218]. The three experiments recorded a total of 25 neutrino interactions in a short burst lasting less than 13 seconds. The attribution of an even earlier signal by the Mont Blanc Liquid Scintillator Detector (LSD) [220] remains controversial. The increase in neutrino intensity was detectable, since timing and arrival direction were known. The discovery prompted hundreds of papers interpreting the signal, for a review see e.g. [250].

Giant water Cherenkov detectors use natural clear lake water, sea water or ice, both as a neutrino target and as a Cherenkov radiator. The Cherenkov light is recorded by photomultiplier tubes with large surface and aperture. The detectors are well shielded against background by immersing them deep under ground level. Such detector systems were first proposed by Soviet physicist Moisey Alexandrovich Markov [158] in the 1960s. Consequently, the first project to take shape was the Baikal Deep Underwater Neutrino Telescope, situated in the Baikal lake at depths around 1.1 km. Construction took place since the early 1990s and was completed with 192 detection modules in 1998 [435]. Because of its limited volume, about $10^6 \, \mathrm{m}^3$, only upper limits on the astrophysical neutrino flux were obtained [379]. In 2015, an upgrade program started to enlarge the instrumented detector volume to a km^3 [792].

The ANTARES detector [522], developed since 2007, uses the water of the Mediterranean sea about 70 km off the coast of Toulon in France in 2.5 km depth. Completed in May 2009, it comprises 885 light detection modules filling an instrumented volume of about $0.1 \, \mathrm{km}^3$. However, as in all such detectors, the effective volume is larger than the instrumented one at very high energies, since they receive Cherenkov light also from the surroundings. ANTARES thus reaches an effective volume of almost a km^3 at PeV energies. Even this size turns out too small to observe a significant neutrino flux of astrophysical origin [777].

The first observation of a significant neutrino flux of astrophysical origin – without using an external trigger event – is due to the IceCube detector in Antarctica [423]. It uses the clear ice deep under the surface as target and Cherenkov radiator. It is the successor of the AMANDA array [383] in the same location near the Amundsen-Scott research station at the South Pole. IceCube instruments an ice volume of about $1 \mathrm{km}^3$ with over 5000 photodetectors at depth between 1.4 and 2.4 km below the surface. The detector is sensitive to neutrino energies above 100 GeV. Its effective volume reaches several hundred km^3 at PeV neutrino energies. The detector set-up is described in Focus Box 7.8. Its methodology to measure muons from high-energy muon neutrino interactions is covered in Focus Box 7.9.

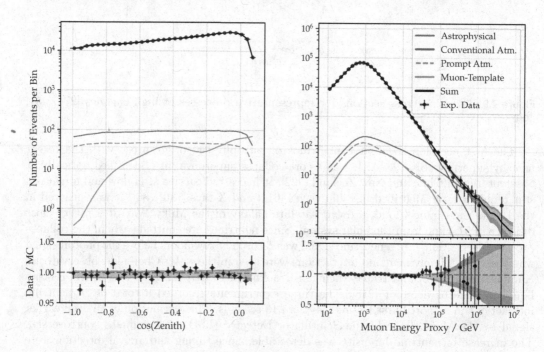

Figure 7.10 Muon zenith angle (left) and energy distribution (right) collected by Ice-Cube [854] with contributions from various sources described in the text.

Figure 7.10 shows the composition of the sample of muons observed by IceCube in nearly ten years of exposition [500, 854], as a function of the angle with respect to the

The neutrino observatory IceCube uses the clear ice deep under the surface at Antarctica as a radiator for Cherenkov light. Over 5000 detector modules, carrying large aperture vacuum photomultipliers and support electronics, have been deployed at depths between 1450 m and 2850 m, by melting holes into the ice with hot water. The schematic on the right shows the configuration completed in 2010. A DeepCore section is densely populated with photomultipliers to allow the detection of low-energy events.

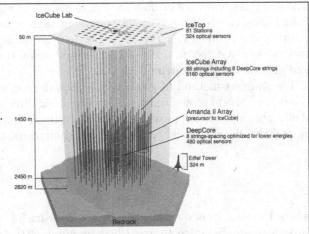

(Credit: IceCube Science Team – Francis Halzen, Department of Physics, University of Wisconsin)

The underground detector is complemented by the surface array IceTop of ice tanks with photomultipliers sensitive to atmospheric showers above the detector [497]. The volume of the detector is large – about $1 \, \text{km}^3$ – and the target mass is substantial – about $10^9 \, \text{t}$. However because of the tiny interaction probability of neutrinos, few interactions occur inside the detector itself. It normally detects Cherenkov light from relativistic muons created by inelastic reactions of muon neutrinos. The neutrinos dominantly come from the decay of charged pions or kaons in atmospheric showers created by cosmic rays. Subdominant contributions come from air shower muons making it to the detection volume, neutrinos from the decays of rare heavy quarks created in air showers and neutrinos from astrophysical sources. At the high-energy tail of the energy spectrum, the latter become dominant.

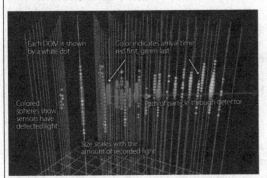

(Credit: IceCube Science Team – Francis Halzen, Department of Physics, University of Wisconsin)

The schematic on the left shows how Ice-Cube and similar observatories estimate energy and direction of the neutrinos. The image shows an event where the muon crosses the detector horizontally. The spheres indicate detectors registering Cherenkov light. The size of the sphere is proportional to light intensity, the grey scale indicates time. Dark detectors fired the earliest, brighter ones the later.

Timing allows to determine the shower direction, and most importantly the incident angle with respect to the zenith. A lower limit for the neutrino energy is determined from the light intensity caused by the muon. Since the spectrum is rapidly falling, this is a fair estimate of the neutrino energy itself. The system is sensitive to neutrinos above 100 GeV. Background is rejected by analysing the shower shape. Significant signals for astrophysical neutrinos are found at energies above 200 TeV and large zenith angles, as shown in Figure 7.10.

Focus Box 7.8: IceCube neutrino observatory

The detectors under ice or under water measure the yield and time of arrival of Cherenkov light pulses caused by a particle traversing their volume. These are used to reconstruct position, direction and energy of the particle. For reactions of high-energy neutrinos in the rock below the detector, $\nu_\mu \mathcal{N} \to \mu X$, straight muon tracks are used to approximate energy and direction of the neutrino.

The simplest method of geometrical track reconstruction uses the positions x_i and the times t_i of the photomultipliers observing a significant signal. Since the muon track is a straight line, the muon position is described by $\vec{x}(t) = \vec{x}_0 + \vec{v}t$. One estimates the position $\vec{x}_0$ and the direction $\vec{v}/|v|$ of the muon by minimising the sum of squared differences:

$$\min_{\vec{x}_0, \vec{v}} \sum_{i=1}^{N} |\vec{x}_i - (\vec{x}_0 + \vec{v}t_i)|^2$$

where the sum runs over all detectors with a useful signal. Detectors far away from the trajectory are discarded to avoid bias by noise [521]. In IceCube, this algorithm results in an angular resolution of about $1°$ and a position accuracy of a few meters.

The muon energy is taken as representative for the neutrino energy. At energies below about 1 TeV, muons stop inside the detector volume and their energy can be determined by their range in the ice [319]. The energy loss by ionisation only grows as the logarithm of the particle energy (see Focus Box 1.2). However, muon energy loss at energies above 1 TeV is dominated by radiative processes: bremsstrahlung, $e^\pm$ pair production and electromagnetic interactions with nuclei as shown in the Figure below.

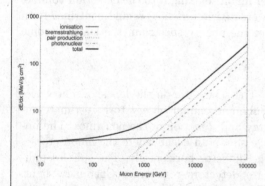

Relativistic charged particles from these processes emit Cherenkov light which is detected. These are stochastic processes which are subject to important fluctuations from event to event. However, the total mean energy loss, $dE_\mu/dx = a + bE_\mu$, is roughly proportional to the muon energy E_μ as shown by the figure on the left.

Experimentally, the number of Cherenkov photons registered by all hit detector elements is a measure of the muon energy loss. The sum of all signals is compared to the expected total signal from a "standard" muon with the same position and direction, which has a unit energy loss of 1 GeV/m. This expected signal is calculated by detailed Monte Carlo studies [496], including the effects of the ice and details of the light detection technology. The energy loss is then determined by the ratio of observed and expected "standard" signal sum. Since the dE_μ/dx distribution is skewed, a substantial fraction of high signals is discarded from the calculation of the sum. This truncated sum is less sensitive to high stochastic energy losses. The resulting resolution is about $\sigma(\log_{10} E/E_\mu) \simeq 0.22$ for energies between 1 TeV and 1 EeV, thus about 160% in energy. Within this resolution, the difference between muon and neutrino energy is insignificant.

Focus Box 7.9: Muon reconstruction in IceCube

zenith direction and the muon energy. Contributions come in several classes:

- The vast majority of neutrinos are emitted by meson decays in atmospheric showers caused by cosmic rays (labelled "Conventional Atm." in the figure). Their contribution is maximum around the zenith direction and at energies around a TeV.

- More than two orders of magnitude rarer are neutrinos emitted by decays of heavy quarks charm and bottom in the same atmospheric showers (labelled "Prompt Atm." in the figure).

- An even rarer category are background muons produced inside atmospheric cosmic ray showers (labelled "Muon-Template" in the figure). They basically only come from above and do not reach energies beyond a few 10 TeV.

- Finally, there are muons from the interactions of astrophysical neutrinos in the rock below the detector (labelled "Astrophysical" in the figure), which become a significant contribution at high energies. They have an almost flat zenith angle distribution.

IceCube has so far identified 35 neutrino events with a high probability to come from astrophysical sources [854]. The sample is restricted to zenith angle above 85° and energies above 200 TeV. Two of the highest energy events in that sample, nicknamed "Ernie" and "Bert" by the researchers, are shown in Figure 7.11, both with energies in the PeV range. In these event displays, the graphics attempts to represent a movie in a static picture: hit photodetectors are represented by spheres; the grey scale indicates the arrival time from early (dark) to late (bright), the size of the spheres is proportional to light intensity. The sample allowed to roughly measure the neutrino spectrum from astrophysical sources, summing over all arrival directions. The observations are compatible with isotropic arrival and a single power law spectrum with spectral index $\gamma_\nu = 2.37 \pm 0.09$.

Recently, an even higher energy event has been added to the sample [801]. Its energy of $(6.05 \pm 0.072)\,\mathrm{PeV}$ and the characteristics of the observed particle shower indicate that it might come from a neutrino interaction at the Glashow resonance, i.e. that a W boson was resonantly produced and decayed hadronically.

On one occasion so far, the IceCube detector has registered a high-energy neutrino event coincident with a flaring blazar [653]. A blazar is an active galactic nucleus (see Section 5.4) with a jet roughly pointing towards Earth. The neutrino interaction had an energy of about 290 TeV and was registered at the same time as a flare of blazar TXS 0506+056, studied with photons in a multi-wavelength campaign by multiple telescopes ranging from radio frequencies to gamma rays. If such coincidences are confirmed, they would indicate that indeed hadrons are accelerated in relativistic jets of active galactic nuclei.

To make neutrinos a systematic tool in astrophysics, even larger detectors are required. Several of those are in planning or construction phases. In the Baikal lake, an extension of the existing detector to gigaton size is under construction since 2015 [792]. Further information is found on the project home page baikalgvd.jinr.ru. In the Mediterranean, the KM3NeT project [586] (see https://www.km3net.org) plans gigaton detectors at two sites. Its ARCA detector will deploy 230 photodetectors off the Sicilian coast and specialise on the detection of astrophysical neutrinos from the southern hemisphere, where the galactic centre is located. The smaller but more densely instrumented ORCA detector will deploy additional 115 photodetectors at the current ANTARES site off the coast at Toulon. With an effective volume of about 400 Gt at PeV energies, KM3NeT will be sensitive to neutrino fluxes thirty times smaller than ANTARES. A similar initiative aims at installing the detector P-ONE of km^3 size in the North Pacific ocean [764], where a cabled research network already exists in the Cascadia Basin.

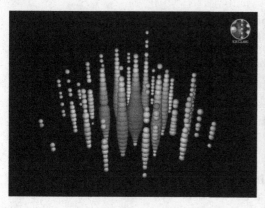 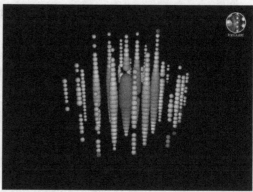

Figure 7.11 Graphical representation of two PeV energy neutrino events observed in Ice-Cube, nicknamed "Ernie" (left) and "Bert" (right) [500]. Spheres represent fired detector modules, with size indicating energy and grey scale indicating time from early (dark) to late (light) (see Focus Box 7.8). The muons come from below, the Cherenkov cone thus moves parallel to the detector strings, from bottom to top. (Credit: IceCube Science Team – Francis Halzen, Department of Physics, University of Wisconsin)

Also IceCube has an ambitious upgrade program called IceCube-Gen2 [802]. With novel photodetection units containing several photomultipliers, it will not only instrument a much larger volume but also improve the surface area for the detection of Cherenkov light. The detector imbedded in Antarctic ice will be complemented by a much enlarged surface array. An array of antennas will cover almost $500\,km^2$, both on the surface and at $200\,m$ depth. It will capture Cherenkov radiation at radio frequencies, based on the promising test experiments RICE [471] and ARA [549]. IceCube-Gen2 will increase the rate of registered cosmic neutrinos by an order of magnitude. It will be sensitive to neutrino sources which are a factor of five fainter and extend the energy reach by several orders of magnitude. More details are discussed in Section 11.3.

FURTHER READING

Antonio Capone, Paolo Lipari and Francesco Vissani, *Multiple Messengers and Challenges in Astroparticle Physics*, Springer (2018)

Antonio Ereditato (Edt.), *The State of the Art in Neutrino Physics*, World Scientific (2018)

Malcolm S. Longair, *High Energy Astrophysics*, 3rd Edition, Cambridge Univ. Press (2011)

Thierry J.-L. Courvoisier, *High Energy Astrophysics: An Introduction*, Springer (2013)

8 In the Heliosphere

This chapter is about phenomena in the solar system and how they influence cosmic rays. We will describe the Sun as a star. We will speak about the solar wind and the embedded Sun's magnetic field. We will learn about solar activity and the 11- and 22-year solar cycles. We will describe the structure of the heliosphere and illustrate how charged particles propagate through it. We will see how the galactic cosmic ray spectra are affected by solar activity and discuss long- and short-term solar modulation. We will also see that the Sun itself accelerates particles during explosive events, to create what is called solar energetic particles. We will explore the Earth's magnetosphere, the near Earth radiation environment, its coupling with the solar wind and its transient phenomena. Finally, we will take a quick look at space weather physics.

8.1 THE SUN

The Sun is a main sequence G2 V star[1], as defined in the Hertzsprung-Russell diagram, with a luminosity of $\simeq 3.8 \times 10^{26}$ W and effective temperature of 5772 K [286]. The mass of the Sun is $\simeq 2 \times 10^{30}$ kg, about 33'300 Earth masses, and its radius is almost $R_\odot \simeq 70 \times 10^4$ km, about 109 times the one of the Earth. The Sun is mostly made of hydrogen (90%) and helium (10%), but light elements such as carbon, nitrogen and oxygen are also present ($\sim 0.1\%$, see Figure 3.2). The enormous gravitational force of the Sun is counterbalanced by the nuclear fusion reactions taking place in the centre of the star, where the temperature is such that the high end of the proton energy spectrum is sufficient to overcome the electrostatic repulsion between protons and to initiate the proton-proton cycle of helium nucleosynthesis, as described in Section 4.2. The total Q value of 26.73 MeV sustains life on Earth and drives the entire heliosphere environment.

A schematic diagram of the Sun is presented in Figure 8.1 where different zones are shown:

1. The **core** is the central zone of the Sun. It extends up to 25% of $R_\odot$ and contains almost half of the solar mass. Here the temperature is about 15×10^6 K and thermonuclear reactions for the helium nucleosynthesis take place. All the energy that sustains the star is produced in the core.

2. The **radiative zone** extends from the end of the core up to about 70% of $R_\odot$. Here the temperature drops rapidly from 7×10^6 to 2×10^6 K. All the energy released from the core is carried outward by the gamma rays produced in the nuclear reactions. In the radiative zone, gamma rays are scattered, absorbed and emitted many times.

3. The **tachocline** is a thin ($\simeq 0.04 R_\odot$) transition region between the Sun's interior, which rotates as a rigid body, and the outer convective zone which exhibits differential rotation typical of fluids. It is currently believed that the solar dynamo mechanism that generates the Sun's magnetic field occurs in the tachocline region.

4. The **convective zone** extends from $0.7 R_\odot$ to $R_\odot$. Here the density and temperature are dramatically reduced, becoming $0.2 \, \text{g/cm}^3$ and 5700 K respectively. In the convective zone, the energy is carried outwards by huge bubbles of plasma that emerge from the inner layers and expand reducing their temperature. Once cooled, the bubbles

---

[1]In this classification, G2 stands for the second hottest stars of the yellow G class and the V represents a main sequence (or dwarf) star, the typical star for this temperature class.

DOI: 10.1201/9781003181385-8

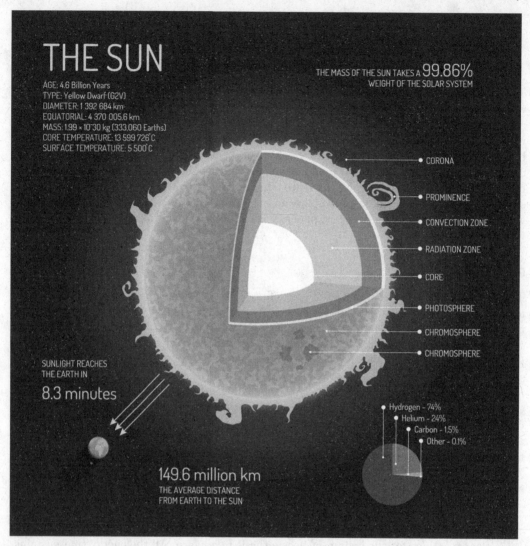

Figure 8.1 Diagram of the Sun drawn to scale. From the inner to the outer: core, radiative zone, tachocline, convective zone, photosphere, and atmosphere (chromosphere, transition layer, corona, and solar wind). On the top left of the figure, the photosphere details of a granule and sunspot are magnified. Transient phenomena like prominences and solar flares are also shown. See text for details on each region. (Credit: Shutterstock)

have higher density than the surrounding material and tend to sink. The global effect is the transfer of heat through the convection mechanism which creates high levels of granularity with small regions (about 1000 km in size) at different temperatures. The convective zone exhibits the differential rotation typical of fluids, where the poles rotate slower and the equator rotates faster.

5. The **photosphere**, about 100 km thick, is the first layer transparent to visible light, and has an equivalent black-body temperature of 5770 K. Here gamma and X-rays are free to travel out into space. Observations of the photosphere at different wavelengths show the presence of *sunspots* which appear darker and cooler compared to their

surroundings. The lower temperature of sunspots is due the high intensity of the magnetic field which inhibits the convection mechanism. Sunspots often come in pairs with opposite magnetic polarity. They are transient and recurrent phenomena; the number of sunspots varies from maximum to minimum every eleven years. More details on sunspots and the solar cycle are given in Section 8.2.

6. The **atmosphere** is the outer region of the Sun. It consists of four different parts:

  - The *chromosphere* with a thickness of 2000 km and a temperature that gradually increases with altitude from 4100 K to 20000 K. The starting temperature is low enough such that the spectrum of the chromosphere is dominated by emission and absorption lines of simple molecules such as carbon monoxide and water. Here the helium passes from a neutral state to a partially ionised one in the upper layers. The chromosphere can be seen in visible and ultraviolet light.

  - The *transition layer*, a thin layer of about 200 km, where the temperature rises rapidly up to 10^6 K. The transition layer is formed above the chromosphere at an altitude that is not well defined, similar to a halo with fuzzy edges. The high temperature causes helium to become completely ionised. From the observational point of view, the transition region can be seen by instruments sensitive to extreme ultraviolet light.

  - The *corona* extends up to a few $R_\odot$ and reaches temperatures up to 20×10^6 K. How the corona is heated is still not well understood but it is known that part of the heating comes from magnetic reconnection (see Focus Box 5.6). The corona finally evolves into the *solar wind* which is a continuous flow of plasma embedded with the Sun's magnetic field.

  - The *heliosphere* is considered to be the final stretch of the Sun's atmosphere extending as far out as the solar wind pressure is predominant with respect to the interstellar medium. Figure 5.5 shows an artist's impression. The heliosphere is evolving with the solar activity and its interaction with cosmic rays affects the lower energy range of their spectra. More details of the heliosphere, with recent spacecraft measurements, will be given in Section 8.5.

The Sun's magnetic field is quite strong. The order of magnitude of its field strength is ~ 1 mT at the polar surface and a tenth of that on average. The field strength is however very variable and even its polarity is not fixed. The Sun's structure is also very dynamic. In addition to its rotational motion, multiple surface phenomena can be observed, such as prominences, which are huge arcs made of denser and cooler plasma that can extend up to tens of thousands of kilometers above the solar surface, solar flares, which are explosive releases of electromagnetic energy in the Sun's corona, as well as surface details such as the Sun's granularity and sunspots on the photosphere. These phenomena will concern us through the next sections.

8.2 SUNSPOTS AND SOLAR CYCLES

Sunspots are dark regions of relative low-temperature plasma (~ 4000 K) with very strong local magnetic fields (up to 30 kG), which are formed on the photosphere close to the equator at latitudes between 20° and 30°. Each sunspot is associated with another one of opposite magnetic polarity (i.e. inward vs. outward pointing magnetic field), together they form a bipolar sunspot. The leading spot of the bipolar field has the same polarity as the solar hemisphere, while the trailing spot is of opposite polarity. Bipolar sunspots may associate to form groups. The Sun's activity is always changing with time and the number

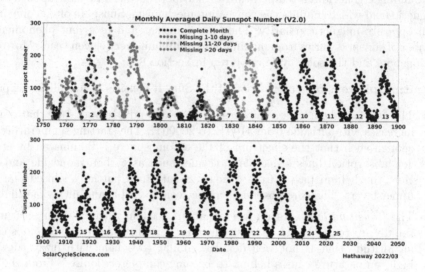

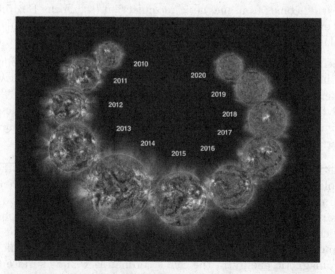

Figure 8.2 Monthly averaged sunspot numbers since 1750. Different symbols indicate the number of missing days in each monthly average, with the black dots representing complete months. (Credit: `http://solarcyclescience.com/solarcycle.html`)

Figure 8.3 Evolution of the Sun in extreme ultraviolet light from 2010 through 2020, as seen from the telescope aboard Europe's PROBA2 spacecraft. (Credit: Dan Seaton/European Space Agency, collage by NOAA/JPL-Caltech)

of sunspots is strongly correlated with the solar activity, being maximum during high levels of solar activity and minimum during quiet solar periods. For this reason, a useful quantity to monitor the solar activity is the Wolf number or universal sunspot number, defined as $W = k(10g + f)$, where f is the number of individual spots while g is the number of recognisable groups of sunspots and k is a correction factor which takes into account the technological evolution of instruments.

Figure 8.4 Schematic representation [272] of the global solar magnetic field reversal according to the Babcock model [160]. N indicates the north pole with magnetic field lines pointing out of the Sun; S indicates the south pole with magnetic field lines pointing into the Sun. (a) Axisymmetric poloidal magnetic field; (b) progressive growing of the toroidal magnetic field component due to the solar differential rotation; (c) formation of sunspot pairs with closed magnetic field lines; (d) end of the cycle: the poloidal magnetic field is reestablished with opposite polarity.

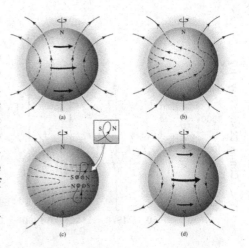

The first historical observation of sunspots was mentioned around 800 BC in China. Later, in the 17th century a systematic registration of their number and time variation was performed thanks to the invention of the telescope in Europe. Figure 8.2 shows the monthly averaged sunspot number recorded since 1750. The number of spots exhibit alternating periods of minimum and maximum which last about eleven years, defining the 11-year solar cycle. Maximum and minimum of the sunspot number vary through each solar cycle, but in general range from zero to a few hundred. The "first" solar cycle is defined as the period from 1755 to 1766. In 2022, we are in the 25th solar cycle. Figure 8.3 shows the time evolution of the Sun's corona from 2010 through 2020 as observed in extreme ultraviolet light by the telescope aboard Europe's PROBA2 spacecraft, demonstrating the approximately 11-year solar activity cycle. The bright images in solar maximum around 2014 show regions of plasma with magnetic loops anchored to the photosphere inside sunspots. The darker images in 2019 and 2020 indicate a quiet Sun during solar minimum.

Another important solar cycle lasts for about twenty-two years and is related to the Sun's magnetic field reversal. The Sun's magnetic field is strongly correlated with the solar activity and the sunspot number. During the solar minimum, the configuration of the Sun's magnetic field is well approximated by a dipole, with north and south hemispheres of opposite polarities. In a dipole configuration, the overall polarity of the Sun is referred to by the number A. If the northern hemisphere magnetic field lines point outward, the polarity is defined positive $(A > 0)$. In the opposite configuration, the polarity is negative $(A < 0)$. During the maximum of a solar cycle, when the sunspot number is at its maximum, the overall polarity of the Sun is not well defined: positive and negative magnetic field lines are organised such that, on average, the total magnetic field is zero. Every solar maximum, the Sun's magnetic field changes configuration reversing its sign with a periodicity of about twenty-two years, defining the 22-year solar cycle.

The exact details of the solar dynamo for magnetic field formation and reversal are still not well understood. In this section, we will give a simple representation of the Sun's magnetic field reversal as described by the Babcock model [160], following the diagram in Figure 8.4. Figure 8.4(a) shows the first stage of the reversal, starting during solar minimum when the solar magnetic field can be described by an axisymmetric dipole with a major poloidal component and a negligible toroidal one. This configuration is not stable because the Sun's surface does not rotate as a rigid body but exhibits differential rotation: elements of fluid near the equator spin faster than ones near the poles, as indicated by the bold arrows. Figure 8.4(b) shows how the differential rotation makes the poloidal magnetic field component progressively less important than the toroidal one. As seen in Figure 8.4(c),

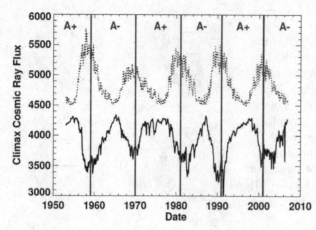

Figure 8.5 Anticorrelation of the cosmic ray flux with sunspot number and the 22-year solar cycle [554]. The solid line shows the monthly averaged counts from the Climax Neutron Monitor; the dotted line is the rescaled sunspot number (multiplied by five and offset by 4500). Vertical lines indicate epochs of different Sun's magnetic field polarity: A^+ are positive and A^- are negative polarities.

the progressive intensification of the toroidal magnetic field causes eruptions of magnetic field line loops on the photosphere which appear as sunspots. Sunspots of different polarities migrate towards opposite directions: leading spots, with same polarity as their own hemisphere, tend to migrate towards the equator, while trailing spots, with opposite polarity, migrate towards the solar pole. At the equator, the leading spots of each hemisphere cancel each other, while at the poles, the trailing spots neutralise the existing toroidal magnetic field in favour of the poloidal component. Progressive formation and migration of sunspots continue until the solar dipole field completely reverses, as shown in Figure 8.4(d), which represents the Sun's magnetic field configuration at the following solar minimum.

A well-known consequence of the 22-years cycle is the difference in the shape of the neutron monitor counts. Neutron monitors are described in Focus Box 8.1. They measure the integral cosmic ray flux at Earth above the local rigidity cut-off. Figure 8.5 shows the monthly average counts of cosmic rays measured by the Climax neutron monitor together with the rescaled sunspot number for six consecutive solar cycles. During periods of $A < 0$ polarity, the neutron monitor counts are sharply peaked, whereas during $A > 0$ periods, they are broadly flat. This is due to the drift motion of galactic cosmic rays in the heliosphere and will be described in detail in Section 8.6. We will see that the Sun's magnetic field direction has important consequences for the propagation of the galactic cosmic rays, especially when particles of opposite charge sign are considered.

8.3   THE SOLAR WIND AND THE SUN'S MAGNETIC FIELD

The solar wind is the extension of the solar corona far away from the Sun. Its plasma is expanding due to the very high pressure difference between the hot plasma at the Sun's surface and the interstellar medium, which sucks material from the Sun. Ludwig Biermann was the first who suggested that a stream of ionised gas was continuously emitted by the Sun, based on his studies on the motion of comets carried out in the 1950s [137]. In the late 1950s, Eugene N. Parker found that the corona could not stay in a simple hydrostatic equilibrium. He showed that the gravitational attraction of the Sun is not sufficient to balance the outward force exerted by the thermal pressure of the plasma, therefore the corona must expand [159]. This continuous outflow of plasma was named solar wind by Parker.

The solar wind initially travels subsonically and rapidly accelerates to supersonic velocities above 250 km/s at a few $R_\odot$. To match the conditions of the interstellar medium, it will eventually decelerate to subsonic velocity through the formation of a termination shock

Since International Geophysical Year 1957/58, a world-wide network of neutron monitors allows to closely follow the time evolution of the cosmic ray flux at Earth. Neutron monitors are relatively simple and stable instruments which can run unattended over long periods and automatically record the flux of galactic cosmic rays impinging on the atmosphere. When energetic enough, starting at about 500 MeV, cosmic rays cause air showers with an appreciable hadronic component, including neutrons. Due to their long lifetime, these can reach the Earth and be registered.

A neutron monitor consists of four principle elements:

- An outer shell of proton-rich material like paraffin or polyethylene, called a *reflector*. It absorbs low-energy neutrons coming from outside environment and reflects those generated on the inside.

- A lead target, called a *producer*, with which high-energy neutrons from the air shower interact, producing multiple low-energy neutrons. These are trapped inside due to the reflector.

- A *moderator* material, again rich in protons, which degrades the energy of neutrons further, so that they can be more easily detected.

- A proportional *counter tube* (see Focus Box 2.3) which registers charged hadrons produced in nuclear reactions of the slow neutrons with an active component of the counter gas. Examples of such reactions are $n\,^{10}B \rightarrow \,^{7}Li\,\alpha$ or $n\,^{3}He \rightarrow \,^{3}H\,p$.

An example of a typical set up [643], the standard NM-64 neutron monitor with six proportional tubes, is shown on the right. A yield function and a correction for atmospheric effects [643] allows to convert the measured count rate into an integral flux of cosmic rays at the location of the neutron monitor.

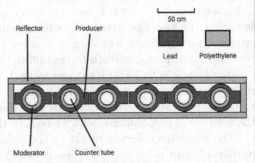

Since neutron monitors count airshowers, they effectively measure the integral flux of cosmic rays above the local geomagnetic cut-off (see Section 8.9). In addition to geomagnetic latitude, this cut-off also depends on the local overburden by the atmosphere [859], especially in the polar region. Using the Earth's magnetic field, a world-wide network with of the order of 50 active neutron monitor sites establishes a giant cosmic ray spectrometer. Results of many stations are recorded in a database sponsored by the European Commission, accessible at https://www.nmdb.eu/nest/. A large fraction of monitors transmits rates in real- or near real-time. They are stored as one-minute and hourly count rates. In addition, the Oulu neutron monitor station in Finland operates the ground level enhancement database https://gle.oulu.fi, making ground level enhancement data from solar energetic particles (see Section 8.8) available from a world-wide network.

Focus Box 8.1: Neutron monitors

at distances where the solar wind pressure is balanced by the pressure of the interstellar medium. More details of the termination shock and the general structure of the heliosphere will be given in Section 8.5.

The high corona temperature makes the solar wind plasma a very good conductor. As a consequence, the Sun's magnetic file lines are frozen in the solar wind while it expands

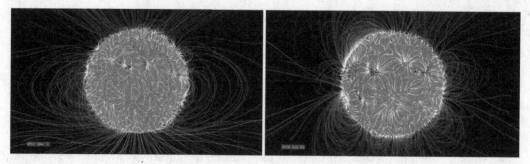

Figure 8.6 Illustration of the Sun's magnetic field during solar minimum in January 2011 (left) and solar maximum in July 2014 (right). Outward and inward magnetic field lines are shown in grey. Closed magnetic field lines are in white. (Credit: NASA Goddard Space Flight Center/Bridgman)

into interplanetary space (see Focus Box 5.2). This is evident from observations of the solar corona. Figure 8.6 (left) shows an illustration of the magnetic field lines close to the Sun during solar minimum in January 2011: at polar latitudes, the magnetic field lines are "open"[2] and streams of fast solar wind, with an average speed of 750 km/sec and originating from colder regions called coronal holes, are free to flow outward. Close to the Sun's equator, instead, the magnetic field lines form closed loops because the inward magnetic pressure is higher than the outward heat pressure, thus inhibiting the escape of the gas. At mid latitudes, the loop-like structures, which connect regions of opposite magnetic polarity, are stretched out by the push of the gas, which eventually reaches the height where the magnetic pressure is not strong enough to contain it, and streams of slow solar wind, with velocities ranging from 300 to 400 km/sec, are formed. Figure 8.6 (right) shows an illustration of the magnetic field lines during solar maximum in July 2014: the coronal holes almost disappear, while the mid latitude streamers extend to higher latitudes, making the slow component of the solar wind predominant with respect to the fast solar wind.

The first direct measurement of the solar wind at different latitudinal positions, and during different solar activity periods, was made by the Ulysses spacecraft during an 18-year mission from October 6, 1990 to June 30, 2009 [248]. Ulysses orbited the Sun with an inclination of 80.2° and was able to measure both solar wind components from the equatorial regions to the poles following an entire solar cycle evolution and Sun's magnetic field reversal. The top of Figure 8.7 shows the distribution of the magnetic field with inward and outward magnetic field lines during solar minimum in Ulysses' first orbit (left) and solar maximum in its second orbit (right), respectively. The intensity of the solar wind is displayed by the radial coordinate at different latitudinal positions. The bottom of Figure 8.7 shows the average monthly and smoothed sunspot number. As shown during solar minimum the field is well organised in a dipolar structure, whereas during solar maximum, the field lines appear more chaotic: the dipolar structure disappears and multipolar structures are formed.

Figure 8.8(A) shows the solar wind speed and corona temperature as a function of the heliographic latitude, measured by Ulysses in the second fast latitude scan, during the maximum in 2001. As can be seen, the solar wind speed is almost latitudinally independent but has a high degree of variability between 400 and 700 km/sec. Figure 8.8(B) shows the solar wind and temperature measured during the minimum in 1995, in the first fast Ulysses' latitude scan. As shown, the slow solar wind extends between −30° and +30°,

---

[2]By "open" we mean that the field lines are only closing far away from their origin, in this case the solar surface.

Figure 8.7 Top: Polar plots of solar wind speed [333] as a function of latitude for Ulysses' first two orbits, during solar minimum (left) and solar maximum (right). Sun's images are from the Solar and Heliospheric Observatory (SOHO). Bottom: Average monthly (white) and smoothed (grey) sunspot number. Vertical white lines delimit when Ulysses was orbiting in the North (N) and South (S) hemispheres.

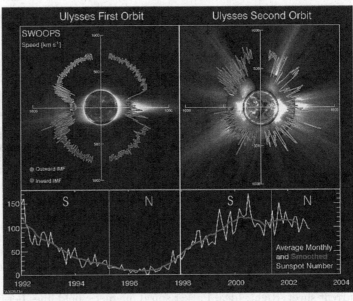

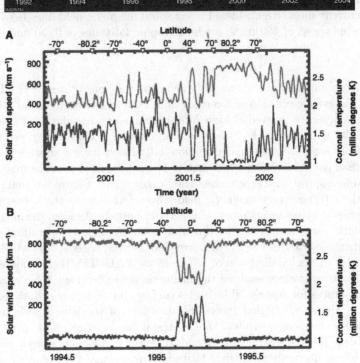

Figure 8.8 Anticorrelation of solar wind speed (upper curve) in units of km/s and coronal temperature at the solar wind source (lower curve) in units of 10^6 K, during (A) solar maximum and (B) solar minimum [334]. Bottom x-axis is time in units of years and top x-axis is Ulysses's heliospheric latitude.

while the fast solar wind, starting at latitudes $> |40°|$, has an almost constant speed of 800 km/sec.

8.4 HELIOSPHERIC MAGNETIC FIELD STRUCTURE

As already mentioned, the Sun's magnetic field lines are frozen in the solar wind while it expands through the heliosphere resulting in magnetic flux tubes which propagate out from the Sun. The locations where the magnetic flux tubes are attached to the Sun are called

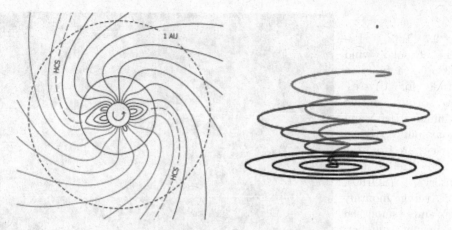

Figure 8.9 Left: Steady-state magnetic field lines in the ecliptic plane [507]. The latitudinal and longitudinal components of the solar wind and magnetic field disappear at the source surface, leaving only a radial expansion for the solar wind, while the heliospheric magnetic field follows the Parker spiral. Dashed lines separate opposite polarities with the heliospheric current sheet. Right: Ideal Parker spiral magnetic field lines between 0 and 25 AU for a solar wind speed of 450 km/s, for heliographic latitudes of 0, 30 and 60 degrees, respectively.

footpoints and are constantly moving as the Sun rotates. A fixed point on the Sun, which is in a specific position at a certain time, emits particles in radial direction. As the Sun rotates, particles are emitted at some later time in a new angular direction. For a given footpoint, the solar wind leaves the Sun at an approximately constant speed; thus, the particles released from some later angular direction will always have a shorter radial distance from the Sun than particles released previously. The net result is that the magnetic field lines are dragged into a spiral structure named the Parker spiral. Figure 8.9 (left) shows the Parker spiral of the interplanetary magnetic field, also referred to as the heliospheric magnetic field, in the ecliptic plane during quiet solar activity periods. As seen, the magnetic field lines are inclined with respect to the radial direction with an angle that gradually increases with increasing distance from the Sun. For normal solar wind speeds, this will lead to an inclination away from the radial direction of 25° to 45° at 1 AU. This inclination is nearly 90° at distances of 10 AU or greater, making the magnetic field effectively form rings around the Sun at great distances for equatorial latitudes corresponding to the ecliptic plane.

Figure 8.9 (right) shows the structure of the heliospheric magnetic field at 25 AU for three different latitudes. Due to the differential rotation of the Sun, outside the ecliptic plane the spiral structure extends upwards in a conical shape, with angular width gradually decreasing with increasing latitude. On average the solar rotation lasts 28 days: at the equator the rotation period is 24.47 days, while at the poles it is around 38 days. Due to the Sun–Earth relative motion, the apparent rotation period of the Sun as viewed from the Earth is around 27 days. A commonly used time interval is the Carrington rotation which lasts 27.2753 days, and is progressivly numbered starting from November 9, 1853. Another commonly used number is the Bartels rotation which lasts exactly 27 days and is numbered starting from February 8, 1832.

The regions above and below the magnetic equator have opposite polarity. Near the magnetic equator, the magnetic field abruptly changes sign passing from one polarity to the other: in one hemisphere, the field points towards the Sun, in the other away from it. A small current density of about 10^{-10} A/m^2 is present at the boundary which is called

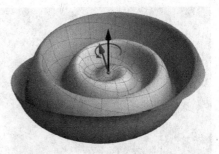

Figure 8.10 Graphical representation of the helio-spheric current sheet [701]. The long arrow represent the Sun's rotation axis, the short one the magnetic dipole direction. The angle between the two is the tilt angle.

the heliospheric current sheet. It defines the magnetic equator and is carried throughout the heliosphere by the solar wind. In addition, the Sun's rotation axis and the magnetic dipole are tilted by an angle. As a consequence, the heliospheric current sheet warps while it expands, assuming a wavy shape similar to a ballerina's skirt, as shown in Figure 8.10. The tilt angle and thus the heliospheric current sheet configuration, varies according to the solar activity. It increases progressively from a value between 4° and 10° during solar minima to values which can exceed 75° during solar maxima. As the solar activity increases, the shape of the heliospheric current sheet becomes increasingly disturbed. At solar maximum the heliospheric current sheet reaches maximum latitudes. After the magnetic field polarity reversal, it then returns to a quiet configuration close to the equatorial plane.

While the Earth moves along its orbit, it traverses sectors with different magnetic field polarity. During solar minimum, only two sector boundary crossings are seen in the ecliptic, whereas, during the maximum, the large dipole axis inclination and the additional multi-poles can lead to more than two boundary crossings. At each sector boundary crossing the Earth's magnetic field and the heliospheric magnetic field may have parallel or antiparallel orientations, or a mixture of the two. The relative orientations of the heliospheric magnetic field and the Earth's magnetic field have important consequences for particles entering the Earth's magnetosphere as will be seen in Section 8.10.

Finally, it is interesting to note that the solar wind is moving at speeds much greater than the magnetosonic speed of the plasma, which results in shocks when objects like the Earth are located within the flow. These shocks are nearly discontinuous increases in the pressure, temperature and density of the solar wind. Important examples are the bow shock of the Earth, and the heliopause at $\sim 120\,\mathrm{AU}$, the point at which the solar wind is stopped by the interstellar medium.

8.5 GLOBAL HELIOSPHERE STRUCTURE

The heliosphere is the region of space where the solar wind extends its influence. As already mentioned, the solar wind accelerates to supersonic speeds very close to the Sun, then it expands steadily until it undergoes a shock transition to subsonic velocities at the termination shock. The region of space beyond the termination shock is called the inner heliosheath. Here, the solar wind expands until its pressure balances the pressure of the interstellar medium at an interface called the heliopause. The region of space beyond the heliopause is called the outer heliosheath, located where the interstellar medium is compressed by the overall motion of the heliosphere and finally forms a bow shock or wave ahead of the heliopause.

Figure 8.11 shows a schematic representation of the global heliosphere with the locations of the termination shock and the heliopause from the direct observations by the Voyager 1 and Voyager 2 spacecrafts [194, 707]. These missions were launched in 1977 and, since then, have been traveling through the solar system following two different trajectories, north and

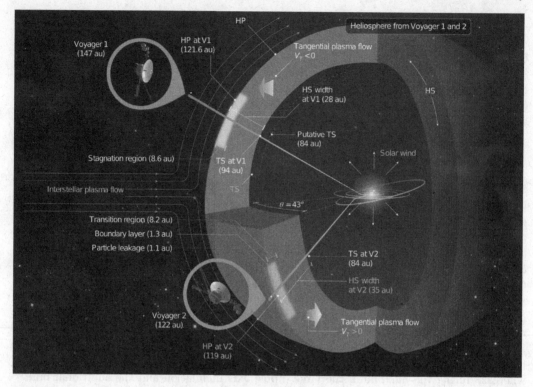

Figure 8.11 Schematic representation [707] of the global heliosphere from Voyager 1 and 2 measurements.

south of the ecliptic plane respectively. They passed through the termination shock and the heliosheath at ∼ 94 AU and ∼ 84 AU respectively. In August 2012, Voyager 1 passed through the heliopause entering the interstellar space at about 121 AU, setting a milestone in the exploration of the universe by humankind. In November 2018, also Voyager 2 passed the heliopause, entering the interstellar space at about 119 AU.

Figure 8.12 shows their measurement of the integral proton rate above 211 MeV and 213 MeV, respectively, as measured by the Low Energy Charged Particle (LECP) instruments [707]. The increases of the proton rate show that the Voyagers have passed into interstellar space where they now measure the Local Interstellar Spectrum (LIS) of galactic cosmic ray at the boundary of the heliosphere. Both spacecrafts are still sending scientific data about their surrounding environment. More recently, the Interstellar Boundary EXplorer (IBEX) found that the velocity of the interstellar flow is too low to form a shock [417]. Thus the outer heliosheath is bounded by a bow wave rather than a shock.

8.6 SOLAR MODULATION

Conditions in the heliosphere significantly affect the galactic cosmic ray spectrum, especially at rigidities below 100 GV. Entering the heliosphere, cosmic rays are subject to convection by the solar wind, drifts along magnetic field gradients and the heliospheric current sheet, diffusion by scattering on irregularities in the heliospheric magnetic field, and adiabatic energy loss due to the solar wind expansion. The overall effect is called solar modulation. The temporal evolution of the galactic cosmic ray spectra and its relation with the solar activity was monitored since the discovery of cosmic rays, and especially after the 1950s

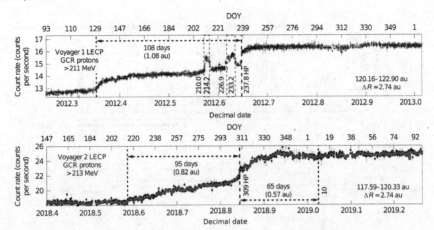

Figure 8.12 Comparison of the cosmic ray proton rates measured by Voyager 1 (top) and Voyager 2 (bottom) over about the same distance scale of 2.74 AU surrounding the respective heliopause crossings [707]. The time axis is shifted so that the heliopause crossings for the two spacecraft coincide.

by neutron monitors on the ground. Multiple balloon missions have been launched into the stratosphere to collect different species of galactic cosmic rays during different phases of the solar cycle over various periods of time ranging from 1 day to 42 days. Spacecraft such as Pioneer 10 and 11 [188], Ulysses [248] and Voyager 1 and 2 [194] have flown in space over long periods of time with the objective to provide a coherent picture of the conditions of the heliosphere and relate them to cosmic ray measurements on Earth. A fleet of satellites near Earth is dedicated to the study of particles, the local magnetic environment and space weather. For example, the Cosmic Ray Isotope Spectrometer (CRIS) instrument on board of the Advanced Composition Explorer (ACE) spacecraft has been functioning since the 1990s, recording galactic cosmic ray spectra and intensities for different nuclei. More recently, a new generation of space experiments like the magnetic spectrometers Payload for Antimatter Matter Exploration and Light nuclei Astrophysics (PAMELA) [630] and Alpha Magnetic Spectrometer (AMS-02) [775], are dedicated to measuring galactic cosmic ray nuclei, electrons and their antiparticles simultaneously. Figure 8.13 gives an overview of the space and balloon experiments measuring the flux of proton, helium and higher charged nuclei, between solar cycles 20 and 24, with the rigidity interval of the instrument. The sunspot number and the epochs of Sun's magnetic field reversal are also given for reference.

Due to the 11-year solar cycle, the intensity at the low-energy end of galactic cosmic rays periodically changes with time inside the heliosphere. Figure 8.14 shows the proton flux around 1 GeV measured by the EPHIN instrument on board of the SOHO mission, PAMELA and AMS-02 with the sunspot number and the number of counts registered by the Oulu neutron monitor during solar cycles 23 and 24. The polarity of the Sun's magnetic field and the periods of fields reversal are also indicated. The anticorrelation between the Sun's activity level and the intensity of galactic cosmic rays is evident. This is considered to be a long-term solar modulation and is related to the changes and evolution of the overall structure of the heliosphere. For a detailed review of solar modulation see for example [511].

Focus Box 8.2 gives a simple account of the diffusive transport of cosmic rays through the heliosphere. The description is similar to the general diffusion equation detailed in Focus Box 6.4, but simpler because inelastic interaction, decay and reacceleration can be neglected. The physical quantity that distinguishes between parallel and perpendicular diffusion

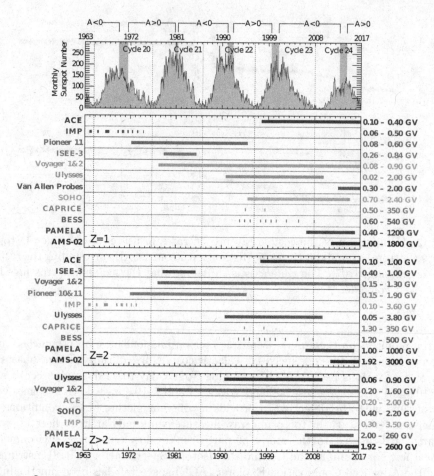

Figure 8.13 Overview of space and balloon experiments [616] measuring the flux of p, He and ions over different solar cycles. The sunspot number is shown at the top of the figure as a proxy of the phase of the solar activity when the data was taken. The vertical bands mark the periods of the solar magnetic field reversal. Only experiments providing multiple measurements in the rigidity range where cosmic rays below the cut-off are negligible ($R \gtrsim$ 0.4 GV) have been included.

coefficients is the pitch angle, which indicates the angle between the direction of motion of a particle and the direction of the magnetic field. The theory that describes the parallel diffusion is called quasi-linear theory. The field lines of the heliospheric magnetic field follow a random walk, so that a particle spiralling around them will be scattered perpendicularly with respect to the nominal background field. This effect is considered in the non-linear guiding centre theory of perpendicular diffusion. All diffusion coefficients can in general depend on position and rigidity. The perpendicular diffusion coefficient is typically modelled with the same rigidity and spatial dependence as the parallel diffusion coefficient, simply rescaled by a factor of a few percent.

Drift motions of particles are caused by gradients and curvature in the global heliospheric magnetic field and by the heliospheric current sheet. Figure 8.15 shows the meridional projection of trajectories followed by a 2 GeV proton in a $qA > 0$ heliospheric configuration, where q is the particle charge. Convection leads to the wavy trajectory around the equatorial

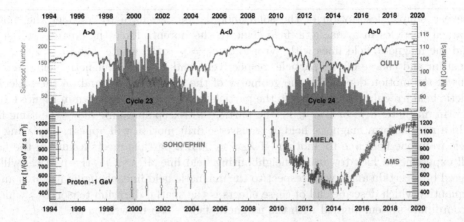

Figure 8.14 Top: Sunspot number and Oulu neutron monitor count rate during solar cycle 23 and 24 (https://www.nmdb.eu/). The shaded areas show the transitions of the Sun's magnetic field polarities. Bottom: galactic cosmic ray proton flux at about 1 GeV measured yearly by SOHO/EPHIN [582], monthly by PAMELA [654] and daily by AMS-02 [776].

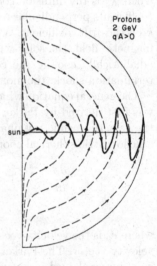

Figure 8.15 Meridional projection of the drift motion of a 2 GeV proton (solid line) in a $qA > 0$ heliospheric configuration [200]. The dashed lines indicate the solar wind flow direction above and below the heliospheric current sheet. The arrows will have opposite direction in a $qA < 0$ configuration.

plane. The heliospheric drift patterns depend on the heliospheric magnetic field polarity: during $A > 0$ cycles, positive charge particles enter the heliosphere mostly from the poles and drift out through the heliospheric current sheet, i.e close to the heliosphere equator. During $A < 0$ cycles, positive charge particles enter the heliosphere mostly through the heliospheric current sheet, and drift out from the poles. The opposite is true for $q < 0$ particles. Since the polarity of the Sun changes every 11 years, the difference in solar modulation between positive and negatively charged particles is particularly visible before and after the Sun's magnetic field polarity reversal. The drift effect explains the shape of the neutron monitor count rate shown in Figure 8.5: sharp peaks are observed in $A < 0$ and broad plateaus in $A > 0$ alternating every 22 years. Neutron monitors measure the integral flux of cosmic rays at all rigidities above the local geomagnetic rigidity cut-off (for the definition of the rigidity cut-off see Section 8.9). The incoming particles interact with the atoms of the atmosphere creating a shower of secondary particles which is detected on the ground. Thus the neutron monitor count rate is dominated by the most abundant species which are positive particles.

To have a simple description of the solar modulation, we have to consider the motion of charged particles in a magnetic field. Due to the Lorentz force the particle will spiral around the magnetic field lines with gyroradius $r_L = p/(qB)$, where p and $q = Ze$ are the momentum and charge of the particle, respectively, and B is the magnetic field strength. Details of the motion depend on the geometry of the magnetic field and on the scale λ of magnetic field irregularities relative to the particle gyroradius r_L as shown in Figure 6.6. In a uniform magnetic field or if $\lambda \ll r_L$ the particle will keep spiralling along the same field line. In a non-uniform magnetic field, a transverse drift motion will appear, displacing the particle from one field line to another. If $\lambda \gg r_L$ and a particle meets a kink in the field, it will experience a kick towards a neighbouring field line. If $\lambda \sim r_L$ the particle will be scattered by a certain angle with respect to the magnetic field line, changing its pitch angle. The quantity which describes all of these effects is the spatial diffusion tensor D_{xx} which – in a coordinate system aligned with the magnetic field – can be written as:

$$D_{xx} = \begin{bmatrix} k_{\parallel} & 0 & 0 \\ 0 & k_{\perp,\theta} & k_A \\ 0 & -k_A & k_{\perp,r} \end{bmatrix}$$

where $k_{\parallel}$ is the diffusion coefficient parallel to the field line, $k_{\perp,\theta}$ and $k_{\perp,r}$ are the diffusion coefficients perpendicular to the field line in the polar and radial directions, respectively, and k_A is the drift coefficient. While the particles are spiralling and drifting along the heliospheric magnetic field, the solar wind expands, causing a decrease in the particle momentum, i.e. adiabatically cooling the cosmic rays. If instead the solar wind compresses, like at the termination shock, the momentum will increase, causing an adiabatic acceleration.

The general equation which models the transport of galactic cosmic rays in the heliosphere was first derived by Parker [165]. According to the Parker equation, the time variation of the galactic cosmic ray phase space density $\rho(\vec{r}, p, t)$ as function of position $\vec{r} = (r, \theta, \phi)$ in a heliocentric spherical coordinate system, momentum p and time t can be written as:

$$\frac{d\rho}{dt} = \underbrace{- \vec{v}_p \nabla \rho}_{\text{convection}} \underbrace{- \langle \vec{v}_d \rangle \nabla \rho}_{\text{drift}} + \underbrace{\nabla (D_{xx} \nabla \rho)}_{\text{diffusion}} + \underbrace{\frac{1}{3} (\nabla \vec{v}_p) \frac{\partial \rho}{\partial \ln p}}_{\text{momentum loss}}$$

where $\vec{v}_p$ is the velocity vector of the solar wind plasma and $\langle \vec{v}_d \rangle$ is the average particle drift velocity vector. The Parker equation is in fact a simplified version of the general diffusion equation explained in Focus Box 6.4.

Focus Box 8.2: Solar modulation and Parker equation

During $A > 0$, positively charged particles mainly propagate in from the poles and follow a faster trajectory than those coming through the heliospheric current sheet. The maximum flux is thus reached faster than during $A < 0$ periods.

PAMELA and AMS-02 precisely measured the time dependence of the differential proton and helium flux during solar cycle 23 and 24. Figure 8.16 shows the PAMELA time dependent helium fluxes together with the proton-to-helium flux ratio (p/He) measured during the minimum of solar cycle 23 and the ascending phase of solar cycle 24. The fluxes mainly change at low energy and the p/He flux ratio shows two different time behaviours: below 1.31 GV the averaged p/He ratio before 2009 is higher than after 2009. Above 1.31 GV the behaviour is reversed; the p/He ratio before 2009 is lower than after 2009.

Figure 8.17 shows the AMS-02 measurement of the differential proton and helium fluxes versus time and rigidity, together with the p/He flux ratio for selected rigidities from 1.92 GV

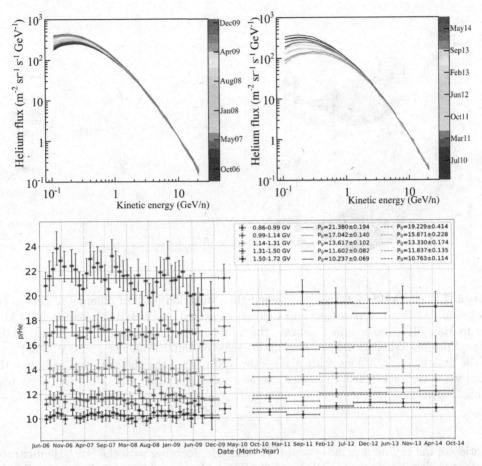

Figure 8.16 Top left: PAMELA helium differential fluxes versus kinetic energy, measured from July 2006 to December 2009 [744]; data are integrated over one Carrington rotation. Top right: PAMELA helium differential fluxes versus kinetic energy, measured from January 2010 to September 2014 [850]; data are integrated over three Carrington rotations, without isotopic composition separation. Bottom: PAMELA proton-to-helium flux ratio measured from June 2006 to October 2014 [850] for five rigidity intervals from 0.86 GV to 1.72 GV. Solid and dashed lines are averaged values of data before and after 2009. After 2009, data are integrated over nine Carrington rotations.

to 22.80 GV. The fluxes were measured every Bartels rotation during solar cycle 24, from May 2011 to May 2017. The low rigidity fluxes were at their minimum around February 2014. The p/He flux ratio gradually decreases after February 2015 at all rigidities below 3.29 GV. Figure 8.18(left) shows the AMS-02 daily helium fluxes for selected rigidity bins from 1.71 GV to 10.10 GV. The measurement covers most of solar cycle 24 from May 2011 to November 2019. Figure 8.18(right) shows the He/p flux ratio versus the helium flux, both averaged over a sliding window of more than one year (14 Bartels rotations) and an increment of one day. The sliding average helps to reduce the daily variability in the fluxes and to underline the long-term solar modulation. A hysteresis was found between the He/p ratio and both the helium and proton fluxes below 2.40 GV, from 2011 to 2014 and from 2014 to 2015. This means that the solar modulation of helium and proton is different before and after the solar maximum in 2014. Figure 8.19 shows the AMS-02 time

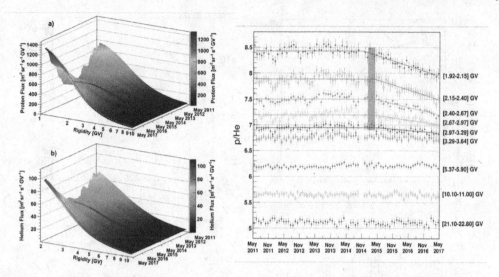

Figure 8.17 Left: AMS-02 three-dimensional time variation [639] of differential proton (a) and helium (b) fluxes versus time and rigidity, measured every Bartels rotation from May 2011 to May 2017. Dashed lines indicate the minimum and maximum of the fluxes. Right: AMS-02 proton-to-helium flux ratio [639] for selected rigidities above 1.92 GV. Solid lines are the best fit of data done with a single break in time; the vertical band indicates the period when the ratio starts to change. No isotopic composition separation was performed in this analysis.

variation of the helium flux for isotopes ^{3}He and ^{4}He, together with the ^{3}He/^{4}He flux ratio for selected rigidities from 2.15 GV to 15.30 GV; the fluxes are integrated over four Bartels rotations from May 2011 to November 2017. Also in this case, a time-dependent behaviour was observed after 2014 for rigidities below ~ 4 GV.

According to the Parker equation, particles with the same rigidity should have the same solar modulation. Differences at low rigidities may be due to differences in the terms containing the velocity, mainly in the diffusion and drift terms. At the same rigidity, proton, helium and their isotopes have different mass-to-charge ratios, so they have different velocities, as long as $\beta < 1$. In addition, the corresponding local interstellar spectra are expected to have different rigidity dependencies [576, 655], inducing differences in the energy loss term of the Parker equation. To understand the relative importance of the different terms better, which causes differences in the solar modulation, sophisticated numerical models based on the Parker equation, are needed [698, 702].

PAMELA and AMS-02 both directly measured the charge-sign dependence of solar modulation [593, 631, 640]. Figure 8.20(top-left) shows the differential electron fluxes measured every six months by PAMELA from 2006 to 2009, together with the local interstellar spectrum. At energies less than ~ 10 GeV the fluxes are largely suppressed compared to their interstellar value. The measurements cover three years of solar cycle 23 when the solar activity was gradually decreasing. Due to the 11-year solar cycle, the PAMELA electron fluxes gradually increased from 2006 to 2009 as shown in Figure 8.20(bottom-left) where the fluxes are normalised to the 2006 flux. Figure 8.20(right) shows the PAMELA positron-to-electron flux ratio (e^+/e^-) measured every three months from 2006 to 2016 and normalised to the 2006 ratio at three different energy intervals between 0.5 GeV and 5 GeV. The measurement covers the period of the Sun's magnetic field polarity reversal from $A < 0$ to $A > 0$. Due

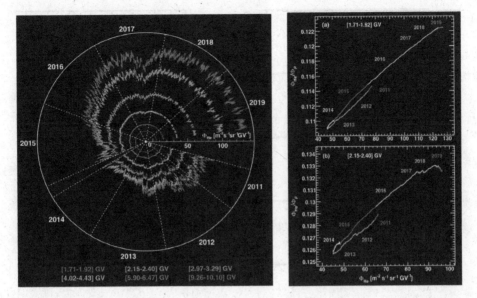

Figure 8.18 Left: AMS-02 daily helium fluxes [837] for selected rigidity bins from 1.71 GV to 10.10 GV measured from May 2011 to November 2019. Right: Sliding average over 14 Bartels rotations with an increment of one day for the AMS-02 He/p flux ratio [837] versus helium flux for (a) [1.71 to 1.92] GV and (b) [2.15 to 2.40] GV.

to the charge sign effect in particle drift, e^+/e^- is increased after the Sun's magnetic field polarity reversal.

Figure 8.21(left) shows the monthly time profiles of the AMS-02 positron and electron fluxes in five energy intervals between 1 GeV and 21 GeV, measured from May 2011 to May 2017. The solar modulation of the opposite charged particles is distinctly different: the fluxes were rescaled to overlap during the period when fluxes were increasing, after April 2015, to simplify the comparison. The charge-sign dependence of solar modulation is evident in Figure 8.21(right) where the AMS-02 positron-to-electron flux ratio is shown. According to the drift effect, during $A < 0$ the e^+ mainly follow the heliospheric current sheet, thus are more suppressed than the e^-. From the polarity reversal to the $A > 0$ period, e^+ are favoured to diffuse from the poles, faster than e^-; thus, the flux ratio is increasing.

8.7 TRANSIENT PHENOMENA IN THE SOLAR WIND

The Sun can emit magnetic disturbances with the solar wind. These can significantly change the Parker spiral structure thus affecting the long-term solar modulation, especially during solar maximum. Such transient phenomena are considered to be a short timescale solar modulation. They can be non-recurrent like coronal mass ejections (CME) or recurrent like corotating interaction regions (CIR). Both phenomena affect the galactic cosmic ray spectra at low rigidities on short timescales and in a major way and lead to the formation of magnetic barrier structures which prevent the diffusion of galactic cosmic rays. This causes a temporary reductions of the galactic cosmic ray spectra at low rigidities, called Forbush decreases. Forbush decreases are characterised by a fast flux suppression on the scale of hours and a long recovery phase lasting from a few days to a month. Their causes and effects will be further described in the next Sections.

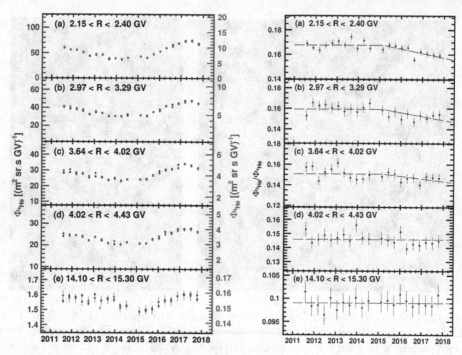

Figure 8.19 AMS-02 helium fluxes for isotopes ^{3}He and ^{4}He (left) and the ^{3}He/^{4}He flux ratio (right) for selected rigidities from 2.15 GV to 15.30 GV, integrated over four Bartels rotations from May 2011 to November 2017 [685].

8.7.1 INTERPLANETARY CORONAL MASS EJECTIONS

Coronal mass ejections are huge clouds of plasma ejected from the Sun at velocities greater than the solar wind speed. The fastest coronal mass ejections can move at more than 2000 km/sec, creating a shock in front of them in which particles can be accelerated (see Section 8.8). A coronal mass ejection propagating outward from the Sun at distances greater than $50R_\odot$ is called an Interplanetary Coronal Mass Ejection (ICME). Upon reaching Earth, these can result in large disturbances of the Earth's geomagnetic field and cause changes in the particle population of the magnetosphere, called magnetic storms (see Chapter 1 and Section 8.10).

Figure 8.22(left) shows the classical structure of an interplanetary coronal mass ejection [464]: a magnetic cloud is emitted out from the Sun during the coronal eruption with embedded and twisted magnetic field lines; the twisted magnetic field is frozen into the plasma cloud in a structure called the flux rope; the high speeds of the interplanetary coronal mass ejection lead to the formation of a shock wave in front of the cloud. Between the shock and the leading edge of the cloud is a particularly turbulent sheath of relatively strong magnetic fields. Each of these features may be important for the effects of an interplanetary coronal mass ejection on the shape and structure of the associated Forbush decrease. In fact, the time variation of the cosmic rays' profile strongly depends on the observer's position. Forbush decreases associated with interplanetary coronal mass ejections may be of three types: those caused by a shock and ejecta, those caused by ejecta only and those caused by a shock only [299]. If the observer is invested by both the shock and the cloud, the measured Forbush decrease will have a sharp drop and a long recovery phase (see Figure 8.22, insert A). If a coronal mass ejection is preceded by a shock, a characteristic two-step

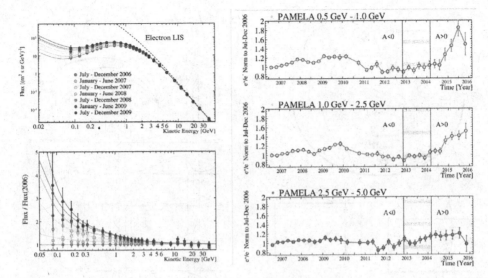

Figure 8.20 Top-left: PAMELA electron fluxes measured every six months from 2006 to 2009 together with the electron local interstellar spectra (dashed line). Bottom-left: The electron fluxes normalised to the 2006 flux. Right: PAMELA positron-to-electron flux ratio (e⁺/e⁻) measured every three months and normalised to the 2006 ratio at three energy intervals between 0.5 GeV and 5 GeV, measured from 2006 to 2016 [631]. (Credit: with kind permission of Società Italiana di Fisica)

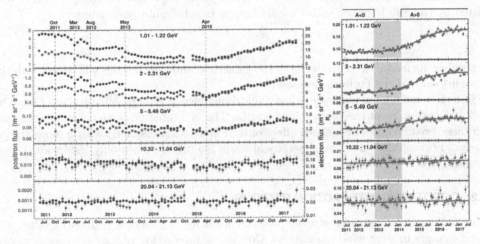

Figure 8.21 Left: Monthly time profiles of the AMS-02 positron and electron fluxes in five energy intervals between 1 GeV and 21 GeV, measured from 2011 to 2017 [640]. Right: Corresponding positron-to-electron flux ratio $R_e = (d\Phi_{e^+}/dE)/(d\Phi_{e^-}/dE)$.

Forbush decrease is expected; the first is a step associated with the shock, the second with the ejecta [299, 442]. The galactic cosmic ray intensity is suppressed inside the ejecta due to the cumulative diffusion across magnetic field enhanced regions. A decrease may even be observed from distant ejecta if it was energetic enough to create an interplanetary shock [552]. The observed depth of a Forbush decrease depends on the trajectory through the interplanetary disturbances [276]. The faster the propagation of the interplanetary disturbance, and

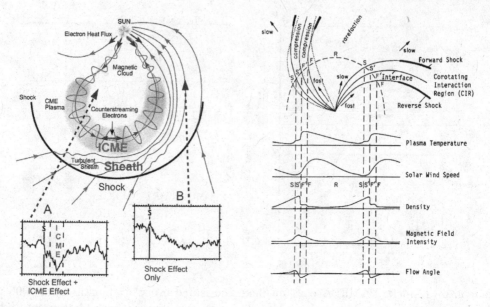

Figure 8.22 Left: Classical structure on an interplanetary coronal mass ejection [464] with cloud (shaded), flux rope (arrow lines) and shock wave (bold black). Inserts show galactic cosmic ray flux as a function of time. Depending on the observer position, the associated Forbush decrease will show (A) a double mark structure with shock and interplanetary coronal mass ejection effect or (B) with shock effect only. Right: Schematic representation of a corotating interaction region [355] formed by two high speed streams of solar wind emitted by the Sun together with typical changes in solar wind parameters.

the stronger its magnetic field, the more the galactic cosmic ray density will decrease during the main phase of the Forbush decrease.

Since forward shocks are wider than the driving ejecta, it is possible to pass through a shock but not intercept the coronal mass ejecta. The measured Forbush decrease associated with these events will show a shock effect only (Figure 8.22, insert B) with a more gradual decrease. Forbush decreases are generally of lesser magnitude when only the forward shock is present [271].

8.7.2 COROTATING INTERACTION REGIONS

Figure 8.22(right) shows two streams out of corotating interaction regions together with typical changes in solar wind parameters. Corotating interaction regions are shocks created at the interface between slow and fast solar wind streams, which corotate with the Sun. When coronal holes form, the Sun's plasma is able to easily escape into the interplanetary magnetic field along open magnetic field lines, resulting in a high speed stream of solar wind emitted from this region. As the high speed stream collides with the normal solar wind, the latter will be accelerated and compressed. The high speed stream, on the other hand, is decelerated and compressed. The region of compressed plasma created by this collision is the corotating interaction region. The compression causes a forward and backward wave propagating through the solar wind, which can become a shock at radial distances approaching 2 AU as the sound speed in the solar wind decreases. Corotating interaction regions have been observed to cause Forbush decreases as they sweep across regions in interplanetary space. Furthermore, because the coronal holes which cause these events can last longer than

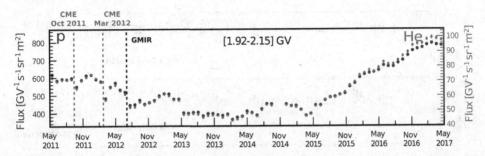

Figure 8.23 AMS-02 monthly proton (black) and helium (grey) fluxes for the rigidity bin [1.92 to 2.15] GV [639]. The black vertical dashed-line indicates a decrease in the fluxes during July 2012, induced by a global merging interaction region formed by two coronal mass ejections emitted by the Sun on October 2011 and March 2012 (vertical dashed-lines) [710].

the rotational period of the Sun, corotating interaction regions can cause recurring Forbush decreases with a period of ~ 27 days given by the solar rotation period.

In the outer heliosphere, these disturbances can coalesce, forming merging interaction regions (MIR), huge structures with compressed magnetic fields and plasma. An accumulation of these can lead to the formation of global merging interaction regions (GMIR), which have a wide latitudinal and longitudinal extent and affect the global structure of the heliospheric magnetic field. The overall effect of these global structures is to create additional barriers for the inward propagation of galactic cosmic rays, further increasing their modulation on a long timescale. Contrary to the Forbush decrease, their flux suppression does not have a recovery phase. Figure 8.23 shows an example of such a global event: the big drop of the AMS-02 monthly fluxes measured in July 2012, was attributed to a global merging interaction region formed by two strong coronal mass ejections emitted by the Sun on October 2011 and March 2012, which had already been responsible for the two preceding Forbush decreases observed in the fluxes [710].

8.7.3 FORBUSH DECREASES

Forbush decreases are temporary decreases in the galactic cosmic ray flux due to heliospheric disturbances, first reported in 1937 by Scott E. Forbush [97]. In the 1950s, John A. Simpson, using neutron monitors, showed that the origin of the Forbush decreases was in the interplanetary medium. Forbush decreases are caused by huge heliospheric magnetic field structures (compared to the size of the Earth) convected by the solar wind that sweep away some of the galactic cosmic rays. Since these field structures are a result of solar activity, a Forbush decrease is fundamentally a heliospheric phenomenon. Forbush decreases are diverse, varying in size, duration, rate of decrease, rate of recovery and overall profile. They can be categorised into recurrent and non-recurrent Forbush decreases. Recurrent Forbush decreases are associated with corotating interaction regions formed at the leading edges of high speed solar wind streams. Non-recurrent Forbush decreases are caused by the passage of transient solar wind structures, interplanetary coronal mass ejections, and/or the shocks that they drive.

Figure 8.24 shows the normalised low-energy proton and helium fluxes from AMS-02 for selected rigidities between 1 GV and 10 GV during three non-recurrent Forbush decrease events in September and October 2011, together with the normalised count rate for neutron monitors at different geomagnetic latitudes, as well as different solar wind parameters. The

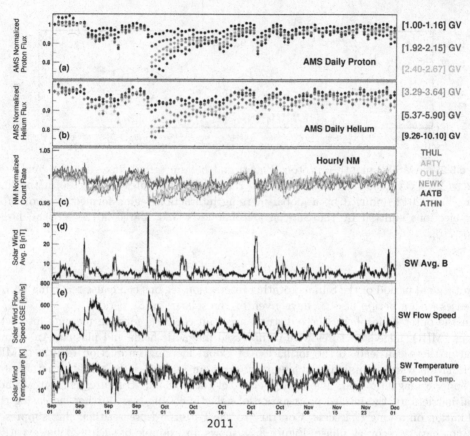

Figure 8.24 Normalised AMS-02 daily (a) proton [776], (b) helium [837] fluxes, relative to the flux on September 1, 2011, for selected rigidity bins. (c) Neutron monitor normalised count rate of Thule-USA (rigidity cut-off 0.30 GV), Apathy-Russia (cut-off 0.65 GV), Oulu-Finland (cut-off 0.81 GV), Newark-USA (cut-off 2.40 GV), Alma-Ata B-Kazakhstan (cut-off 5.90 GV) (data from https://www.nmdb.eu/). (d) Solar wind average magnetic field, (e) flow speed, (f) temperature and expected temperature (data from https://omniweb.gsfc.nasa.gov). All data have been taken from September 1 to December 1, 2011.

start of each Forbush decrease on September 9, September 27, and October 25, 2011 is accompanied by a sudden increase in each of the solar wind parameters, indicating the arrival of an interplanetary coronal mass ejection shock. The temperature on September 27–29 is lower than the expected temperature, indicating the presence of a magnetic cloud at Earth. Neutron monitor stations at different latitudes are sensitive to the integral cosmic ray spectra above their local rigidity cut-off, as can be seen from the shape of the ratios where broader dips correspond to stations at lower rigidity cut-offs. It is evident how the changes in the solar wind conditions can influence the fluxes measured at or near Earth, differently for each AMS-02 rigidity bin and each neutron monitor station. The lower the energy, the deeper is the dip in the spectrum.

Figure 8.25(left) shows the normalised PAMELA proton, helium and electron fluxes in the rigidity interval from 1 GV to 2 GV during the Forbush decrease in December 13, 2006 [668]. Due to high statistics, the proton flux was measured with a time resolution of

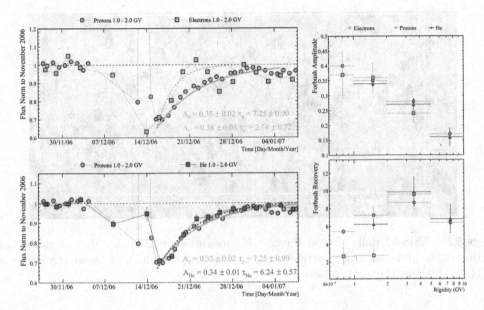

Figure 8.25 Left: PAMELA proton, helium and electron fluxes in the rigidity interval from 1 to 2 GV during the Forbush decrease on December 13, 2006. Right: Amplitudes and recovery times of protons, helium and electrons, in nine rigidity intervals between 0.4 and 20 GV [668].

three to six hours up to 5 GV, and one day above 5 GV. Helium and electron fluxes were measured with a two-day time resolution. The increase measured in the proton and helium fluxes on December 14 is associated with a solar energetic particle event (see Section 8.8). Data were fitted with an exponential function to study the amplitude and recovery time of the Forbush decrease event. Figure 8.25(right) shows the resulting amplitudes and recovery times in nine rigidity intervals between 0.4 and 20 GV. The amplitude of the decrease is found to be similar for all particles. The recovery times of proton and helium nuclei show similarities as well. On the contrary, electrons below 2 GV exhibit a faster recovery time than nuclei. This behaviour can be interpreted as a charge-sign dependence effect due to the different global drift pattern between positively and negatively charged particles in a $A < 0$ solar period.

Historically, neutron monitors have been used to measure recurrent Forbush decreases in the galactic cosmic ray count rate associated with corotating interaction regions. These Forbush decreases recur at the solar rotation period of 27 days. Shorter periods of ~ 13.5 and ~ 9 days were also observed. These periodicities are also observed from space. As an example, Figure 8.26 [774] shows the AMS-02 daily proton fluxes measured in 2016 for three rigidity bins, with Bartels rotations also indicated. Similar results where obtained with an analysis of the AMS-02 daily helium fluxes [837].

8.8 SOLAR FLARES AND SOLAR ENERGETIC PARTICLES

Solar flares are explosive events in the Sun's corona associated with a release of electromagnetic energy. The emitted photons extend across the electromagnetic spectrum from radio waves, through X-rays to gamma rays. The classification of solar flares is denoted with letters A, B, C, M and X according to the peak intensity in watts per square meter of

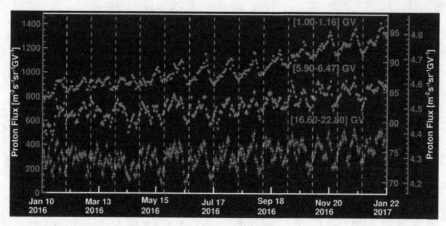

Figure 8.26 AMS-02 daily proton fluxes [776] measured in 2016 for three rigidity bins. Vertical dashed lines separate Bartels rotations. Double-peak and triple-peak structures are visible in different Bartels rotations.

X-rays in the wavelength range from 1 to 8 Ångstrom (1 to 8×10^{-10} m). The classes are delimited by a 10-fold increase in strength, with a subdivision from 1 to 9 within each class: A with intensity $< 10^{-7}\,\mathrm{W/m^2}$; B with $[10^{-7} - 10^{-6}]\,\mathrm{W/m^2}$; C with $[10^{-6} - 10^{-5}]\,\mathrm{W/m^2}$; M with $[10^{-5} - 10^{-4}]\,\mathrm{W/m^2}$; and X with $> 10^{-4}\,\mathrm{W/m^2}$.

The magnetic fields in active regions are large loops with footpoints anchored inside of sunspots at the photosphere. The footpoints move around due to convection and may cause the loop to interact with itself or another loop in such a way that two lines of opposite polarity meet [310]. This creates a cross-like configuration of magnetic field lines, with zero magnetic field at the crossing and a sudden release of the magnetic tension in the loop, initiating a highly energetic process called magnetic reconnection, described in Focus Box 5.6. Figure 8.27(a) displays a schematic of the flare geometry, showing the reconnection site and the generation of a turbulent fast shock which accelerates particles in the region above and below. Energised electrons stream along the closed and open magnetic field lines within the flare geometry, emitting electromagnetic radiation defining the characteristics of the flare. Electrons accelerated near the reconnection site stream along closed magnetic field lines back towards the corona, producing soft X-rays via bremsstrahlung. A hard X-ray source is visible where the particles energised near the reconnection site collide with the denser top loop. Likewise, electrons moving downwards inside of the loop will collide with the chromosphere below, creating two hard X-ray sources at the footpoints of the magnetic loop.

Solar flares are often, but not always, accompanied by coronal mass ejections. When magnetic reconnection occurs, the magnetic field lines are significantly curved and magnetic tension can cause the material from the corona to erupt and expand into space, forming a coronal mass ejection. This ball of plasma and magnetic fields continues to expand and propagate outwards as an interplanetary coronal mass ejection.

During these explosive events, Solar Energetic Particles (SEP) may be accelerated to energies higher than the solar wind, from a few hundred keV to a few GeV, escaping their acceleration site, and moving through the heliosphere, generally following the heliospheric magnetic field. In contrast to the solar wind, they thus qualify as cosmic rays. There are two potential acceleration sites for solar energetic particles: magnetic reconnection as shown in Figure 8.27(a), and the shocks driven by interplanetary coronal mass ejections shown in

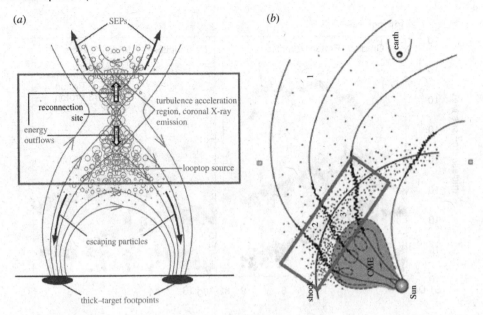

Figure 8.27 The two potential acceleration mechanisms for solar energetic particles [700]. (a) Acceleration during the magnetic reconnection which drives a coronal eruption. The thin solid lines are closed coronal loops and open magnetic field lines. The foam represents turbulence. The bold arrows are the directions traveled by particles escaping the reconnection site. (b) Fermi acceleration on a shock driven by a coronal mass ejection, delimited by the box.

Figure 8.27(b). In the magnetic reconnection region particles are subject to direct current electric field acceleration, stochastic acceleration or shock acceleration. In contrast to that, at coronal mass ejection driven shocks the particles are subject to first order Fermi acceleration only. Historically, solar energetic particles were divided in two main categories: gradual events that are the result of coronal shock waves, and impulsive events that are the result of flares. This classification, however, is an oversimplification of the effective observations which may show overlapping features.

Gradual events are dominated by protons and tend to have a solar wind-like composition, as shown in Figure 8.28(a). They have a particle flux increase extended in time which implies continuous acceleration, like in the shock front of an outward propagating coronal mass ejection. They originate from all longitudes across the solar disk, indicating that the acceleration source has access to a broad set of magnetic field lines [296]. This is in agreement with an expanding coronal mass ejection which connects to an increasing volume of magnetic field lines as it moves into interplanetary space. Gradual events display a wide range of time profiles that can be explained via connectivity to the coronal mass ejection shock front and its relationship with the Parker spiral.

Impulsive events show a quick impulsive increase followed by a rapid decay as shown in Figure 8.28(b). Their composition differs from the solar wind in that they contain high charge states, are enriched with anomalously high abundances of ^{3}He, and display high electron-to-proton ratios. Impulsive events come from source regions which are magnetically well connected to the observer, implying that the acceleration regions are small and only have access to a limited range of open magnetic field lines [296].

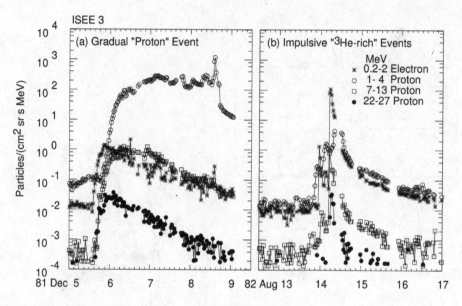

Figure 8.28 Intensity profiles as a function of time [296] of protons and electrons for "pure" gradual and "pure" impulsive solar energetic particle events measured by NASA's International Sun-Earth Explorer-3 (ISEE-3). (a) Gradual events show a time profile with a quick increase followed by a gradual decline in intensity over many hours to days. This event on December 5, 1981 was associated with an erupting filament as part of a coronal mass ejection, but with no accompanying flare. (b) Impulsive events show a quick increase followed by a rapid decay. These events on August 13 and 14, 1982 followed a series of flares with no evidence of an accompanying coronal mass ejections.

Lower energy solar energetic particle events can only be observed in space. Very high-energy solar energetic particles, instead, can also be detected on Earth by neutron monitors. These have enough energy, greater than about 500 MeV, to cause a shower of secondary particles when they strike the atmosphere, so as to be measured on the ground. These events are called Ground Level Enhancements (GLE).

Different sources of acceleration may be studied by taking at look at the energy spectra of the event. Many models were developed to explain the different features of the solar energetic particle spectra considering both acceleration and propagation processes. Figure 8.29 shows the September 10, 2017 time-integrated energy spectra for protons. The spectra were fitted by two simple functions: a double power law with an energy break typically associated with the limits of shock acceleration; and a combined function, which is like the former, but modulated by an exponential cut-off. The cut-off energy is then attributed to particles escaping the shock. This results in a steeper decrease at the high-energy end, which fits the data slightly better. The precise measurements from space spectrometers like PAMELA and AMS-02 help to constraining the high-energy range further. However, these detectors are not always exposed to solar energetic particles due to their orbital motion in the Earth's magnetic field where they cross different geomagnetic regions with different cut-off rigidities. Figure 8.30 shows the integrated proton energy spectra of all solar energetic particle events measured by PAMELA from 2006 until the end of the mission in 2014 together with a fit using a single power law modulated by an exponential cut-off [649]. The fits show that an exponential cut-off is indeed required.

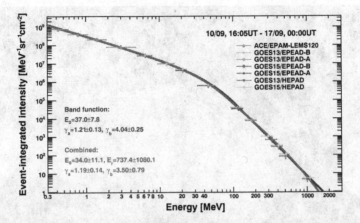

Figure 8.29 Time-integrated differential flux for combined measurements of protons by ACE and GOES on September 10, 2017 [696]. The upper curve is a fit done with a double power law. The upper insert quotes the two spectral indices and the break energy. The lower curve is fitted with a an additional exponential cut-off. The lower insert gives the parameters including the cut-off energy.

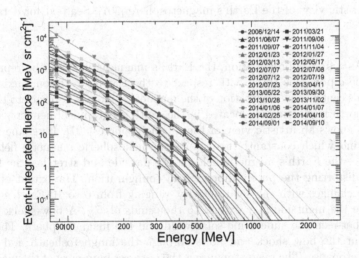

Figure 8.30 Time-integrated differential flux of protons from solar energetic particle events measured by PAMELA from 2006 to 2014, together with a fit using a single power law modulated by an exponential cut-off.

8.9 THE EARTH'S MAGNETOSPHERE

We now step down in distance scale to discuss what happens to cosmic rays close enough to Earth to feel the influence of the terrestrial magnetic field. At the same time, the order of magnitude of magnetic fields changes. The Earth's magnetic field is much weaker than that of the Sun, varying between about $25\,\mu\mathrm{T}$ at the equator and $65\,\mu\mathrm{T}$ at the poles. The magnetosphere is the region of space around the Earth where this magnetic field extends its influence. The Earth's magnetic field is generated by electric currents induced by the convection currents of the molten iron and nickel mantle that flows on the surface of the

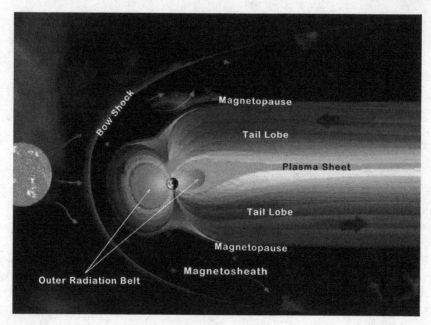

Figure 8.31 Artistic view of the Earth's magnetosphere [307]. See text for details.

Earth's core. As a first approximation, the Earth's magnetic field can be represented as a dipole, with the dipole axis tilted with respect to the Earth's rotational axis and shifted from the Earth's centre. The orientation of the magnetic dipole is opposite to the rotational axis with the north magnetic pole located in the south geographical hemisphere.

Figure 8.31 shows an artistic view of the magnetosphere [307]. Its shape is influenced by the solar wind which constantly transports the heliospheric magnetic field. The solar wind compresses the Earth's magnetic field on the dayside and stretches the field lines on the nightside, distorting the purely dipolar field configuration. The extent of the dayside magnetosphere changes with the solar activity, ranging from 6 to 10 Earth's radii ($R_E = 6371$ km), while the nightside may extend to thousands of R_E. A bow-shock is formed at the boundary between the supersonic solar wind and the magnetosphere. The solar wind is decelerated at the bow shock and flows through the magnetosheath and tangentially along the magnetopause. The magnetopause is the outmost boundary of the magnetosphere, which partially prevents the solar wind plasma from penetrating the inner regions of the magnetosphere. At the interface between the day- and nightsides, close to the magnetic poles, there are two regions called the polar cusps where the magnetosheath plasma extends deeper into the dense region of the atmosphere. On the nightside, the magnetic field lines are highly stretched and become "open" forming the magnetotail. In this region, the plasma mantle coming from the magnetosheath extends almost parallel to the magnetic field lines in the tail lobes. The division between the north and south regions of the magnetotail is characterised by magnetic field lines of opposite polarity which come very close to each other, forming a neutral current sheet layer. The inner region of the magnetosphere is characterised by a dipolar magnetic field with closed field lines.

Trajectories of charged particles are considerably bent when entering the magnetosphere. For this reason, the directions of the galactic cosmic rays measured by an observer near Earth are different from the incoming direction that the particles had before entering the magnetosphere. This effect increases with decreasing particle rigidity, until a real selection

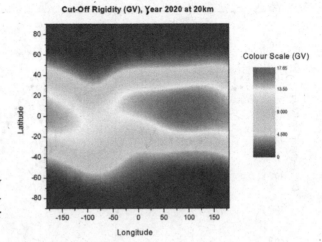

Figure 8.32 The vertical rigidity cut-off calculated at 20 km altitude using IGRF-12 for the year 2020 [796].

in rigidities occurs: particles below a particular rigidity, called the geomagnetic rigidity cut-off [368, 427], are excluded from penetrating some regions of the magnetosphere. The rigidity cut-off depends on altitude, latitude and the direction of the incoming particle. Figure 8.32 shows an example of the rigidity cut-off calculated for vertical incoming directions at an altitude of 20 km, using the 12th generation of the International Geomagnetic Reference Field (IGRF-12) for the year 2020 [796]. The rigidity cut-off varies from zero at the geomagnetic poles to ~ 15 GV at the equator.

The effect of the geomagnetic cut-off on the spectra of galactic cosmic rays is demonstrated in Figure 8.33 with differential proton fluxes measured by AMS-01 at various geomagnetic latitudes [324]. The spectra have two components: the high-energy galactic cosmic rays start out above the latitude dependent geomagnetic rigidity cut-off. The flux observed below the cut-off can only come from protons due to backsplash, when galactic cosmic rays interact with the upper layers of the Earth's atmosphere. When energy and emission angle allow, such protons can be trapped in the Earth magnetic field as explained in the next Section. Consequently, in Low Earth Orbit, particles below the cut-off come from the zenith as well as the nadir direction, while above they only come from the zenith hemisphere [324].

8.9.1 RADIATION BELTS

Trapping of charged particles in the Earth's magnetic field is possible everywhere, provided that the pitch angle with respect to the field and the particle rigidity allow it. However, there are two regions around Earth which are much more densely populated with trapped particles than others. These are located in the inner magnetosphere and called Van Allen radiation belts. The two are divided by a less populated slot region. The inner belt extends from about 1600 to 13000 km above the equator. The outer belt extends from about 19000 to 40000 km altitude. Figure 8.34 shows a cutaway model of the near-Earth radiation belts with the two Van Allen Probe satellites and others flying at different altitudes through the radiation belts. In Low Earth Orbits at altitudes between 100 and 1000 km, where many space missions are operating, trapped particles are also present, as first observed by AMS-01 [324]. Particles of energies up to few GeV/nucleon were observed below the geomagnetic cut-off as shown in Figure 8.33. These particles are secondaries produced by the interaction of galactic cosmic rays with the upper layers of the atmosphere.

The Low Earth Orbit intersects the South Atlantic Anomaly (SAA), a region of space where the geomagnetic field is particularly weak due to the misalignment of magnetic and rotational axis. Because of this, charged particles can penetrate deeper into the

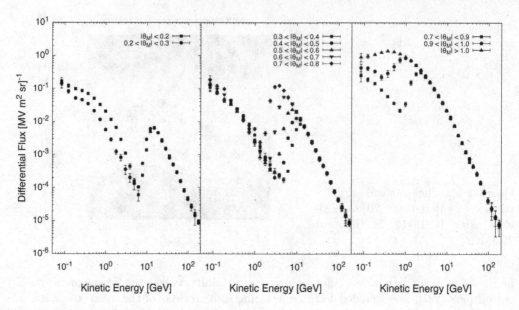

Figure 8.33 Differential proton flux measured by AMS-01 [324] in bins of geomagnetic latitude θ_M. The geomagnetic cut-off is clearly visible at energies varying with latitude. Below the cut-off, the observed particle flux can only come from trapped protons.

magnetosphere. At higher altitudes, from 1000 to 3600 km, the Medium Earth Orbits (MEO) intersects the inner radiation belt. It is populated mainly by protons in the energy range from 10 to 100 MeV; in addition, there are also less energetic electrons. The population of this region is quite stable but can vary with the solar activity. The particles mostly come from the decay of neutrons, generated by the collision of cosmic rays with the atoms of the atmosphere. For example, a proton with an energy of 5 GeV produces a shower with about seven neutrons, a fraction of which diffuse outward into the magnetosphere where they decay into protons and electrons. The decay products, in turn, become trapped particles [332].

At an altitude of 36000 km, there is the Geostationary Orbit (GSO) which is used for communication satellites. This orbit intersects the outer radiation belt which contains mainly electrons of energies up to ~ 10 MeV. The majority of these electrons come from magnetic storms, so the population density of the outer magnetic belts fluctuates greatly with solar activity.

Particles in the radiation belts have rigidities below the cut-off and are trapped in the Earth's magnetic field like in a magnetic bottle (see Focus Box 5.1). The motion of a trapped particle in the geomagnetic field can be decomposed into three components as shown in Figure 8.35. The first component is the gyro motion of the particle around the magnetic field line following a helical orbit. The second component is a bounce motion due to gradients in the latitudinal direction of the magnetic field. While it gyrates along the magnetic field line, the particle reaches higher latitudes where the magnetic field is stronger. This causes a reflection of the particle by a magnetic mirror, bouncing back-and-forth between mirror points as demonstrated in Focus Box 5.1. The third component is a slow drift motion around the Earth that comes from perpendicular magnetic field gradients. The direction of the drift motion depends on the sign of the particle; it is anticlockwise (eastward) for electrons and clockwise for ions. The drift motion is responsible for the Earth's ring current (see Section 8.9.2). The gyro motion typically takes from milliseconds to seconds, the bounce

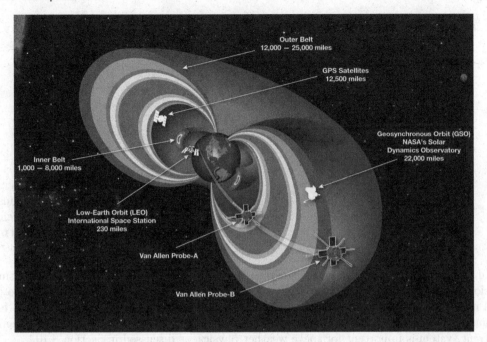

Figure 8.34 Cutaway model of the near Earth radiation belts with the two Van Allen Probes and other satellites flying through them. (Credit: NASA)

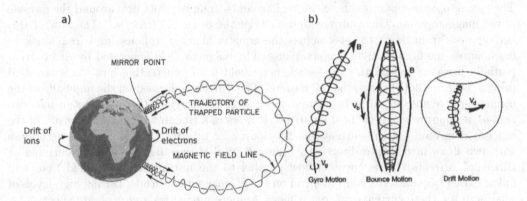

Figure 8.35 a) Trajectories of trapped particles in the magnetic field lines around the Earth [229]. b) Decomposition of the trapped particle motion into gyro, bounce and drift motion [268].

motion from seconds to minutes and the drift motion from minutes to hours. Particles can thus be trapped in the Earth magnetic field over long timescales.

8.9.2 MAGNETOSPHERIC CURRENT SYSTEMS

Magnetospheric currents are produced by the motion of charged particles at the boundaries between the flow of different plasmas in the Earth's magnetosphere. Moving charged particles, in turn, produce magnetic fields which couple with the Earth's intrinsic magnetic

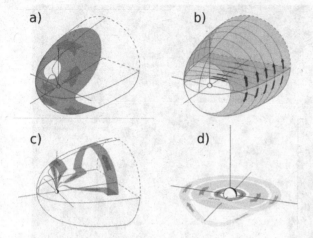

Figure 8.36 Magnetospheric current systems [661]: a) magnetopause current; b) tail current; c) field-aligned currents; d) ring currents.

field to form the topology of the magnetosphere. The currents at the interface of plasma regions are largely affected by the activity of the Sun, especially during interplanetary coronal mass ejection events which may lead to magnetic storms at Earth (see Chapter 1 for examples). Therefore, understanding and monitoring the variations in the magnetospheric current system is important for space weather physics, as discussed in Section 8.10.

There are five principal current systems in the magnetosphere [661]. Figure 8.36(a) shows the magnetopause current or Chapman-Ferraro current. It is generated at the interface between the plasma flowing in the magnetosheath and the plasma inside the magnetosphere. The magnetopause separates these two regions and strong currents flow around the dayside of the magnetopause. Their current density is of the order of $20\,\mathrm{nA/m^2}$. The flow of this current goes from dawn to dusk across the equatorial magnetopause and from dusk to dawn across the high-latitude magnetopause. The magnetic field generated by this current partially prevents the Earth's dipole magnetic field from penetrating into the solar wind plasma. Figure 8.36(b) shows the tail current. This current is formed on the nightside of the magnetosphere and includes the current sheet layer that divides the magnetotail into two regions of opposite magnetic field polarities. It flows in a ribbon with current density of the order of $30\,\mathrm{mA/m}$. The tail current closes its loop via a return current on the magnetopause with two flows in opposite directions. Figure 8.36(c) shows the field-aligned currents or Birkeland currents. These currents flow parallel to the Earth's magnetic field lines and follow closed loop patterns which depend on the geomagnetic latitude. During high levels of solar activity, these currents can reach mega-Ampere intensities and generate spectacular aurorae. Figure 8.36(d) shows the ring currents. This system is made of currents flowing around the Earth, an internal current in the anticlockwise direction, an external current in the clockwise direction, and a cut ring current on the dayside. The external ring current generates a magnetic field opposite to the Earth's dipole magnetic field. These currents are mainly due to the drift motion of the trapped charged particles already discussed in Section 8.9.1 and are thus much weaker, with densities of a few $\mathrm{nA/m^2}$.

8.10 SPACE WEATHER PHYSICS

In this section, we will give you a glimpse into space weather and magnetic storm and substorm phenomena [392]. Space weather is a branch of space physics that concerns the electromagnetic and radiation conditions on the ground, near Earth and in the solar system. The magnetic field carried by fast interplanetary plasma may go through a magnetic reconnection process when it encounters the Earth's geomagnetic field. During this process, magnetic energy is converted to kinetic energy, which may be spent for particle acceleration. Magnetic reconnection may also lead to magnetic storms in the Earth's magnetosphere. These can cause damage to spacecraft by electrical charging of circuitry or by depositing energy when penetrating sensitive electronics. Electromagnetic induction may occur in long conductors on Earth, causing power station damage and component failures in long electrical wires. Electromagnetic interference from the magnetic storms may lead to disruption of radio or satellite communications. Examples have been given in Chapter 1.

Solar flares and coronal mass ejections are accelerators for solar energetic particles. These are a risk to astronauts and aircrews of transpolar flights, who are subject to rapid increases in the absorbed dose of ionising radiation (see Focus Box 1.1) over short time periods. In addition, solar energetic particles lead to significant degradation of electronic devices in space and on the ground.

In recent years, the awareness of danger and risks posed by explosive events coming from the Sun has lead to an increased interest in space weather physics which seeks to detect and eventually predict space weather events.

8.10.1 MAGNETIC STORMS AND SUBSTORMS

Magnetic storms are long lasting perturbations of the magnetosphere caused by the interaction of the magnetic field carried by interplanetary coronal mass ejections with the Earth's magnetic field. When the shock front of a coronal mass ejection hits the Earth, the magnetosphere is compressed. This compression leads to a rapid increase of the magnetopause current and the horizontal component of the magnetic field on the Earth's surface. This component is monitored using different geomagnetic activity indices defined in Focus Box 8.3. This initial phase of the magnetic storm may last from two to eight hours and is called the sudden impulse. If the magnetic field carried by the coronal mass ejection has a long lasting component in the south direction with respect to the Earth's geographical poles, it will point in the opposite direction of the Earth's magnetic field. This configuration allows a magnetic reconnection process on the dayside of the magnetosphere. Extended periods of southward interplanetary magnetic field lead to the main phase of the magnetic storm, during which the solar wind can freely flow into the magnetosphere enhancing charged particle diffusion into the inner magnetosphere. This injection of charged particles causes an increase in the ring current with a consequent decrease of the horizontal component of the surface magnetic field. The change in the magnetic field can be quantified as a decrease in the so-called disturbance storm-time index, D_{st}. The magnetic flux transfer to the geomagnetic tail is also increased during this phase. There, further reconnection processes can occur, leading to additional energy transfer to the inner magnetosphere. A schematic of the entire process is shown in Figure 8.37 where the boxes indicate the magnetic reconnection regions. Gradually, the southward component of the interplanetary magnetic field will decrease and eventually disappear, the reconnection process will stop and no additional particles can diffuse into the inner magnetosphere. Progressively, the ring current will decrease to its typical values. This phase of the magnetic storm is called the recovery phase and can last several days.

If the interplanetary magnetic field passes through periods of alternating southward and northward directions, the magnetic disturbance is called a magnetospheric substorm. A series of substorms is characterised by short periods, each lasting for about an hour, when the ring current increases. A series of substorms has recursive impulsive phases that disturb the main phase of the storm. During the periods when the interplanetary magnetic field points southwards, the consequent reconnection processes accumulate energy in the magnetotail. This continuous process of energy storage in the magnetotail is called the

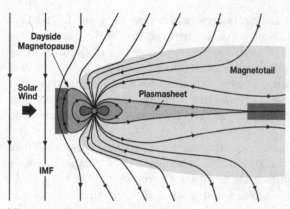

Figure 8.37 Schematic of the reconnection sites in the magnetosphere during periods of southward interplanetary magnetic field. The boxes indicate the regions of magnetic reconnection on the dayside and in the magnetotail.

(Credit: NASA, https://mms.gsfc.nasa.gov/science.html)

Geomagnetic activity indices are all based on ground measurements of the Earth's magnetic field recorded in magnetograms. A magnetogram is a measurement of temporal variation in the local strength and direction of the geomagnetic field. To characterise the level of activity, indices are established using these time-dependent measurements integrated over specific time intervals. In general, an average value that corresponds to a reference level of geomagnetic activity is subtracted from the measurement.

A widely used index is the K-index or the corresponding planetary K_p-index, where the latter is based on multiple measurements performed world wide. The K-index quantifies the disturbance in the horizontal component of the Earth's magnetic field. Since the high magnetic latitude magnetograms are subject to higher levels of variations, this index is mostly sensitive to auroral zone activity. It is given in an almost logarithmic scale, from 0 to 9 with subdivisions of 1/3, such that 28 values are possible. The A-index and the planetary A_p-index have similar meanings but different numerical values.

The auroral electrojet indices are defined in order to quantify the strengths of electrojets measured in the auroral regions based on variations of the horizontal magnetic field component. Electrojets are Hall currents carried primarily by electrons at altitudes from 100 to 150 km. Only high magnetic latitude magnetograms are considered in these indices, therefore they are uncontaminated by any ring current effects. There are four auroral electrojet indices: the auroral upper index (AU) defined as the maximum positive disturbance; the auroral lower index (AL) defined as the lowest negative disturbance; the AE index defined as the difference between AU and AL; and the AO defined as the average of AU and AL.

The disturbance storm-time index, D_{st}, measures the strength of the ring current. This index is also based on measurements of the horizontal component of the Earth's magnetic field, but only considers low magnetic latitude magnetograms.

Detailed definitions and time variations of these and others geomagnetic activity indices are available at https://www.ngdc.noaa.gov/geomag/indices/. Magnetograms have existed since Victorian times, and the British Geological Survey has preserved records from the 1850s onward.

Focus Box 8.3: Geomagnetic activity indices

---

growth phase of the substorm. If too much energy is stored in the tail, it becomes unstable and the energy is released. This is called the substorm onset and immediately leads to the so-called expansion phase, when particles are accelerated towards the geomagnetic poles causing aurorae. The expansion phase typically lasts from minutes to a halfhour, during which the auroral brightness is increased and large-amplitude magnetic disturbances are measured. Also, the ionospheric current flow is enhanced during this phase of the substorm, and electrojets are generated that flow through the field-aligned current from the magnetic reconnection site at the magnetotail to the poles. The substorm recovery phase begins about an hour after the substorm onset. It may last a few hours and may overlap with the growth phase of the next substorm.

9 In the Milky Way

In this chapter, cosmic rays with energies up to 10^6 GeV are discussed. These represent the bulk of cosmic rays and are produced by sources residing in our own galaxy. Direct energy measurements and identification of individual cosmic ray particles *in situ* are currently limited to this energy range. Galactic cosmic rays probably reach energies beyond this limit and up to what is called the knees of their spectrum, as discussed in Chapter 10, but the experimental methods radically change. We will discuss here how direct observations are made, concentrating on recent space experiments. The impact of the results on cosmic ray modelling and searches for signals from antimatter, dark matter and exotic forms of matter are discussed in the final sections of this chapter.

9.1 MODELS CONFRONTING OBSERVATIONS

The bulk of cosmic rays, up to energies around 10^9 GeV, is believed to originate in our own Galaxy. In Chapter 5, we have delineated the principles of the current understanding of cosmic ray acceleration and of astrophysical objects where such mechanisms can occur. The current generally accepted model of galactic cosmic rays is based on the so-called supernova paradigm in which cosmic rays come from supernova explosions in the disk of our galaxy. They acquire energy through diffusive acceleration by the shock wave following the explosion. Subsequently they are released into the interstellar medium and propagate diffusively in the turbulent magnetised galactic halo, as sketched in Chapter 6. They are confined for a long period until they reach the boundary of the galaxy and eventually escape.

The supernova paradigm is a framework, which was built on available observations of cosmic photons and charged cosmic rays, as well as astrophysical observations and theoretical investigations. In this scenario, the spectra of primary cosmic rays – such as protons and nuclei directly coming from their sources – are predicted to follow a universal power law in rigidity $d\Phi/dR \sim R^{-2.7}$ when they arrive near Earth. We argue this result of standard diffusion transport in Focus Box 9.1. The energy spectrum of electrons, which are also primary cosmic rays, is expected to be softer than that of nuclei because of the larger energy losses suffered by leptons with respect to hadrons; it is predicted to be $\sim E^{-3}$. These spectra are expected to follow a featureless power law up to the knees, at energy $\sim 10^9$ GeV.

Antimatter counterparts, such as positrons e^+, antiprotons $\bar{p}$ and antideuterons $^2\overline{H}$, are instead thought to be produced in interactions of primary cosmic rays with the interstellar medium, as discussed in Section 4.5. The observed relative abundance of the different species reflects their production mechanisms. Protons are the most abundant component, $\sim 89\%$, followed by helium nuclei, $\sim 9\%$. Heavier nuclei and electrons account for $\sim 1\%$. The abundance of positrons and antiprotons relative to protons is 1/1000 and 1/10000 respectively (see Sections 9.6 and 9.4). To date no antideuteron nor heavier antinucleus observation has been firmly established in cosmic rays.

It is important to understand that there is no complete top-down model of cosmic rays, which would describe their complete life cycle, from the formation of stars and their nuclear processes, via the acceleration and propagation of cosmic rays to their arrival in the heliosphere. Such a model would predict the abundances spectra of cosmic rays as they enter, traverse and exit the interstellar medium. While awaiting such a break-through, bottom-up models guide the way. They start by postulating a certain functional form of cosmic rays spectra upon injection into the interstellar medium, with adjustable parameters like the abundance and the spectral index for each species. They then follow the transformation

DOI: 10.1201/9781003181385-9

In commonly adopted diffusion models, the cosmic ray diffusion equation can be simplified without loss of generality by assuming that cosmic rays effectively propagate only in the direction perpendicular to the galactic disk. One can then solve the diffusion equation under the general assumption that particles at the boundary freely escape. Here we follow the treatment in [638] to give a summary of predictions from standard diffusion models for primary cosmic ray nuclei spectra observed near Earth, and show how they relate to the parameters of the diffusion equation with the aim of connecting experimental observation with theory expectations.

Proton spectrum: At high energies, where ionisation losses and advection effects can be neglected, the cosmic ray proton spectrum as a function of rigidity R observed near Earth can be expressed as

$$\frac{d\Phi_H}{dR} = \frac{d\Phi_H^{SN}}{dR}\frac{N_{SN}}{2\pi R_d^2 H}\frac{H^2}{D(R)}$$

where $d\Phi_H^{SN}/dR \propto R^{-\gamma}$ is the average proton spectrum injected in the ISM by cosmic ray sources, generally assumed to be supernovae, N_{SN} is the rate of supernovae explosion in the galaxy, a few per century. H and R_d are the galactic halo size and disk radius, respectively, and $D(R)$ is the diffusion coefficient (see Focus Box 6.4). The differential flux of protons observed near Earth thus results from the product of the total injection rate over the entire galaxy and the cosmic ray confinement time in the galaxy, $\tau_{esc}(R) = H^2/D(R)$. For a diffusion coefficient $D(R) \propto R^\delta$ the above equation results in :

$$\frac{d\Phi_H}{dR} \propto R^{-\gamma-\delta}$$

The spectrum observed near Earth is thus softer than the injected spectrum by a factor equal to the rigidity dependence of the diffusion coefficient.

Primary nuclei spectra: For nuclei heavier than hydrogen, energy losses due to spallation on the ISM cannot be neglected, but one can assume that spallation occurs in an infinitely thin disk. Assuming also that these nuclei are only produced by sources (i.e. neglecting spallation of heavier nuclei for mostly primary nuclei such as O, Si and Fe), and again neglecting advection, the diffusion equation can be simplified and solved to obtain the propagated spectrum of species p:

$$\frac{d\Phi_p}{dR} = \frac{d\Phi_p^{SN}}{dR}\frac{N_{SN}}{2\pi R_d^2 H}\frac{H^2}{D(R)}\frac{1}{1+X(R)/X_p^{crit}}$$

where $X(R) = v\tau_{esc}(R)\langle M \rangle n$ is the cosmic ray grammage, with $v \simeq c$ and and $\tau_{esc}(R)$ the cosmic ray velocity and confinement time in the galaxy respectively, $\langle M \rangle \simeq 1.4\ m$ the average ISM mass, with m the mass of the proton, and n the average particle density experienced by cosmic rays while propagating in the rarefied galactic halo or in the more dense disk which depends on the galactic halo size H and the galactic disk thickness; $X_p^{crit} = \langle M \rangle/(\xi\sigma_p)$ is the critical grammage for a nucleus of type p, with σ_p its spallation cross section on the ISM and ξ an energy-independent correction factor that takes into account that spallation reactions approximately conserve kinetic energy per nucleon and not rigidity [572, 638]. The grammage $X(R)$ has the units of a column density, i.e. mass × length^{-2}, and is independent of the nucleus type. Since $X(R) \propto H/D(R)$ is a decreasing function of rigidity, at high rigidities cosmic rays traverse a small grammage. When $X(R) \ll X_p^{crit}$ the relation between the injected and the propagated spectrum we gave above for protons is recovered.

Focus Box 9.1: Standard diffusion models: primary cosmic ray nuclei

of abundances and spectra by the transport through the interstellar plasma and the helio-sphere, again with adjustable parameters like the diffusion coefficient and Alfvén speed, as well as properties of the galaxy like its halo size, the thickness and radius of the galactic disk and the column density. Experimental data and theory on cross sections obviously also enter into the game. This way, conditions at the beginning of the cosmic ray life cycle are related to observations at the end. And observations can be used to test hypotheses and tune injection and transport parameters. We give an idea of how this is done in Focus Boxes 9.1, 9.2 and 9.11.

Understanding cosmic ray propagation from source to detection is of crucial importance to establish not only a model for cosmic rays themselves, but also a benchmark to search for new physics, such as dark matter and antimatter signals in cosmic ray spectra. It is equally important for the interpretation of signals from other cosmic messengers. The analysis of the elemental composition of cosmic rays is an important tool to improve our knowledge about this part of a future top-down model.

As we have already noticed in Chapter 3, cosmic ray elemental abundances differ in interesting ways from those found in stellar matter. When they arrive near Earth, all nu-clear species are a mixture of those coming from their sources unaltered, so-called primaries, and those produced in interactions with the interstellar medium, so-called secondaries. The fractions depend on the properties of each species. Those abundantly produced in stellar nucleosynthesis and moreover tightly bound, like carbon or oxygen, are dominated by pri-maries (see Chapter 4). If we assume that these represent a "standard" abundance, hydrogen and helium nuclei are significantly underrepresented in cosmic rays, as seen in Figure 3.2. We summarise the properties of their spectra in Section 9.3.1.

There are also a few nuclear species, such as lithium, beryllium, boron, fluorine and the sub-iron elements, which are not abundantly produced by stellar nucleosynthesis. Li, Be and B are intermediate stages on the way to the synthesis of heavier and more stable elements, as described in Chapter 4. However, when comparing cosmic ray and solar system abundances they are found to be much more abundant in cosmic rays, as already shown in Figure 3.2. Such nuclei are mainly produced by spallation of heavier primary cosmic ray nuclei in interactions with the interstellar medium. Their properties are thus particularly important to derive some of the parameters involved in propagation modelling. Their spectra result from the convolution of energy spectrum and fragmentation cross section of the primary progenitors. The average amount of matter traversed by cosmic rays in their journey through the galaxy before they interact with an atom of the interstellar medium, the so-called grammage, also enters into the game. Hence, in cosmic ray diffusion models the ratio of secondary to primary spectra is inversely proportional to the diffusion coefficient. The spectrum remains a power law, but with a modified spectral index. We discuss this prediction in Focus Box 9.2. Because of their production mechanism, secondary nuclei are less abundant than primaries neighbours in mass: for instance boron nuclei, the most abundant light secondary nuclei, are found in a ratio 1:3 with the neighbouring primary oxygen nuclei.

Accurate measurements of secondary-to-primary ratios are crucial experimental tools to improve cosmic ray propagation models. Abundant and stable light nuclei, such as carbon or oxygen, represent primary ones. Lithium, beryllium or boron are good representatives for light secondary ones. We describe their features in Section 9.3.2. For heavier nuclei covered in Section 9.3.3, silicon or iron are good candidates for a primary nucleus, fluorine and sub-iron elements for secondaries.

Another important diagnostic tool for cosmic ray propagation are radioactive secondary isotopes with lifetimes of the order of the cosmic ray confinement time in the galaxy, like e.g. ^{10}Be. The degeneracy between the diffusion coefficient, the grammage and the size of the galactic halo can be broken by measuring the spectra of such unstable secondary nuclei and of their stable counterpart. Indeed, the energy spectrum of unstable secondaries

Secondary nuclei are produced mostly by spallation of heavier primary nuclei on the ISM. Using the same approach and general assumptions as in Focus Box 9.1, and taking into account that a secondary nucleus can be also destroyed in spallation reactions, the spectrum of a stable secondary nucleus s observed at Earth can be expressed as:

$$\frac{d\Phi_s}{dR} = \frac{X(R)}{1 + X(R)/X_s^{crit}} \sum_p \frac{d\Phi_p}{dR} \frac{1}{X_p^{crit}}$$

where X and X^{crit} are the grammages defined in Focus Box 9.1, and the p index denotes a progenitor species, with the sum running over all cosmic ray isotopes fragmenting into species s. Beside secondary production from isotopes of primary nuclei, also the so-called tertiary production is taken into account, that is the case when the progenitor isotope is of secondary origin. Similarly, when computing the spectra of a given secondary nuclear species the calculation for all its isotopes are summed together. Some species can be also produced by radioactive decay of unstable species, which we have neglected because its contribution is in general small compared to spallation. Moreover, the above expression only applies to stable secondary species, since we have not taken into account destruction of species s by radioactive decay. We will discuss unstable secondary nuclei in Focus Box 9.11.

The above expression is often simplified assuming that a given secondary nuclear species is mostly generated by a single progenitor species. This simplification is based on the observation that the highest contribution comes indeed from the fragmentation of the closest primary species and that nuclear fragmentation cross sections for different projectile nuclei differ only in their absolute normalisation, and on the assumption that primary nuclei have all the same spectral shape. Under this approximation, the above equation leads to:

$$\frac{d\Phi_B}{dR} = \frac{X(R)}{1 + X(R)/X_B^{crit}} \frac{d\Phi_C}{dR} \frac{1}{X_C^{crit}}$$

At sufficiently high rigidities ($R \gtrsim 100\,\text{GV}$), where $X(R) \ll X_B^{crit}$, one finds the scaling law:

$$\frac{d\Phi_B}{dR} \bigg/ \frac{d\Phi_C}{dR} \propto X(R)$$

used to extract the cosmic ray grammage – and indirectly the rigidity dependence of the diffusion coefficient – from the secondary-to-primary B/C ratio measured near Earth. It is worth noting that the assumption that the carbon nuclei spectrum observed at Earth has negligible contribution from spallation of heavier nuclei (such as nitrogen and oxygen) is not a good approximation. A better "standard" primary nucleus is oxygen, for which this approximation holds better. However, before AMS-02 published a precision oxygen spectrum [627], no accurate measurements of the oxygen spectrum were available, and the carbon spectrum was used instead. Using the relation between the grammage and the diffusion coefficient given in Focus Box 9.1, the above expression gives the relation between the rigidity dependence of the spectrum of a secondary species s to the spectrum of its primary progenitor p:

$$\frac{d\Phi_s}{dR} \propto \frac{d\Phi_p}{dR} \cdot \frac{1}{D(R)} = \frac{d\Phi_p}{dR} R^{-\delta}$$

Hence the spectral index of the secondary species is softer than the spectral index of its primary progenitor by a factor $-\delta$, where δ is the spectral index of the diffusion coefficient.

Focus Box 9.2: Standard diffusion models: secondary cosmic ray nuclei

which decay before reaching the boundary of the galaxy, such as ^{10}Be, is only sensitive to the diffusion coefficient. The energy spectrum of their stable counterpart, such as ^{9}Be, which can escape the galaxy, is sensitive to both the diffusion coefficient and the size of the galactic halo. Therefore, the ratio of unstable secondaries to their stable counterpart allows to determine the diffusion coefficient and the size of the galactic halo separately. These unstable secondaries are called radioactive clocks because they allow to measure the time cosmic rays spend travelling in the galaxy by looking at their surviving fraction measured near Earth. Other cosmic ray radioactive clocks are the heavier ^{26}Al, ^{36}Cl and ^{54}Mn which decay to ^{26}Mg, ^{36}Ar and ^{54}Fe respectively. A direct measurement of the isotopic composition of Al, Cl and Mn is challenging, and at energies above about $10\,\text{GeV/n}$ impossible with current instruments. However, an indirect measurement of the surviving fraction of heavy nuclei can be extracted from the mother to daughter ratios Al/Mg, Cl/Ar and Mn/Fe. Similarly, the measurement of the ^{10}Be surviving fraction at higher energy can be extracted from the Be/B ratio as detailed in Focus Box 9.11. We summarise experimental results on isotopes in Section 9.3.4.

In recent years, the deployment of sophisticated cosmic ray instruments in space has allowed cosmic ray physics to enter a new era of precision measurements which have revealed unexpected features in cosmic ray spectra. They are thus challenging the current supernovae paradigm and stimulate theoretical investigations. In the next section, we will review such cosmic ray instruments. In the rest of the chapter, we will review the latest experimental results on charged galactic cosmic rays pointing out possible interpretations.

9.2 SPACE DETECTORS

As we have shown in Chapter 2, cosmic ray instruments have equipped rockets and satellites since the beginning of space flight itself. The International Space Station provides an ideal platform for the long-term exposure of sophisticated particle detectors. The year 2020 marked the twentieth anniversary of constant manning for the ISS, 2021 the tenth year of data taking by the Alpha Magnetic Spectrometer. New cosmic ray calorimeters have been deployed in the meantime.

The power law shape of the cosmic ray all-particle spectrum has driven a two-fold experimental approach to the detection of cosmic rays. Space-borne and balloon detectors are employed to measure cosmic ray spectra and elemental composition at energies approaching $10^6\,\text{GeV}$. Above this energy, the particle flux summed over a whole hemisphere is less than a few particles per square meter per year and rapidly decreases to about 100 particles per square kilometer per year at $10^9\,\text{GeV}$ (see Section 3.2), thus requiring kilometer-size observatories which can only be deployed on the Earth's surface. As discussed in Section 3.1, space-born detectors have direct access to cosmic ray particles before they enter the Earth's atmosphere and cause air showers. The accurate identification of the incoming cosmic ray particle is thus experimentally feasible.

Both spectrometry-based and calorimetry-based cosmic ray observatories are currently active in space missions. As discussed in Focus Box 3.2, the primary observable of spectrometers is the magnetic rigidity, i.e. the signed ratio of particle momentum and electric charge. For all astrophysical processes involving magnetic fields, like in acceleration and transport of cosmic rays, rigidity is indeed the relevant quantity. Spectrometers are generally equipped with additional detectors for particle identification, to enhance their capability to assess particle charge and distinguish light from heavy particles. Cosmic ray antiparticles can only be distinguished from particles by direct detection with a magnetic spectrometer, since there is no way to measure the sign of the electric charge by calorimetric means. Magnetic spectrometers equipped with velocity measuring devices also have the ability of measuring the isotopic composition of light nuclei. However, with current state-of-the-art technology, isotopic composition can be measured only at energies below $10\,\mathrm{GeV/n}$, as we will discuss in Section 9.3.4. Therefore the nuclei spectra we will discuss in Sections 9.3.1 to 9.3.3 are summed over isotopes.

Pioneers in the direct measurement of the spectra for individual cosmic ray species – as described in Section 2.6 – have been the balloon-borne magnetic spectrometer BESS shown in Figure 2.16, and the satellite experiment PAMELA shown in Figure 2.17.

Figure 9.1 Left: Artists impression of the AMS-02 detector (Credit: NASA). Right: The AMS-02 instrument installed on the main truss of the International Space Station with an astronaut working in a nearby site (Credit: NASA).

Calorimetric detectors, on the other hand, are more compact and can thus cover a large solid angle, but do not distinguish between particles and antiparticles. Their primary observable is the kinetic energy of the incident particle, as discussed in Focus Box 3.4. The energy resolution of a calorimeter generally improves with energy, while the resolution of magnetic spectrometers worsens with rigidity. When combined with devices for the measurement of particle charge, calorimeters deliver the energy spectrum of individual cosmic ray species, summing over particles and antiparticles as well as isotopes.

The spectra are measured as a function of magnetic rigidity by spectrometers, as a function of kinetic energy by calorimeters. Conversion from one quantity to the other requires an assumption about the isotopic composition of the observed elements based on available measurements (see Section 9.3.4).

When we describe modern detectors below, we insist particularly on two features. The first is their ability to establish and monitor the rigidity or energy scale of the measurement. This is important, since an error on the absolute energy or rigidity scale results in an incorrect flux normalisation. We will also multiply differential fluxes by a power of the rigidity or energy to make spectral features visible. In this chapter, the appropriate scale factor is $(d\Phi/dR) \cdot R^{2.7}$ for nuclei and $(d\Phi/dR) \cdot R^3$ for electrons and positrons. A scale error in rigidity or energy thus translates into a shift in the scaled flux. We discuss in Focus Box 9.9 how the rigidity or energy scale is calibrated and monitored in-flight in modern space detectors. The second important feature concerns the identification of rare species, like positrons or antiprotons. Since their matter counterparts, protons and electrons, are orders of magnitude more abundant, experiments need strong rejection power on the dominant species to determine their flux. So in experiments, light particles like $e^{\pm}$ must be well separated from heavy particles of like absolute charge, such as protons. In spectrometers, this especially applies to like-sign pairs, where a rare species must be distinguished from a dominant one. Examples are positrons and protons, as well as antiprotons and electrons.

The calculation of spectra at Earth for both primary and secondary species requires the knowledge of spallation cross sections, σ_p, which enter in the definition of the critical grammage X_p^{crit} for a given cosmic ray isotope p as explained in Focus Box 9.1 and Focus Box 9.2. Isotopic production cross sections determine for each species, how many primaries are destroyed by fragmentation while they propagate through the interstellar medium, and how many are produced as secondaries when heavier nuclei interact.

Currently, the accuracy of cosmic ray data, of the order of few percent up to multi-TV rigidities, is not matched by precision isotopic production cross section measurements, preventing to fully exploit cosmic ray data. Cross sections data are available for a limited set of isotopes and few energies of order GeV/n. They are affected by large uncertainties, 20% to 50%, and often data from different experiments disagree. There are several databases collecting isotopic cross sections measurements as the EXFOR [679] and the LANL `https://nucleardata.lanl.gov/` databases, though their collections might not be exhaustive. A summary of the current status of isotopic cross sections measurements and implications for the modelling of cosmic ray transport are discussed in reference [657]. To face the lack of data, semi-empirical parametrisations of isotopic cross sections were developed already in the early stages of cosmic ray physics [170], and later refined and constantly improved as more data were became available in the 1990s [253]. Alternative data-driven empirical formulae were developed by Webber and collaborators and updated in the early 2000s [340]. The use of semi-empirical parametrisations and parametric formulae – normalised to available data and to nuclear reaction calculations wherever possible – is still the approach of current propagation tools as GALPROP [657]. Similarly, studies attempting to interpret cosmic ray data or extract information on cosmic ray propagation, origin or acceleration use parametrisations of spallation cross sections normalised to the sparse available data [713, 797, 820].

Lately, efforts are being made to measure cross sections relevant for cosmic rays at accelerators, as for instance at CERN by the NA61/SHINE and COMPASS++/AMBER experiments at the SPS, as well as the LHCb and ALICE experiments at the LHC [650, 695, 726, 730].

Focus Box 9.3: Nuclear cross sections: implications for cosmic ray modelling

There is a third recurring problem which we will encounter not only for the measurement of elemental spectra but also for their interpretation in terms of cosmic ray propagation. It is our poor knowledge of cross sections for nuclear reactions. Such reactions can transform a heavier element into a lighter one by inelastic interaction.

The target can be interstellar matter, mostly protons and helium nuclei, met before the cosmic ray reaches us. The transformation probability must then be taken into account in modelling spectra and composition of cosmic rays. We comment on the implications of nuclear cross sections for propagation models in Focus Box 9.3. The extraction of propagation parameters strongly relies on the knowledge of spallation cross sections. Moreover, if one could firmly establish the probability for secondary production, this would serve as benchmark to identify excesses or deficits in observed cosmic ray spectra and pin down their origin, acceleration mechanism and propagation history. As we have seen in Section 4.4, the identification of signals from dark matter or primordial antimatter requires a precise determination of the background from secondary production of positrons, antiprotons and anti-nuclei.

The target for nuclear reactions can also be met inside the detector itself, the loss and gain of the initial and final species must then be corrected for. The poor knowledge of nuclear fragmentation cross sections thus affects also cosmic ray measurements themselves and contributes a sizeable systematic error. In particular, in direct detection the

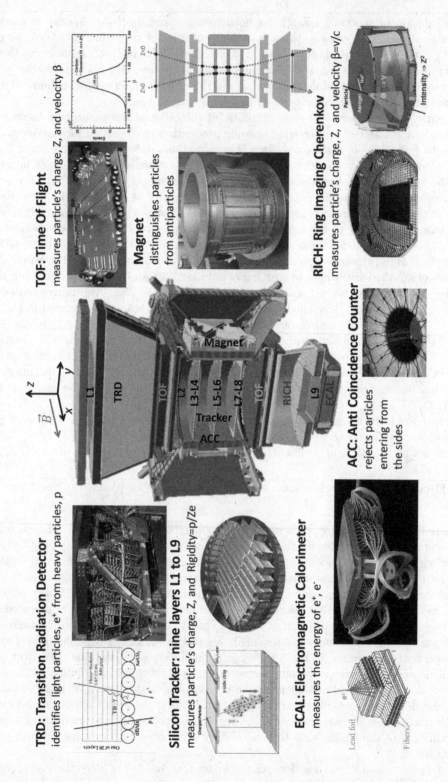

Figure 9.2 Schematic cut through the AMS-02 detector (centre) showing its components, TRD, TOF, Tracker, Magnet, ECAL, ACC, and RICH, with their main functions (adapted from [775]). Cosmic rays are detected entering the instrument from the top.

The AMS-02 detector and its components are shown in Figure 9.2. Particles enter the detector from the top, opposite the z-direction in the coordinate system indicated. We describe the components in the order in which they are encountered:

TRD: The **Transition Radiation Detector** has twenty layers. Each layer is composed of a 20 mm thick fleece radiator and a layer of proportional tubes for X-ray detection. The TRD distinguishes light from heavy particles, by measuring the transition radiation emitted by highly relativistic particles (see Focus Box 2.11). It identifies one positron in a background of more than 1000 protons at 90% positron detection efficiency.

TOF: The **Time-Of-Flight** system consists of two double layers of scintillation detectors (see Focus Box 2.6) placed above and below the magnet. It measures the time a particle takes to traverse from the top to the bottom layer with a resolution ranging from 160 ps for singly charged ($Z = 1$) particles down to 50 ps for heavier ($Z > 6$) nuclei. The TOF measures the cosmic ray arrival direction and velocity, and gives the **trigger** for charged particles to the overall data acquisition system.

ACC: Sixteen **Anticoincidence Counters** surround the inner bore of the magnet to reject particles entering the detector from the sides with an efficiency of 0.99999.

Tracker and Magnet: The heart of the detector is a **magnetic spectrometer** (see Focus Box 3.2) composed of a cylindrical permanent magnet, and a tracker. The magnet generates a dipolar field with the main component directed along the x-axis. The tracker has nine layers of double-sided silicon micro-strip detectors. Layers L3 to L8 are inside the magnet bore, L2 above the magnet, L1 on top of the TRD and L9 just above the ECAL. The total lever arm from L1 to L9 is 3 m. At each tracker layer the x- and y- coordinates of the particle impact point are measured with accuracies of 13 to 20 μm and 5 to 10 μm respectively. The bending of the particle trajectory inside the magnetic field gives the rigidity, $R = p/(Ze)$, and its direction allows to distinguish positively charged particles ($Z > 0$) from negatively charged particles ($Z < 0$).

RICH: The **Ring Imaging Cherenkov** counter is composed by a radiator plane made of NaF and silica aerogel, a conical mirror on the sides, and a photo-detection plane at the bottom. The particle's velocity is obtained from the aperture angle of the Cherenkov light cone (see Focus Box 2.11) with a relative resolution better than 0.1%, allowing to measure isotopic composition of light nuclei up to $\sim 10 \, \text{GeV/n}$.

The **particle charge** squared, Z^2, is measured independently from the intensity of the emitted light in the RICH, and from the energy deposited in the active detector materials of TRD, TOF, Tracker and ECAL.

ECAL: The 3D sampling **Electromagnetic Calorimeter** is made of nine multi-layered lead and scintillating fibre sandwiches for a total of $17X_0$ (see Focus Box 3.4). The nine sandwiches are stacked such that the fibres run alternatively along the x-coordinate (five layers) and the y-coordinate (four layers). ECAL measures the energy of electrons and positrons with few percent accuracy. It also gives the **trigger** to the overall data acquisition for photons and measures their energy. The 3D reconstruction of the particle shower in ECAL, and the matching of the energy measured in ECAL with the momentum measured in the tracker give an additional discrimination power to separate leptons ($e^{\pm}$) from hadrons (p and $\bar{p}$) better than 1 positron over 10'000 protons at 90% positron detection efficiency, independent of the TRD. The ECAL also allows to calibrate the rigidity scale of the spectrometer *in situ* using cosmic rays electrons and positrons.

The complete detector in its flight configuration has been carefully calibrated in particle beams at CERN just before launch. Calibration and alignment of components are constantly updated and refined during data taking on the ISS [775, Sec. 1.].

Focus Box 9.4: The Alpha Magnetic Spectrometer AMS-02 [775]

knowledge of the amount of incoming cosmic ray nuclei fragmenting in the detector material is of paramount importance to correctly assess the overall flux normalisation. There are two principle approaches to this problem. One can estimate the survival probability of a given nucleus using simulation. The systematic error is then estimated comparing different nuclear interaction models or comparing Monte Carlo simulations to test beam data [784]. An alternative and more robust approach is to compare the simulation to data collected with the detector itself [542, 763, 775]. At energies below a few hundred GeV/n, this can be done using test beam data. At high energy, cosmic ray data can be used to measure survival probabilities in the detector, provided the detector design allows to clearly identify incoming cosmic ray nuclei before they start to fragment, to use a portion of the detector as target, and to measure the amount of fragmentation products on the downstream side. With this approach, the AMS-02 detector has measured charge-changing nuclear fragmentation cross sections on carbon target for the most abundant primary cosmic ray nuclei from helium to iron of rigidities from few GV up to TV [771, 774]. In Focus Box 9.6, we will discuss how this issue is dealt with by recent space-born cosmic ray experiments. In indirect detection measurements by ground experiments, nuclear fragmentation cross sections enter in the modelling of extensive air showers. Recently, attempts to probe high-energy hadronic interactions with extensive air showers are being carried out at ground-based experiments [688] to complement measurements of relevant reactions at accelerators [726].

The only magnetic spectrometer currently acquiring cosmic ray data is the Alpha Magnetic Spectrometer AMS-02 shown in Figure 9.1. We briefly describe its principle components and their performance in Focus Box 9.4; more details are available in Reference [775]. The AMS-02 components, shown in Figure 9.2, have been designed to perform simultaneous and accurate measurements of the individual spectra of positrons, electrons, antiprotons, protons and nuclei up to the nickel region ($Z \sim 30$). The spectrometer has a reach in magnetic rigidity from a fraction of a GV up to a few TV.

Particles enter the detector from the top, in the negative z-direction in the coordinate system indicated in Figure 9.2. The heart of the detector is a spectrometer comprising a cylindrical permanent magnet and a series of silicon sensors tracking the particle trajectory. The dipolar field is generated by 64 NdFeB blocks arranged in a Halbach array, reaching $0.15\,T$ in the centre. It points in the x-direction, bending the trajectory in y. The trajectory is located by nine layers of silicon strip detectors. The tracking layers are arranged with six layers inside the magnet bore and three outside. The rigidity measurement thus results from a mixture of sagitta and bend-angle measurements as described in Focus Box 3.2. The relative rigidity resolution is about $\Delta R/R \simeq 0.1$ for $R < 20\,GV$ and increases with rigidity. An important figure of merit is the so-called maximum detectable rigidity, i.e. the magnetic rigidity where the measurement error reaches 100%. For AMS-02, it varies between $2.0\,TV$ for protons and $3.7\,TV$ for iron nuclei. The spectrometer is completed by a large set of partially redundant devices identifying particles and measuring their kinetic energy, as described in Figure 9.2 and Focus Box 9.4. AMS-02 is installed on the International Space Station since May 2011. In its first ten years of exposure, it has collected over 180 billion cosmic ray particles, arguably the largest sample since the discovery of cosmic rays. The collaboration plans to continue operating the observatory until the end of the ISS lifetime, through at least 2030.

Since 2015, the energy region beyond a few TeV is being explored with large-size calorimeters, such as DAMPE [602] on a free-flying Chinese satellite, and CALET [725] on the International Space Station. This pushes the frontier of current direct cosmic ray measurements to $10^5\,GeV$ and, in the case of CALET, promises to progress towards knee energies as more data are collected. The DAMPE and CALET missions both started in 2015 and are currently operational.

The Calorimetric Electron Telescope (CALET) is a modern calorimetric detector measuring the flux of $e^{\pm}$ as well as nuclei from the TeV to the PeV region. On the right is a schematic side view of the instrument with a 1 TeV electron traversing the apparatus, simulated by Monte Carlo [725]. Cosmic rays entering from the top are measured in the following three detectors:

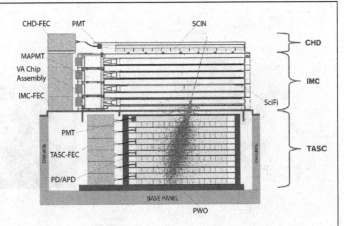

CHD: The **Charge Detector** is made of a double layer of scintillators (see Focus Box 2.6) read by photomultiplier tubes (PMT). The signal of each CHD layer is used as input to trigger the overall data acquisition. The CHD measures the absolute value of the particle charge from the energy deposited in the scintillator. Its large dynamic range allows to identify nuclei up to $|Z| \sim 40$.

IMC: The **Imaging Calorimeter** is a 3D sampling calorimeter with a total depth equivalent to $3X_0$ (see Focus Box 3.4). It is made of seven layers of tungsten alternated with two layers of scintillating fibres (SciFi) arranged orthogonally, and capped by additional SciFi double layers for a total of 16 active layers. The fibres are read individually by multi-anode photomultiplier tubes (MAPMT). The IMC provides a 3D reconstruction of the early phase of the incoming particle shower allowing to determine the shower starting point and the cosmic ray arrival direction with angular resolutions of 0.14° for electrons and 0.24° for photons. The IMC also measures the absolute value of the charge for nuclei up to silicon ($Z = 14$) from the energy deposited in the SciFi fibres. For each IMC double layer the signals from the two SciFi layers are combined to generate input to the trigger system.

TASC: The **Total Absorption Calorimeter** is a homogenous calorimeter equivalent to 27 X_0 (see Focus Box 3.4). It consists of twelve layers of lead tungstate (PWO), each composed of 16 PWO logs. The layers are arranged alternatively with the logs running along orthogonal directions to allow 3D reconstruction of the particle shower. The PWO logs of the first layer are read individually by a photomultiplier tube (PMT) to provide additional input to the trigger system. The remaining eleven layers are read by silicon photodiodes and silicon avalanche photodiodes (Dual PD/APD). The read-out system is configured to provide enough dynamic range to measure the showers induced by a 1000 TeV proton. The TASC measures the energy of electrons with a resolution better than 2% above 20 GeV. IMC and TASC together represent a thickness of $1.2\lambda_I$ for hadrons (see Focus Box 2.9).

The **trigger** for the data acquisition is obtained combining the signals from the CHD layers, the IMC layers, and the first TASC layer [598]. Leptons ($e^{\pm}$) are distinguished from protons and nuclei by comparing the 3D reconstruction of their shower in the IMC and TASC. The discrimination power is of the order of 1 electron over a background of 10'000 protons at 80% electron detection efficiency. CALET components have been calibrated before launch in particle beams at CERN, the calibration is constantly updated using flight data [598].

Focus Box 9.5: The Calorimetric Electron Telescope (CALET) [725]

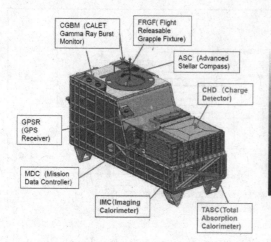

Figure 9.3 Left: Principle components of the CALET calorimetric detector [728] with the charge measurement device (CHD) followed by an imaging calorimeter (IMC) and a total absorption calorimeter (TASC) for electromagnetic showers. The CALET payload is also equipped with a gamma-ray burst monitor (CGBM). Right: The CALET instrument (ellipse) installed on the Exposed Facility of the Japanese Experiment Module on the International Space Station (Credit: JAXA/NASA).

The Calorimetric Electron Telescope (CALET) has been developed to measure the cosmic ray $e^{\pm}$ spectrum in the kinetic energy range from 1 GeV to 20 TeV, the gamma-ray spectrum up to 10 TeV and the proton spectrum from 50 GeV to 1000 TeV. The individual spectra of nuclei can be measured in the kinetic energy range from 10 GeV to 1000 TeV, for nuclei up to iron as well as trans-iron nuclei up to $Z \sim 40$ [725]. The CALET set-up and its position on the ISS are shown in Figure 9.3. We describe the components of the CALET payload in more detail in Focus Box 9.5.

The Dark Matter Particle Explorer (DAMPE) measures the spectra of photons and $e^{\pm}$ in the energy range 5 GeV to 10 TeV, and the spectra of protons and nuclei up to iron ($Z \leq 26$) in the kinetic energy range from 50 GeV to 100 TeV. The principle components of the detector are shown in Figure 9.4.

9.3   COSMIC NUCLEI

In this section, we discuss the latest experimental results on the differential fluxes of cosmic ray nuclei. Proton and nuclei spectra are usually plotted multiplied by $R^{2.7}$, for measurements from magnetic spectrometers, or by $E^{2.7}$, for calorimetric instruments, to put in evidence deviations from the featureless power law predicted by traditional cosmic ray models. We start with the lightest and most abundant species, hydrogen and helium nuclei and then continue upwards in nuclear mass.

9.3.1   PROTONS AND HELIUM

The most abundant species in cosmic rays are hydrogen nuclei (see Figure 3.2), i.e. $Z = +1$ nuclei including protons and deuterons, often simply called protons in the literature. The tritium life time is too short to contribute significantly. We will see in Section 9.3.4 that indeed protons make up the bulk of hydrogen nuclei, the ratio ^{2}H/^{1}H being around 2% at 1 GeV/n. Likewise, the spectra for $Z = +2$ include ^{3}He and ^{4}He, with predominance of ^{4}He.

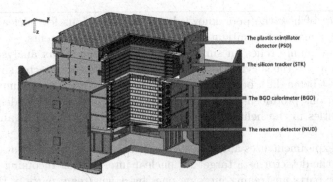

The plastic scintillator detector (PSD)

The silicon tracker (STK)

The BGO calorimeter (BGO)

The neutron detector (NUD)

Figure 9.4 Principle components of the DAMPE [602] free-flying calorimetric cosmic ray detector, composed of a Plastic Scintillator Detector (PSD) and a Silicon-Tungsten tracker-converter (STK) for charge measurement, a BGO imaging calorimeter of $32X_0$ depth for energy measurement, and a Neutron Detector (NUD), which together with the calorimeter provides lepton/hadron identification. The DAMPE satellite was launched in December 2015, the instrument is currently in operation. Components have been calibrated before launch in particle beams at CERN, the calibration is constantly updated using flight data.

The hydrogen and helium spectra have been measured by many instruments, both calorimeters and magnetic spectrometers (see Figures 2.18 and 2.21 in Chapter 2). Some of the pioneering calorimetric experiments on high mountains [233] and balloons already reported a decreasing hydrogen-to-helium flux ratio as energy increased over a wide range from few TeV to hundreds of TeV [230, 252, 259, 287, 414]. They also all reported a feature-less power law behaviour for both the hydrogen and helium spectra over the entire energy range, while spectral indices differed between experiments. However, the limited statistical and systematic accuracy of those pioneering experiments was not sufficient to come to a definite assessment of the behaviour of the proton and helium spectra. At the 23d International Cosmic Ray Conference in 1993 [255], the need for high-precision measurements of both hydrogen and helium spectra in the region above 100 GeV/n was advocated.

More than a decade later, a first hint appeared of a deviation from a featureless power law for both the hydrogen and helium spectra at energies around 200 GeV/n. This was indirectly inferred in a study [429] combining high-energy measurement from CREAM (2.5 TeV to 250 TeV) with lower energy measurements from AMS-01 (0.2 GeV/n to 200 GeV/n). The measurements of these and other pioneering experiments are plotted in Figure 2.18.

A clear assessment of the actual energy dependence of the proton flux could only come from experiments able to connect the GV region to the TV region. The first such instrument has been the PAMELA magnetic spectrometer, which measured the proton and helium spectra in the rigidity range from 1 GV to 1.2 TV [457, 631]. PAMELA observed for the first time a spectral hardening in both the proton and helium spectra at rigidities in the range 230 GV to 240 GV for protons and helium (similar to the value reported in [429]). PAMELA observed that the proton-to-helium flux ratio was a smooth decreasing function of the rigidity over the entire rigidity range of their measurements, finally assessing that hydrogen and helium nuclei have different spectral shapes at modest energies. And neither spectrum follows a featureless power law with constant spectral index.

Few years later the AMS-02 magnetic spectrometer [542, 560] measured the proton and helium spectra in the rigidity ranges 1 GV to 1.8 TV and 1.9 GV to 3 TV, respectively, with unprecedented statistics and systematic accuracy. The calorimeters CALET [722, 778] and DAMPE [693, 784] extended the energy range, although with larger uncertainties. The

We give examples of how the poor knowledge of cross sections for nuclear fragmentation inside the detector material is mitigated by current space experiments.

The first example mainly relies on simulation. In their helium flux analysis, the DAMPE collaboration compares the survival probability predicted by the detector simulation to that obtained from test beam data below 300 GeV. At higher energies, they compare predictions from two simulations produced using different hadronic interaction models [784]. The resulting uncertainties in the helium nuclei flux are 13% below 300 GeV and between 13% and 15% above.

The AMS-02 experiment uses a data-driven approach using in-orbit data. The idea is to use a portion of the detector as a target for nuclear interactions. Incoming nuclei are identified by detector parts upstream, outgoing ones by downstream parts of the set-up. Thus nuclear fragmentation probabilities in the detector material are directly measured for the most abundant primary nuclei from helium to iron, and from a few GV up to TV rigidities. The results are then used to calibrate the nuclear cross sections in the simulation. Remaining uncertainties on the flux measurements result from the residual difference between data and simulation. Nuclei are identified by dE/dx in the AMS-02 tracker layers on top, inside the magnet and just before the electromagnetic calorimeter on the bottom. Comparing charge from layers encountered before the TRD and the upper TOF layers (see Focus Box 9.4) and downstream of the two, determines the survival probability in their accumulated material. For helium nuclei, a special dataset is used in addition. When the ISS accommodates incoming space craft, its orientation is such that the AMS axis points horizontally. Thus high-energy particles can enter from the bottom and the lower tracker layers can be used to identify the incoming nuclei. Different combinations of upstream and downstream detectors can then be used to obtain the overall survival probability in the AMS-02 detector [771, 775]. All these data then calibrate the nuclear model of the simulation in the rigidity range were fluxes are measured [771, 775]. With this method, the flux uncertainties arising from the hadronic interactions model are a few percent: 1% up to 100 GV rising smoothly to 2% at 3 TV for helium [542]; < 2.2% below 100 GV rising smoothly to 3% at 3 TV for carbon [627]; < 2.7% below 100 GV rising smoothly to 3.5% at 3 TV for oxygen [627]; and < 4% below 100 GV rising smoothly to 4.5% at 3 TV for iron [774].

A similar approach is adopted by the CALET experiment. Incoming nuclei are identified by the charge measured in the CHD. The nuclear charge is then remeasured along the trajectory at different depths in the IMC (see Focus Box 9.5), using dE/dx in pairs of adjacent scintillating fibres, and compared to the incoming charge to identify surviving nuclei. The measured survival probabilities are compared to the prediction of the nuclear model in the simulation to estimate the resulting uncertainties on the flux. The latter are < 1% for both carbon and oxygen fluxes in the energy range from 10 GeV/n to 2.2 TeV/n [763].

Focus Box 9.6: Nuclear cross sections: systematics for flux measurements

results of modern spectrometric and calorimetric measurements of the differential spectra for these dominating elements are shown in Figure 9.5. The spectra are well described by a double broken power law [756, 797, 816, 848]:

$$\frac{d\Phi}{dR} = c_{30} \left(\frac{R}{30}\right)^{\gamma} \left[1 + \left(\frac{R}{R_l}\right)^{\Delta\gamma_l/s}\right]^{s} \left[1 + \left(\frac{R}{R_h}\right)^{\Delta\gamma_h/s}\right]^{-s}$$

where c_{30} is a normalisation constant at the starting rigidity of 30 GV, chosen such that effects of solar modulation are negligible. The first term with spectral index γ describes a general power law. It is broken at two rigidities, R_l and R_h, by the amounts $\Delta\gamma_l$ and $\Delta\gamma_h$, respectively, described by the two remaining terms. The parameter s describes the

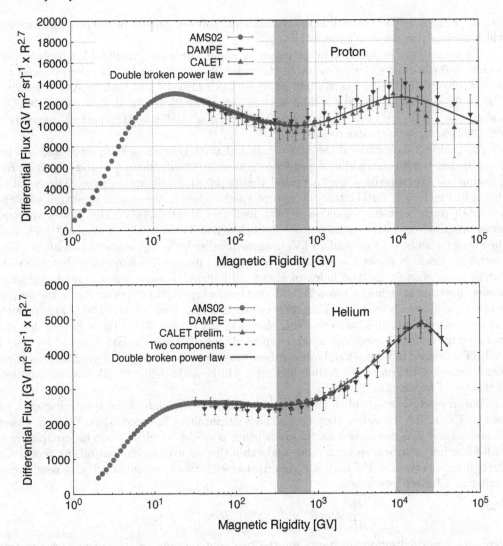

Figure 9.5 The proton and helium differential spectra multiplied by $R^{2.7}$ as measured by the recent experiments AMS-02 (full dots) [775], CALET (upward triangles) [778, 840] and DAMPE (downward triangles) [693, 784]. Conversion of kinetic energy to magnetic rigidity assumes dominance of the ^{1}H and ^{4}He isotopes. The error bars correspond to the quadratic sum of statistical and systematic errors. The full lines are the result of the fit to a double broken power law, the dashed line to the two-component fit mentioned in the text. The vertical bands indicate the position of the two transition regions.

smoothness of the two transitions; it is fixed to the same value at both breaks, since the high rigidity data have little sensitivity to it. Parameters obtained from a fit of this function to all data in Figure 9.5 are given in Focus Box 9.10. The rigidity ranges where the two transitions occur are indicated in the figure by the vertical bands, a few hundred GV for the first and a few tens of TV for the second. CALET proton data with an extended energy range [840] confirm the turn-over of the spectrum at high energies seen by DAMPE. The general spectral index for protons, $\gamma_H = -2.815 \pm 0.004$, is softer than the one for helium,

$\gamma_{He} = -2.728 \pm 0.003$. This is due to an additional soft component in the proton spectrum which is clearly visible in Figure 9.5.

The transitions in spectral index occur at about the same rigidities for protons and helium, and the indices change by about the same amount. In our fit, the spectral index for protons changes by $\Delta\gamma_l = +0.26 \pm 0.07$ at $R_l = (648 \pm 116)$ GV and back by $\Delta\gamma_h = -0.29 \pm 0.07$ at $R_h = (10 \pm 5)$ TV. For helium we find $\Delta\gamma_l = +0.25 \pm 0.02$ at $R_l = (799 \pm 96)$ GV and $\Delta\gamma_h = -0.51 \pm 0.14$ at $R_h = (19 \pm 3)$ TV. Similar studies of the spectral shapes (see e.g. [756]) have given similar results.

The AMS-02 data allowed the first detailed characterisation of the proton and helium spectral shapes with the rigidity dependence of their spectral indices, $\gamma = d(\log\Phi)/d(\log R)$. Calorimetric experiments allow to extend this approach to higher energies. The result is shown in Figure 9.6. Rather than an abrupt spectral break, the proton spectrum exhibits a smooth progressive hardening as rigidity increases above 200 GV. The proton spectral index progressively increases from $\gamma = -2.8$ below 200 GV to $\gamma = -2.6$ at about 1 TV and then back to about -2.8 around 20 TV, in agreement with the fit results quoted above. The spectrum of helium nuclei also exhibits a smooth and progressive hardening above 200 GV but with an offset in spectral index of about $+0.1$ units. It starts out with a significantly harder spectrum at rigidities below 200 GV and reaches spectral indices similar to the proton values only above about 1 TV. This convergence has first been observed by AMS through the proton-to-helium flux ratio Φ_H/Φ_{He} shown in Figure 9.7 [775]. The tendency for the two spectral indices to converge is also supported by higher energy data from DAMPE and CALET as shown in Figure 9.6, however these two experiments have not yet provided their measurement of the proton-to-helium flux ratio which would definitely allow a confirmation of this trend beyond the TV.

Broken power laws are obviously a good heuristic description of the spectral shapes, as seen in Figure 9.5. However, they do not have an intuitive interpretation. What is easier to understand is a description of the data using a model which has several components, each following their own spectral index and exhibiting an exponential cut-off. As shown for the helium spectrum in Figure 9.5, a description with two components gives a result very similar to a broken power law:

$$\frac{d\Phi}{dR} = c_1 R^{\gamma_1} e^{-R/R_1} + c_2 R^{\gamma_2} e^{-R/R_2}$$

Here c_i are normalisation constants for the two contributions, γ_i the respective spectral indices and R_i their cut-off rigidities. The similarity is not surprising, since the functional form of the double broken power law can be seen as an approximation to the two-component function for modest variations in spectral index. A fit to the helium spectrum gives spectral indices of $\gamma_1 = -2.83 \pm 0.01$ and $\gamma_2 = -2.33 \pm 0.02$, with corresponding cut-offs of $R_1 = (178 \pm 28)$ GV and $R_2 = (51 \pm 7)$ TV. It is obvious that models of galactic cosmic rays which predict a single constant spectral index for all species up to knee energies are inconsistent with these precision data.

In the proton spectrum, on the other hand, a third softer component is clearly visible below about 1 TV. This has first been observed through the AMS measurement of the flux ratio Φ_H/Φ_{He} [775], where a good part of the systematic errors cancels. The ratio as a function of rigidity is shown in Figure 9.7. It steadily decreases as a function of rigidity and follows a function [775]:

$$\frac{d\Phi_H}{dR}\Big/\frac{d\Phi_{He}}{dR} = a + c\left(\frac{R}{3.5\,\text{GV}}\right)^{\Delta\gamma_s}$$

The lower limit of 3.5 GV has been chosen such that the effect of solar modulation is negligible (see Chapter 8). The two normalisation constants are similar, with $a = 3.15 \pm 0.07$

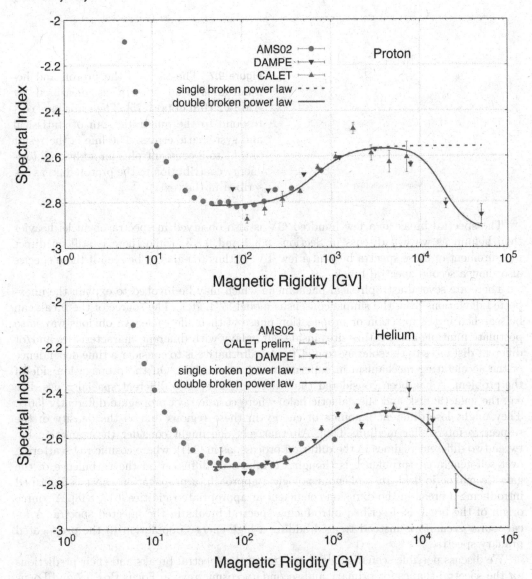

Figure 9.6 The spectral indices for proton (top) and helium (bottom) as a function of magnetic rigidity, derived from the data of the recent experiments AMS-02 (full dots) [775], CALET (upward triangles) [722, 778] and DAMPE (downward triangles) [693, 784]. The curves correspond to the indices $\tilde{\gamma}_1(R)$ and $\tilde{\gamma}_2(R)$ for the broken power laws of Figure 9.5 as explained in Focus Box 9.10. Since the behaviour in the TV range is similar, we use the average of helium and proton parameters for the high-energy break. Conversion of kinetic energy to magnetic rigidity assumes dominance of the ^{1}H and ^{4}He isotopes.

and $c = 3.30 \pm 0.07$. The difference in spectral indices $\Delta\gamma_s = -0.30 \pm 0.01$ is close to other index changes seen in the spectra of protons and light nuclei. The soft proton component evolves with $\gamma_s = \gamma_1 + \Delta\gamma_s$ and dies out somewhere above 1 TV, where the proton and helium spectra start to have the same shape.

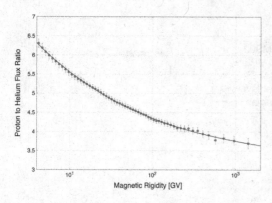

Figure 9.7 The ratio of the proton and helium differential spectra as measured by AMS-02 (full dots) [775]. The error bars correspond to the quadratic sum of statistical and systematic errors. The line is the result of a fit to a constant plus an additional low-energy contribution to the proton flux, as described in the text.

The spectral break at a few hundred GV is also observed in spectra of nuclei heavier than helium as we will discuss in Sections 9.3.2 and 9.3.3, while there is still no direct measurement of these spectra beyond a few TV. It thus remains to be seen if these species also show a second spectral break.

There are several astrophysical phenomena which may be invoked to explain the unexpected deviations from the simple consensus model [674, 705]. The source of the break can be searched in the injection or propagation phase of their life cycle. An obvious way is to postulate multiple accelerating sites inside our galaxy with different characteristics and/or different distances to the solar system [470]. An alternative is to consider a time dependence of the accelerating mechanism in supernovae remnants [791]. On the propagation side of the problem, it has been considered that there might be two different spacial zones, like e.g. the galactic disk and the galactic halo, where cosmic rays propagate differently [567]. They might lose different amounts of energy in these regions due to the density of the respective interstellar medium [663]. Alternatively one might consider the transition between two different regimes in the diffusion process, as in [541], where cosmic ray scattering over self-generated turbulence is thought to take over diffusion on the turbulence of the galactic magnetic field. In a phenomenological approach, a propagation effect is described introducing a break in the diffusion coefficient at appropriate rigidities [617, 848]. A source origin of the break is described introducing spectral breaks in the injected spectra. A local source scenario is instead treated adding a high-energy population to the propagated primary spectra.

We discuss possible scenarios for the origin of the spectral breaks and their predictions for the spectral shapes for primary and secondary cosmic rays in Focus Box 9.7 and Focus Box 9.8. In particular, the hypothesis that there is a break already present in the particle spectra which come out of cosmic accelerators, often called injection spectra, would predict that the spectral change $\Delta\gamma_l$ of primaries and secondaries at the break should be the same. The propagation scenario, where the spectral break arises from the diffusive transport mechanism, predicts a spectral hardening for secondaries twice that of their primary progenitors. The local source scenario, on the other hand, in most of its realisations where the secondary spectra are not affected by the local source, would lead to a hardening only in primary spectra. From this discussion is then evident that the key to ascertain the origin of the spectral breaks is in the comparison of the measured secondary and primary nuclei spectra and their ratios. We will discuss the experimental results on light primary and secondary nuclei, from lithium to oxygen, in Section 9.3.2, together with their impact on cosmic ray modelling and our current understanding on the origin of the spectral break. In Section 9.3.3, we will discuss recent experimental results on nuclei beyond oxygen.

The observation of breaks, first in the proton and helium nuclei spectra, and subsequently also in the carbon and oxygen nuclei spectra, aroused a lively debate among cosmic ray theorists on the physical phenomenon originating the change of spectral indices all occurring at around few hundred GV. Three classes of theoretical interpretations were proposed each tracing back the break to a different stage of the cosmic ray spectrum formation [564, 674]:

Break at injection: A first class of models traces the break to the shape of the spectrum emitted by cosmic ray sources. In this scenario, the change in the spectral index from $-\gamma$ to $-\gamma+\Delta\gamma_l$ at rigidity R_l is already present in the spectrum of primary cosmic rays injected by the source in the interstellar medium, $d\Phi_p^{SN}/dR$. After propagation the primary cosmic ray spectrum, $d\Phi_p/dR$, acquires a softening by $R^{-\delta} \propto 1/D(R)$ from the diffusion mechanism. The spectra of secondary cosmic rays produced by the spallation of this primary, $d\Phi_s/dR$, will exhibit a spectral break at the same rigidity R_l and by the same spectral index change $\Delta\gamma_l$ as their primary progenitor, in addition to the overall -2δ softening expected from the diffusion mechanism.

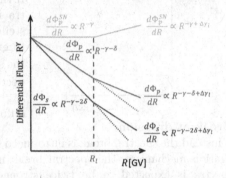

Break at diffusion: In the second class of models, the spectral break is introduced by the propagation mechanism and represented as a change in the diffusion coefficient from R^{δ} to $R^{\delta-\Delta\gamma_l}$ at rigidity R_l (see Focus Box 9.10). In this scenario, the spectrum of the primary

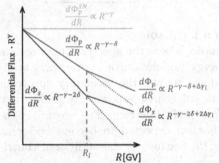

cosmic ray at the source, $d\Phi_p^{SN}/dR$, is a featureless power law proportional to $R^{-\gamma}$, while the propagated spectrum, $d\Phi_p/dR$, exhibits a break at R_l, where its spectral index changes from $-\gamma - \delta$ to $-\gamma - \delta + \Delta\gamma_l$ reflecting the break in the diffusion coefficient. The spectral index of its secondaries will then change from $-\gamma - 2\delta$ to $-\gamma-2\delta+2\Delta\gamma_l$ at rigidity R_l resulting in a spectral break twice stronger than the one of its primary progenitor.

Break due to a local source: A third class of models explains the spectral break as the appearance of a high-energy population of cosmic rays originating from a local source. In this scenario, the spectrum of the bulk of primary cosmic rays injected at the source, $d\Phi_p^{SN}$, is also a featureless power law, since the additional high-energy cosmic rays from the local sources will only appear in a restricted region of the galaxy because of energy losses and interactions. Consequently the secondary spectra, $d\Phi_s/dR$, which originate from the bulk spectrum, will not exhibit any spectral break. Only the propagated local primary cosmic ray spectra will exhibit a spectral break, most likely with a rigidity-dependent break $\Delta\gamma_l(R)$ for rigidities above R_l depending on the shape of the contribution of the local source to the primary species.

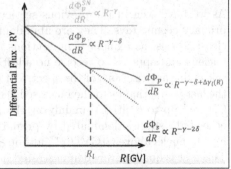

Focus Box 9.7: Theoretical interpretation of the spectral breaks

Secondary cosmic ray spectra as well as secondary-to-primary cosmic ray spectrum ratio are a well recognised experimental probe to understand cosmic ray propagation. From the discussion in Focus Box 9.7 is also evident that accurate measurements of secondary cosmic ray spectra, such as lithium, beryllium and boron spectra, and of their ratio to the spectra of their primary progenitors, as carbon and oxygen, can be used to disentangle among the three classes of theoretical interpretations of the spectral breaks observed in cosmic ray nuclei at few hundreds of GV. Indeed, the three scenarios predict distinct rigidity dependencies for the secondary spectra and for the secondary-to-primary ratios.

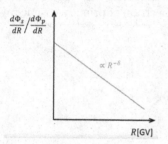

Break at injection:
If the spectral break is a feature of the primary spectra injected by sources of cosmic rays in the interstellar medium, the secondary spectra are expected to exhibit the same spectral break as primaries, and the secondary-to-primary spectrum ratio $\frac{d\Phi_s/dR}{d\Phi_p/dR}$ is expected to follow a featureless power law in rigidity $R^{-\delta}$, with constant spectral index δ coming from the diffusion coefficient (see Focus Box 9.10).

Break at diffusion:
If instead the spectral break is introduced by the propagation mechanism, the spectral break in secondary spectra is expected to be twice stronger than the spectral break of primaries. Thus the secondary-to-primary ratio is expected to exhibit a spectral hardening of $+\Delta\gamma_l$ at rigidity R_l due to the corresponding $-\Delta\gamma_l$ change in the spectral index of the diffusion coefficient (see Focus Box 9.10).

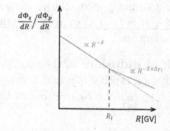

Break due to a local source:
In the last scenario, where the spectral break results from a high-energy population of cosmic rays originating from a local source, the secondary spectrum is expected to be a featureless power law, since they mainly come from bulk injection. The secondary-to-primary ratio is then expected to become softer at rigidities above R_l, because of the additional contribution to primaries.

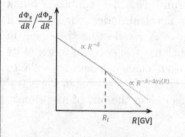

Focus Box 9.8: Secondary-to-primary spectrum ratios as probes of spectral breaks

9.3.2 LIGHT NUCLEI

As we have seen in the previous section, helium nuclei are the first example of so-called primary cosmic rays. They are abundantly produced both in primordial and in stellar nucleosynthesis, as explained in Sections 4.1 and 4.2. Helium is also a very tightly bound nucleus and appears to receive no additional soft contribution in contrast to protons. Detailed characterisations of the spectra of nuclei heavier than helium have become possible in the last years, thanks to the increased accuracy of direct measurements beyond few hundred GV and up to multi-TV mainly by the AMS-02 spectrometer. Carbon and oxygen nuclei, other light species dominated by primaries, all show roughly the same power spectrum as helium nuclei above 60 GV [627], although their abundance is one order of magnitude lower. This fact is qualitatively demonstrated in Figure 9.8 which compares the differential spectra of helium, carbon and oxygen, all measured by AMS-02, and roughly scaled to the helium flux at modest rigidities. A quantitative characterisation of their spectral shape is given in the original AMS-02 paper [627] and in Focus Box 9.10.

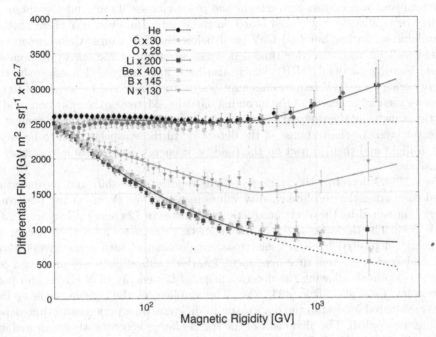

Figure 9.8 Differential flux (times $R^{2.7}$) as a function of magnetic rigidity for primary nuclei (He, C, O) and secondary products of interactions with interstellar matter (Li, Be, B) as measured by AMS-02 [775]. Nitrogen (N) consists of a roughly 50:50 mixture of secondaries and primaries, so its spectral shape lies between the two. The fluxes are scaled so that they roughly match the helium flux at 30 GV. The dashed lines indicate the dependency expected with a constant spectral index. The solid lines result from a fit to the spectra of the three groups with a broken power law as used in Reference [775]. The dashed line shows what the spectra would look like if the spectral indices were constant.

There are as yet no (eagerly awaited) extensions at energies beyond few TeV/n from DAMPE and CALET for none of the nuclei heavier than helium. CALET has published carbon and oxygen spectra up to a maximum kinetic energy per nucleon of 2.2 TeV/n, corresponding to rigidity of about 4.4 TV [763] (see Figure 3.3), far below the high-energy break seen in the helium spectrum.

The CALET measurements of the carbon and oxygen spectra confirm the spectral hardening observed by AMS-02, but they do not agree with the absolute flux normalisation, which is lower by 27%. The reason of the disagreement between these two measurements is still unclear. Possible explanations for this disagreement could be a lack of knowledge of nuclear fragmentation occurring inside the detector itself (see Focus Box 9.6) or a systematic shift of the absolute energy scale [763]. We briefly discuss in Focus Box 9.9 how the energy scale is calibrated in recent space experiments.

In the following, we adopt oxygen as our prototype for light primary nuclei. The reason is that it has little contribution from spallation of heavier cosmic ray nuclei, hence can be considered a pure primary, and through nuclear reactions with the interstellar medium it is, together with carbon, the main progenitor of light secondary nuclei lithium, beryllium and boron. These three nuclei serve as raw materials and intermediate products in the breeding of heavier nuclei inside stars, and are therefore strongly underrepresented in stellar matter

The correct estimation of the energy or rigidity scale and in-flight check of its stability is crucial for an unbiased measurement of absolute fluxes. All detectors undergo tests at particle beams at accelerators before flight and the responses of the subdetectors measuring the energy or rigidity are calibrated based on these data. However, test beam particles are only available up to few hundred GeV, well below the TeV range. Calorimeters are also calibrated in-flight to correct for time-dependent variation of the energy responses using minimum ionising particles (MIP), which also have low energies. Therefore the check of the energy scale up to TeV requires assumptions on the functional behaviour of the energy response for increasing energy of the incoming particles. Moreover the vibrations and shocks during the launch and the thermal environment in space might change the alignment among the detector layers or the response of the detector. All these aspects need to be accurately studied in-flight and their impact on the rigidity or energy scale taken into account in the data analysis.

In the AMS-02 detector, the tracker in-flight rigidity scale shift and its uncertainty is obtained comparing the absolute rigidity values measured for electrons and positrons with the energy measured by the electromagnetic calorimeter in 72 energy points from 2 GeV to 300 GeV [596]. The thickness of the electromagnetic calorimeter allows a 75% containment of the energy deposited by 1 TeV electrons, single channel saturation effects have been studied and corrected, and time-dependent in-orbit calibration using minimum ionising protons is performed, allowing the determination of the energy of electrons and positrons with accuracy 1.4% at 1 TeV [493, 623]. The stability of the rigidity scale as function of time is evaluated analysing the data taken in different time periods and time-dependent corrections are applied. The alignment of the tracker layers is constantly monitored in-flight and corrections are applied. Taking into account both the impact of the uncertainty on the alignment and on the rigidity scale shift, the rigidity scale is determined with an accuracy of 3% at 1 TV [775].

In CALET the accuracy of the absolute energy calibration for nuclei is obtained in-flight checking the consistency of the energy response for proton and helium minimum ionising particles at equivalent rigidity cutoffs. The helium minimum ionising particle peak responses from flight data as function of the rigidity cutoff are compared to Monte Carlo simulated events, where the solar modulation is obtained from data taken as function of geomagnetic latitudes by AMS-01 and checked with BESS data. However, this method leads to inaccuracies of the solar modulation model because these experiment ran in different time periods [598]. The energy scale is determined with 3% accuracy for protons [722] and 2% for carbon, oxygen and iron [763, 779] mostly arising from the accuracy of the beam test calibration.

In DAMPE the absolute energy scale uncertainty was estimated in-flight at 13 GeV using the geomagnetic cutoff for electrons and positrons and resulted in an about 1.3% accuracy [648, 784].

A recent study by Adriani and collaborators [852] identified subtle effects in the light yield of inorganic scintillators at high ionisation densities. They could account for systematic shifts in the calibrated energy scale of space calorimeters of the order of several percent, depending on energy, material properties and calorimeter design.

Focus Box 9.9: Energy and rigidity scale

as already shown in Figure 3.2. They are found three orders of magnitude more often in cosmic rays, thanks to the splitting of heavier nuclei by spallation on the way from their sources to us. As secondary spallation products, on the other hand, they have a significantly softer spectrum (see Focus Box 9.2) as shown in Figure 9.8 with measurements from AMS-02. The unprecedented accuracy of the AMS-02 measurements has revealed for the first time that the light secondary Li, Be and B have similar spectra above 30 GV, and that they exhibit a spectral break at a few hundred GV, as their primary progenitors C and O, but with a spectral index change which is twice stronger than for primaries. Hence the secondary-to-primary ratios Li/O, B/O, Be/O, shown in Figure 9.9, and the analogous ratios to carbon also exhibit a spectral hardening [665]. These observations clearly point to the scenario where the breaks at few hundred GV are originated by the diffusion mechanism as explained in Focus Box 9.7 and Focus Box 9.8.

The difference observed in the Be and B spectra below 30 GV reveals the effect of ^{10}Be decaying to ^{10}B. Instead the Li and B spectral shapes are found to be identical in the wider rigidity range from 7 GV to 3 TV. The GALPROP-HelMod team has found that the calculation of the lithium flux obtained using transport parameters tuned on these AMS-02 data show an excess in lithium above 4 GV, much larger than what can be possibly accomodated in their estimated errors, suggesting that some primary ^{7}Li might be created in novae [742]. The existence of this excess is still under debate, a definite answer can be given by measurements of the lithium isotopic composition since the production of ^{7}Li in novae would lead to a predominace of this isotope over ^{6}Li at high energy. We will discuss isotopic composition measurements in Section 9.3.4.

In cosmic ray diffusion models, the spectral index of secondary-to-primary ratios directly measures the spectral index of the diffusion coefficient δ (Focus Box 9.2). Using the AMS-02 data and in lack of data beyond multi-TV rigidities, we can only determine the general spectral index γ as well as the position R_l and amount $\Delta\gamma_l$ for the first break in the power law. The results are reported in Focus Box 9.10, where we have introduced a spectral break $\Delta\gamma_l$ in the diffusion coefficient at rigidity R_l corresponding to the observed shape of secondary-to-primary ratios.

One observes that all the three lightest primary nuclei – He, C and O – show compatible spectral indices and break parameters, as already obvious from Figure 9.8. Since we have no damping of the spectral index by the parameters of the high-energy break, both the position R_l and the amount $\Delta\gamma_l$ numerically come out lower than in the double break fit quoted above in Section 9.3.1, while the functional form is very much the same. The function corresponding to the average functional form of the three primary species is shown as the solid line overlapping He, C and O spectra in Figure 9.8. The dashed line shows what the spectra would look like if the spectral indices were constant.

All the three lightest secondary species – Li, Be and B – have also similar spectral shapes, which are however softer than the light primaries He, C and O as shown in Figure 9.8. This is expected since secondaries of a reaction naturally have a lower energy than primaries, as discussed in Focus Box 9.2. Light secondaries also exhibit a spectral hardening at few hundreds of GV. The origin of the relative spectral hardening can be ascertained looking at the secondary-to-primary ratios as discussed in Focus Box 9.8. The ratios of the secondary Li, Be and B spectra to the dominantly primary oxygen spectrum from the AMS-02 results are shown in Figure 9.9. The ratios all have a power law shape, as expected from diffusion models (see Focus Box 9.2), but with a spectral break. The spectral index of the ratios Li/O, Be/O and B/O is about −0.4 below few hundred GV, and changes by 0.14 above, i.e. the ratios becomes harder. The spectral hardening of light secondaries is about twice as important as for light primaries, as seen from the results of the individual spectra fits shown in Focus Box 9.10. This is expected when the spectral break for all nuclei is due to a propagation effect where the diffusion coefficient $D(R)$ introduced in Section 6.2 changes at a given rigidity R_l. As an effect of propagation, the change of spectral index will first apply to the primary spectra, then again to the secondary spectra as explained in Focus Boxes 9.1, 9.2 and 9.7. One would thus expect the spectral index change to be twice as large for secondaries compared to primaries. Indeed this is what is observed.

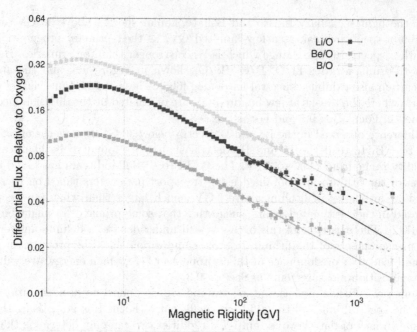

Figure 9.9 Flux ratio of mostly secondary nuclei (Li, Be, B) to mostly primary oxygen (O), as measured by AMS-02 [775]. The solid lines fit the ratios by a power law from 60 GV to 200 GV, the dashed lines above that range.

The measured cosmic ray spectra are used to tune and check analytical and numerical codes connecting spectra at the sources, where cosmic rays are injected into the interstellar medium of the Milk Way, to the spectra observed inside the heliosphere. The principles of this connection were discussed in Chapter 6, and in Reference [813]. As an example, here we briefly summarise how this calculation is carried out in the GALPROP code [393] initiated by Igor Moskalenko and Andrew Strong in the 1990s [288, 289] and ever since continously updated [742, 743, 813] (for the current status and results see https://galprop.stanford.edu). The diffusive transport equation is solved iteratively starting from the heaviest primary nucleus, for instance ^{58}Ni, to calculate the propagated primary spectra as well as the source spectra of the secondaries resulting by its spallation on the ISM, which are in turn propagated to take into account also secondary and tertiary spallation production. The iteration proceeds species by species down to the lightest particles, i.e. protons, secondary electrons and positrons, and antiprotons. A further iteration is then needed to include the effect of radioactive decay, as for instance the contribution to ^{10}B from the ^{10}Be beta decay. The output of these calculations are the spectra in the local interstellar medium (LIS), before they get distorted by solar modulation. In order to directly compare the local interstellar spectra to the measured spectra near Earth, one needs to take into account the effect of solar modulation we discussed in Chapter 8, as in the GALPROP-HelMod framework [612]. The solar modulation is nowadays obtained comparing cosmic ray spectra measured near-Earth with measurements in the local interstellar medium by the Voyager spacecraft [575], as discussed in Chapter 8.

The inputs to the local interstellar spectra calculations are the injection spectra, the diffusion coefficient, and the isotopic spallation cross sections. As we discuss in Focus Box 9.3, the lack of data on isotopic fragmentation cross sections greatly impacts the accuracy of such calculations [657]. However, from the cosmic ray propagation model side, what is unknown or uncertain and can be tuned are in particular the injection spectra and the diffusion coefficient. The experimental information used to do this is generally contained in the fluxes of primary nuclei $d\Phi_p/dR$, which are used to tune the injected spectra, and in the flux ratios of secondary-to-primary nuclei, $(d\Phi_s/dR)/(d\Phi_p/dR)$, used to tune the diffusion coefficient, as discussed in Focus Boxes 9.1 and 9.2. When forming such primary-to-secondary ratios, a good part of the systematic uncertainties cancel. Before the AMS measurements, there were no high accuracy data on our preferred proxy for primaries, which is oxygen. Thus, the flux ratio of B/C as a function of rigidity was traditionally used for tuning transport parameters. The carbon spectrum suffers, however, from a stronger contamination by secondary carbon nuclei, especially those with an oxygen progenitor.

Nitrogen is a good example of a species where neither primary nor secondary origin dominates. It is an important player in stellar nucleosynthesis. As an odd-odd nucleus, it is about a factor three to four less abundant than its neighbours carbon and oxygen, but equally so in stellar matter and in cosmic rays. One can thus conjecture that it represents a mixture of primary and secondary origins. And indeed, its spectral shape lies in between those of our prototype primary and secondary nuclei – oxygen and boron – as seen in Figure 9.8. To reinforce the argument, Figure 9.10 shows that its differential spectrum is very well described by a mixture of an oxygen-like primary and a boron-like secondary spectrum [775, Sec. 12], in about equal overall proportions, but with the secondary fraction decreasing with increasing rigidity. The primary fraction is found to be 56% at 100 GV increasing to 77% at 2 TV. Likewise, the ratio N/O for the species of mixed origin has about half the index values compared to the mostly secondary species, both below and above the break (see Focus Box 9.10).

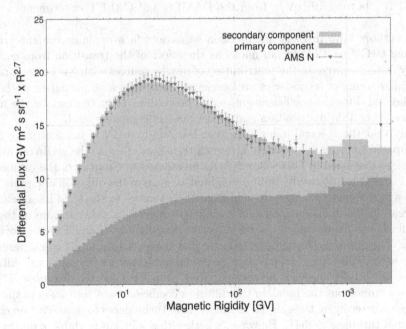

Figure 9.10 The differential spectrum of nitrogen measured by AMS-02, compared to a weighted sum of oxygen (primary) and boron (secondary) spectra measured by the same experiment. The weights are (0.092 ± 0.002) for oxygen and (0.61 ± 0.02) for boron [775]. The secondary fraction steadily decreases with increasing rigidity, from ∼70% at few GV to ∼25% at few TV.

The AMS-02 data on light cosmic ray primary and secondary nuclei provide crucial insights on cosmic ray propagation and have triggered many studies to quantitatively understand cosmic ray propagation and indeed progress has been made by using the rich and high-precision data from AMS-02 [674, 711, 713, 743]. However, the origin of the spectral breaks at few hundred GV observed for all nuclei up to at least silicon and the break at TV observed in protons and helium by DAMPE and CALET is still unclear.

In particular, the GALPROP-HelMod authors have tested the two hypotheses mentioned in Focus Box 9.7, the injection and the propagation scenario [742, 743]. They have randomly sampled the parameter space of injection and propagation to come up with an optimum solution. It turns out that both scenarios – injection with variable spectral index and propagation with variable diffusion coefficients – can fit the data. The resulting description of measured secondary spectra from AMS-02 with the result of their GALPROP tuning [742] is shown in Figure 9.11. Due to the tuning, their spectra follow the rigidity dependence of the data faithfully, without exhibiting the regularities of secondary spectra which we commented on above. The spectral breaks at few hundred GV are well reproduced by both scenarios, but the propagation scenario is preferred since it requires less free parameters and moreover it is consistent with measurements of cosmic ray anisotropy. On the other hand, the TV breaks observed on proton and helium would be more naturally explained by the injection scenario. However, to draw more consistent conclusions data on heavier nuclei beyond 10 TeV/n are needed to see whether these breaks are observed also for heavier species and whether they appear at the same rigidity. Therefore, high-energy extensions of the measurements of the light primary and secondary spectra – C and O and Li, Be and B – beyond 10 TeV/n from the DAMPE and CALET experiments are eagerly awaited.

Other authors have invoked an injection scenario which explains the spectral indices change from 60 GV to 3 TV for all nuclei as the effect of the transition from contribution from many distant sources to the contribution of nearby sources with harder spectra [819]. In this particular scenario, secondaries are accelerated together with primaries at the sources, and the classical diffusion coefficient gets extra power-law terms to describe the high inhomogeneities of the ISM distribution and of the magnetic galactic field.

The fact that the spectral index change at few hundred GV is twice as large for secondaries compared to primaries strongly favours a transport effect as the origin of this spectral break [674], as discussed in Focus Box 9.7. As mentioned in Chapter 6, the diffusion coefficient (see Focus Box 6.4) results from a weighted average of the diffusion of cosmic rays over turbulences in the galactic magnetic field from their sources to the local interstellar region. Thus a break in the diffusion coefficient is indeed revealing a richer nature of the galactic magnetic field turbulence, which can no longer be described by a single average over the entire galaxy. The diffusion through the galactic disk could, for example, run differently than in the halo of the Milky Way [489]. Other spatial inhomogeneities in cosmic ray diffusion can also be invoked [612, 655], since heavier particles propagate shorter distances [577]. Other authors have pinned out the break in the diffusion coefficient as a transition of the diffusion process from scattering on the galactic magnetic field turbulences to scattering on cosmic ray self-generated turbulences [541]. However, whether this solution is viable requires a deeper understanding of the magnetohydrodynamics of the interstellar medium [856]. Work is being done in the direction of performing numerical simulations where the cosmic ray equations of motion are solved in a realisation of the turbulent magnetic field, see reference [761] for a review on test particle simulations of cosmic rays including practical examples.

Very recently, a local reacceleration-propagation model has been proposed by Malkov and Moskalenko [821, 857] to explain the spectral hardening at few hundred GV observed so far for all nuclei up to silicon and iron (see Section 9.3.3), and the TV spectral softening observed on protons and helium nuclei, which would together make a bump between

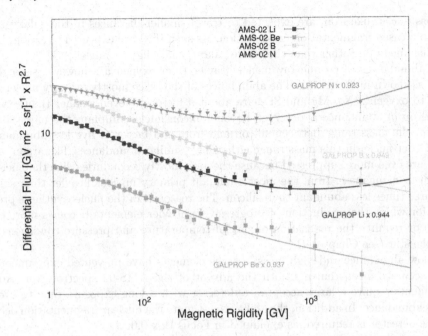

Figure 9.11 Differential fluxes of mostly secondary nuclei (Li, Be, B) as well as nitrogen (N) as measured by AMS-02 [775], compared to the result of GALPROP/HelMod, based on tuning the propagation parameters to AMS light secondary-to-primary flux ratios [743]. To better fit the data, the normalisation of the simulation results has been adjusted by the factors given in the legend.

500 GV and 50 TV. In this model, injection spectra are still featureless power-laws, while a local shock generated the bump accelerating cosmic rays as it passed through the local interstellar medium. Particles below 500 GV were carried away with the interstellar medium, thus they could not reach the Sun and this created the observed bump in all nuclei, both primaries and secondaries. They derive the shock parameters using exclusively the proton data, from AMS-02, CALET and DAMPE, and reproduce the spectra of helium and carbon and the B/C ratio [857]. They indicate as the possible origin of the local shock a passing star, such as Epsilon Indi or Epsilon Eridani. In this scenario, the position of the spectral break at few hundred GV is expected to change on a time-scale of years as the shock gets further away from the Sun [821].

9.3.3 HEAVY NUCLEI

Nuclei heavier than oxygen also contribute information on cosmic ray origin, acceleration and propagation, complementary to the one coming from light nuclei. In the mass range from oxygen to iron several mostly secondary species are found, such as fluorine and the sub-iron elements, like scandium and vanadium, as well as the radioactive isotopes ^{26}Al, ^{36}Cl, and ^{54}Mn. While silicon and iron nuclei are almost pure primary nuclei, other species, such as neon, magnesium, sulphur or calcium, have also a sizeable contribution from spallation of heavier nuclei with different fractions.

In calculating cosmic ray propagation, these heavy nuclei need to be taken into account as progenitors of lower mass secondaries by subsequent spallation processes. This is especially true when searching for subtle effects, as for instance the identification of a possible primordial component in ^{7}Li, or including the modelling of γ-ray emission. In addition, very high-mass nuclei, like iron, are expected to travel smaller distances in the galaxy because of their increased fragmentation cross section. As such they are expected to probe the local interstellar medium, rather than the entire galaxy, as do lighter nuclei [577].

The abundances of cosmic ray nuclei heavier than oxygen are however strongly suppressed, as shown in Figure 3.2. The abundances of the three mostly primary nuclei nearest in mass to oxygen – Ne, Mg and Si – are about a factor of seven smaller than oxygen. A further drop in abundance is observed beyond silicon nuclei. Sulphur, the most abundant nucleus in the mass range between silicon and iron, is a factor of five less abundant than silicon. Other nuclei in this mass range either have similar abundances, like argon and calcium, or are even more suppressed because they are mostly secondaries, like the nuclei from scandium to manganese. Iron, the most abundant primary nucleus heavier than silicon, is about three times less abundant than silicon. The reason is in the nucleosynthesis processes in stars for which the production of progressively heavier elements becomes more energy-intensive or requires the realisation of special temperature and pressure conditions in the core of the star (see Chapter 4).

The low abundance and high mass of these elements have prevented experiments from making accurate measurements until the advent of the AMS-02 spectrometer. Accurate measurements require a large acceptance detector with a long exposure time to overcome their low abundance. In addition, the ability to estimate the nuclear fragmentation occurring inside the detector is required, as explained in Focus Box 9.6.

Pioneer balloon and space experiments, all based on calorimetric measurements, like HEAO, CREAM and TRACER, have measured neon, magnesium and silicon spectra as a function of kinetic energy per nucleon from low energies up to $\sim 10\,\mathrm{TeV/n}$, as shown in Figure 3.3. However, these low-statistics measurements suffered from large uncertainties, of the order of 20% at $50\,\mathrm{GeV/n}$ (i.e. $\sim 100\,\mathrm{GV}$ rigidity) and more at higher energies where they collected only few events. Therefore, the quality of the experimental data did not allow a detailed study of the spectral shape of even these medium-mass nuclei, until the recent AMS-02 measurements spanning a rigidity range from $2.15\,\mathrm{GV}$ to $3\,\mathrm{TV}$, with the unprecedented accuracy of 5% at $100\,\mathrm{GV}$ [731].

These three mostly primary nuclei are found to share the same spectral shape at rigidities above $86.5\,\mathrm{GV}$ as shown in Figure 9.12. The neon and magnesium spectra are already identical from $3.65\,\mathrm{GV}$ onward, most likely because they have a sizeable secondary contribution from spallation while silicon nuclei are predominantly primary. The spectral indices of Ne, Mg and Si follow the same rigidity dependence with the same hardening above $\sim 200\,\mathrm{GV}$. However, comparison with the oxygen spectrum, by looking at the flux ratios Ne/O, Mg/O and Si/O shown in Figure 9.13, reveals that these ratios are not constant. Above $86.5\,\mathrm{GV}$ these primary-primary ratios follow a power law in rigidity with a spectral index of -0.045 ± 0.08. Hence, heavy primary nuclei – Ne, Mg and Si – have a different spectral shape than light primary nuclei – He, C and O – exhibiting a slightly weaker spectral hardening with respect to the light primaries as shown in Focus Box 9.10.

The observed difference between the spectra of Ne-Mg-Si and He-C-O has been interpreted in the framework of the injection scenario, invoking either a local source or the superposition of several sources. For instance, in [746], a local source injecting lower relative abundances of Ne-Mg-Si with respect to He-C-O than the average injected by distant sources, can reproduce both the spectral breaks and the difference between light and medium-mass primaries. However, this scenario does not predict breaks in secondaries, as demonstrated in Focus Box 9.2. This issue is overcome in [858] by invoking a hybrid propagation-injection scenario where the spectral breaks come from both a break in the diffusion coefficient and in the injected spectra at different rigidities for different groups of primary nuclei. An injection scenario with two groups of sources, distant and nearby, and secondaries accelerated at sources [819] seems to fit well also the Ne-Mg-Si data, together

In the scenario where the spectral breaks arise from the propagation mechanism, the diffusion coefficient (see Focus Box 6.4) no longer follows a simple power law, but has an analogous break (see Focus Box 9.7):

$$D(R) = \beta^\eta K \left(\frac{R}{R_0}\right)^\delta \left[1 + \left(\frac{R}{R_l}\right)^{\Delta\gamma_l/s}\right]^{-s}$$

The corresponding power spectrum exhibit a hardening break described by the following function:

$$\frac{d\Phi}{dR} = cR^{-\gamma}\left[1 + \left(\frac{R}{R_l}\right)^{\Delta\gamma_l/s}\right]^{s} = R^{-\tilde\gamma_1(R)} \quad ; \quad \tilde\gamma_i(R) = \frac{d(\log\Phi)}{d(\log R)} = \alpha(Z) - \frac{d(\log D(R))}{d(\log R)}$$

When one fits this function to the AMS-02 data shown in Figures 9.5, 9.8 and 9.12, the results for $45\,\mathrm{GV} < R < 3\,\mathrm{TV}$ are:

Element	$c\,[\mathrm{m^{-2}sr^{-1}s^{-1}GV^{-1}}]$	γ	$\Delta\gamma_l$	$R_l\,[\mathrm{GV}]$	s
H	$(4216 \pm 5) \times 10^{-4}$	2.815 ± 0.004	0.175 ± 0.028	369 ± 54	0.066 ± 0.026
He	$(907.7 \pm 2.2) \times 10^{-4}$	2.728 ± 0.003	0.126 ± 0.013	364 ± 28	0.026 ± 0.011
C	$(29.49 \pm 0.32) \times 10^{-4}$	2.748 ± 0.022	0.162 ± 0.045	277 ± 53	0.051 ± 0.051
O	$(30.69 \pm 0.10) \times 10^{-4}$	2.696 ± 0.003	0.140 ± 0.025	574 ± 74	0.013 ± 0.020
$\langle\mathrm{He, C, O}\rangle$		2.712 ± 0.002	0.131 ± 0.011	459 ± 26	0.023 ± 0.009
Li	$(3.63 \pm 0.06) \times 10^{-4}$	3.09 ± 0.02	0.43 ± 0.09		
Be	$(1.88 \pm 0.03) \times 10^{-4}$	3.09 ± 0.02	0.23 ± 0.11		
B	$(5.13 \pm 0.05) \times 10^{-4}$	3.08 ± 0.01	0.30 ± 0.06		
$\langle\mathrm{Li, Be, B}\rangle$		3.08 ± 0.01	0.32 ± 0.05		
N	$(5.96 \pm 0.06) \times 10^{-4}$	2.88 ± 0.01	0.32 ± 0.05		
Ne	$(5.20 \pm 0.06) \times 10^{-4}$	2.75 ± 0.01	0.21 ± 0.03		
Si	$(5.29 \pm 2.41) \times 10^{-4}$	2.71 ± 0.20	0.01 ± 0.41		
Mg	$(6.23 \pm 0.04) \times 10^{-4}$	2.74 ± 0.01	0.12 ± 0.03		
$\langle\mathrm{Ne, Si, Mg}\rangle$		2.74 ± 0.01	0.17 ± 0.02		

Note that the spectra of secondaries have little sensitivity to R_l and s_l, we therefore fixed them to the values obtained for $\langle\mathrm{He, C, O}\rangle$.

The high-energy proton and helium data from CALET and DAMPE indicate a second, softening break. Introducing a diffusion coefficient with two breaks:

$$D(R) = \beta^\eta K \left(\frac{R}{R_0}\right)^\delta \left[1 + \left(\frac{R}{R_l}\right)^{\Delta\gamma_l/s_l}\right]^{-s_l} \left[1 + \left(\frac{R}{R_h}\right)^{\Delta\gamma_h/s_h}\right]^{s_h}$$

The nuclei spectrum function reads:

$$\frac{d\Phi}{dR} = cR^{-\gamma}\left[1 + \left(\frac{R}{R_l}\right)^{\Delta\gamma_l/s_l}\right]^{s_l}\left[1 + \left(\frac{R}{R_h}\right)^{\Delta\gamma_h/s_h}\right]^{-s_h} = R^{-\tilde\gamma_2(R)}$$

Fits in the range $30\,\mathrm{GV} < R < 40\,\mathrm{TV}$ using data from AMS, CALET and DAMPE shown in Figure 9.5 yield the following parameters:

Element	c	γ	$\Delta\gamma_l$	$R_l\,[\mathrm{GV}]$	$\Delta\gamma_h$	$R_h\,[\mathrm{TV}]$
H	12777 ± 29	2.711 ± 0.005	0.263 ± 0.072	654 ± 207	-0.29 ± 0.07	10 ± 5
He	2634 ± 8	2.702 ± 0.003	0.252 ± 0.020	799 ± 96	-0.51 ± 0.14	19 ± 3

The data show little sensitivity to s_h, therefore we set $s_h = s_l = s$ and obtain $s = 0.122 \pm 0.042$ for both H and He spectra.

Focus Box 9.10: Spectral index and diffusion

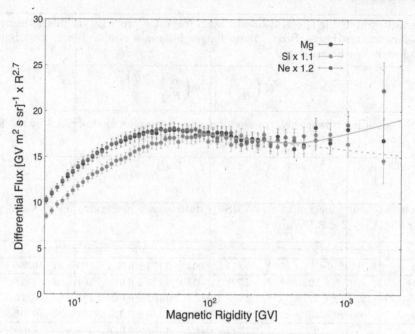

Figure 9.12 Differential fluxes of medium-mass primary nuclei, neon, magnesium and silicon as a function of rigidity measured by AMS-02 [731], and multiplied by $R^{2.7}$. The three spectra follow the same rigidity dependence above ~ 90 GV. A spectral hardening is observed above 200 GV. The solid lines result from a fit to the spectra of the three groups with a broken power law as used in [731], with parameters given in Focus Box 9.10. The dashed line show the spectral shape expected for constant spectral indices.

with light primaries and boron. There are as yet no calculations of Ne-Mg-Si spectra with the local reacceleration-propagation model of Malkov and Moskalenko [821, 857].

Neon, magnesium and silicon are the progenitors of the medium-mass secondary fluorine nuclei. In the mass range between oxygen and silicon, fluorine is less abundant than silicon by a factor of seven and it is the only one considered of pure secondary origin. It is suspected that it might have some primary contribution, but low because fluorine is easily destroyed in stars. However, before the advent of AMS-02, calculations of its primary component were inconclusive because of the lack of accurate data on the fluorine spectrum [742], in addition to the uncertainties on spallation cross sections. Indeed, measurements of the fluorine spectrum performed by pioneering experiments were limited to energies below $\sim$ 100 GeV/n and suffered from uncertainties exceeding 100% at 50 GeV/n. The recent results from the AMS-02 magnetic spectrometer over the rigidity range from 2.15 GV to 2.9 TV with a total uncertainty of 5.9% at 100 GV has changed radically this situation [773]. The fluorine spectral shape has now been characterised in detail. The measurement of its spectral index as a function of rigidity shows that it also hardens above 200 GV as observed for all the other nuclei. However, when compared to the light secondary nuclei, for which we use the boron spectrum as template, the fluorine spectrum does not follow the same rigidity dependence below 150 GV as shown in Figure 9.14.

It is instructive to compare secondary-to-primary ratios for light (B/O) and medium-mass nuclei (F/Si), as shown in Figure 9.15(left). The spectral index change of F/Si before and after the spectral hardening at few hundred GV is found to be 0.15 ± 0.07, which is compatible with the value observed for the B/O of 0.140 ± 0.025. However, the two ratios

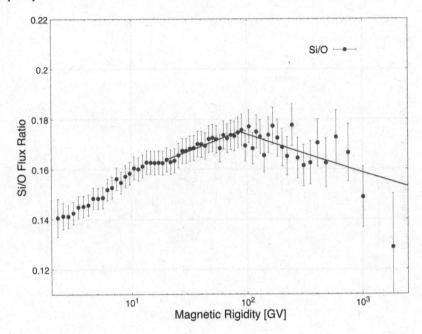

Figure 9.13 Flux ratio of medium-mass primary nuclei Si to the light primary oxygen O, as measured by AMS-02 [775]. The solid lines fit the ratios with a power law from 20 GV to 86.5 GV, and a different spectral index above that range.

do not follow the same shape, in contrast to commonly used diffusion models. Instead the double ratio (F/Si)/(B/O) follows a power law $\propto R^{0.052\pm0.007}$, as seen in Figure 9.15(right).

As discussed in Focus Box 9.2, in diffusion models secondary-to-primary ratios map the grammage (or the inverse of the diffusion coefficient) which is assumed to be the same for all species (Focus Box 9.1) provided they have similar origin and travel similar galactic distances. The GALPROP-HelMod team has analysed the above results in a recent paper [849]. A rough estimate of the effective propagation distances x_A [577], assuming that nuclear cross sections depend on the mass number A as $A^{2/3}$, gives

$$x_A \sim 2.7\,\text{kpc} \left(\frac{A}{12}\right)^{-1/3} \left(\frac{R}{4\,\text{GV}}\right)^{\delta/2}$$

This yields an effective propagation distance for silicon about 10% smaller than that of oxygen and neon, which are very close in mass and hence are expected to probe similar galactic volumes. In their study, they calculate the F/Si and F/Ne ratios using the diffusion coefficient from B/C and find consistent results with the AMS-02 data above 10 GV. Therefore they conclude that B/O and F/Si practically probe the same galactic volume and that is unlikely that the observed difference is related to different diffusion properties of light and medium-mass nuclei. They instead explain the discrepancy below 10 GV as due to a primary fluorine component. However, they still retain a different distribution of the sources of light and medium-mass cosmic rays as a possible explanation. This latter hypothesis is also supported by the observed difference in the spectra of light – He-C-O – and medium-mass – Ne-Mg-Si – primary nuclei and can be tested looking at other species in the same mass range, such as Na and Al, but also at heavier secondary-to-primary ratios, like the scandium-to-iron and the vanadium-to-iron ratios, which would probe cosmic ray transport in the local galaxy.

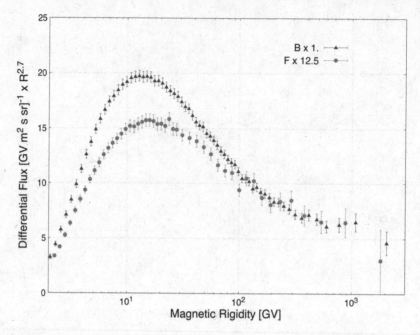

Figure 9.14 Differential fluxes of medium-mass secondary fluorine nuclei (F) compared to the flux of the light secondary boron nuclei (B) as measured by AMS-02 [773]. The fluxes have been multiplied by $R^{2.7}$ and rescaled by the factors indicated in the picture. The two spectral shapes are the same only at rigidities above $\sim 150\,\mathrm{GV}$.

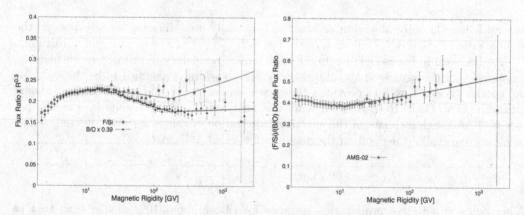

Figure 9.15 Medium-mass (F/Si) and light (B/O) secondary-to-primary ratios multiplied by $R^{0.3}$ (left) and their ratio (right) as a function of rigidity as measured by AMS-02 [773]. The B/O ratio has been rescaled by 0.39 for comparison. Both ratios harden above 200 GV with the same spectral index change, as found fitting the data with two different spectral indices below and above the break (left plot, solid lines). However, their rigidity dependence is different, the double ratio increases with increasing rigidity following a power law (right plot, solid line).

A similar combined study of the light and medium-mass secondary-to-primary ratios has been conducted with the USINE propagation tool [797]. The authors found that the F/Si calculated using the transport parameters from the light secondary-to-primary ratios overestimates the data by 15%, which is within the uncertainties of the set of spallation cross sections they used for the production of secondary fluorine. They therefore conclude that light and medium-mass nuclei can be described by the the same transport parameters and that the fluorine data can be described without the need of introducing a primary component. A recent update with more up-to-date cross sections finds a smaller overestimate [845]. This is yet another example of the need for accurate measurements of isotopic fragmentation cross sections to interpret current cosmic ray data (see Focus Box 9.3).

Sodium and aluminium are expected to be mixtures of primary and secondary nuclei, like nitrogen (see Figure 9.10). Measurements of the sodium and aluminium spectra were recently released by the AMS-02 collaboration [772]. These measurements span the rigidity range from 2.15 GV to 2.15 TV and have uncertainties of 5% and 4.8% at 100 GV for sodium and aluminium respectively, largely improving on the 50% uncertainties of previous measurements. Sodium and aluminium nuclei, as nitrogen, though being produced in stars, have also a sizeable contribution from spallation of heavier cosmic ray nuclei, mostly silicon and also magnesium in the case of sodium. They are about a factor four less abundant than silicon nuclei.

The spectral shape of sodium nuclei is quite similar to nitrogen below 100 GV, while above it is significantly softer. On the contrary the aluminium spectrum follows the one of nitrogen only above 100 GV. Overall their spectral shape is situated between the spectral shapes of primary – He-C-O and Ne-Mg-Si – and secondary nuclei – Li-Be-B and F –, similar to what is observed for nitrogen (see Figure 9.8). These spectra are indeed also well described by a linear combination of a primary and a secondary components, defined by silicon and fluorine spectra respectively. In this way, the primary fraction of sodium is found to be 35% at 100 GV increasing to 62% at 2 TV, while the aluminium primary fraction is higher: 67% at 100 GV increasing to 78% at 2 TV.

The GALPROP-HelMod group has recently used these data to study the origin of sodium and aluminium in cosmic rays finding an excess of aluminium at low rigidities [847], which could have the same origin as similar excesses they had previously found for the lithium and fluorine, i.e. a contribution from novae [742, 849].

Cosmic ray iron nuclei are the only mostly primary nuclei beyond silicon and are found in a ratio of 1:3 with respect to silicon. Because of their high mass ($A = 56$), their effective propagation distance is about 30% smaller than that of oxygen. Therefore iron nuclei probe the local interstellar medium. There have been many measurements of the iron spectrum by pioneering experiments, though with large uncertainties (20% at 50 GeV/n corresponding to 100 GV). The AMS-02 spectrometer and the CALET calorimeter have recently provided more accurate measurements of the iron differential flux. The AMS-02 measurement [774] is made as a function of rigidity from 2.65 GV to 3 TV with uncertainties of 4.8% at 100 GV and 9% at 1 TV. The CALET measurement [779] is made as a function of kinetic energy per nucleon from 10 GeV/n to 2 TeV/n with uncertainties of 9.8% at 50 GeV/n and 10% at 500 GeV/n (corresponding to 1 TV). These recent measurements are shown in Figure 9.16. The AMS-02 and CALET measurements agree in shape but not in normalisation: CALET finds a 20% lower flux than AMS-02. A possible source of this discrepancy can be the hadronic interaction model (see Focus Box 9.6) or a systematic error in the energy scale determination (see Focus Box 9.9). In their publication, the CALET collaboration also indicates the trigger efficiency as a possible cause, which they plan to investigate further [779].

The spectral shape of the Fe flux, when compared to other primary nuclei, follows the shape of light primaries – He-C-O – as shown in Figure 9.17 with measurements from AMS-02 and detailed in reference [774].

The AMS-02 data on iron nuclei have been analysed together with results from Voyager-1 and ACE-CRIS by the GALPROP-HelMod team to update their calculation of the local interstellar spectrum of iron in the range 1 MeV/n to 10 TeV/n. They find an excess of iron at low-energy hinting to a past supernova activity in the solar neighbourhood [803].

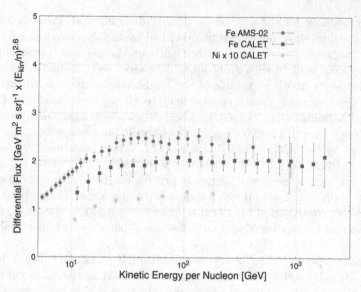

Figure 9.16 Iron differential flux as measured by AMS-02 [774] and CALET [779] as a function of kinetic energy per nucleon and multiplied by $(E_{kin}/n)^{2.7}$. The AMS-02 measurement has been converted from rigidity to kinetic energy per nucleon assuming iron is all ^{56}Fe. The CALET and AMS-02 measurements agree in shape but have a 20% difference in absolute normalisation [779]. Also shown is the measurement of the nickel differential flux by CALET [839], multiplied by a factor of 10.

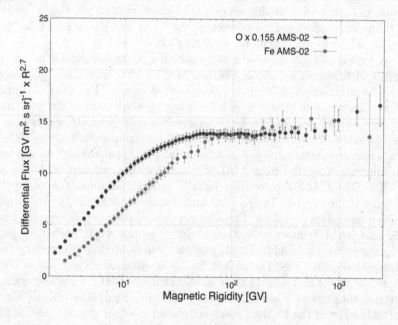

Figure 9.17 Iron and oxygen differential flux as a function of rigidity as measured by AMS-02 [774, 775]. The spectral shape of the high-mass primary nuclei – Fe – is similar to that of the light primaries – He-C-O – above ~ 80 GV.

Accurate measurements of nuclei between silicon and iron require large exposure time, because of their low abundance. Some of them are challenging to measure because of higher fragmentation and contamination from heavier neighbouring nuclei as e.g. for the sub-iron elements for which until now no individual flux measurements exist.

While there are measurements of nuclei in this mass range by pioneering experiments [199, 235, 247, 494], suffering by large uncertainties, there are yet no published results by recent experiments, i.e. AMS-02, CALET or DAMPE. Their results are expected in the coming years as discussed at recent conferences.[1] The AMS-02 collaboration has presented a preliminary measurement of the sulphur spectrum from few GV up to few TV, which seems to align with the medium-mass spectra – Ne-Mg-Si. While lacking more accurate data on nuclei heavier than silicon, current calculations of the nuclei spectra are tuned mostly on the HEAO3-C2 measurements, which however suffer from undocumented systematic uncertainties. The low-energy data from ACE-CRIS, taken in the inner heliosphere, and Voyager-1, taken in its travel through the boundary of the heliosphere, are used to compute the local interstellar spectra, as done for instance by the GALPROP-HelMod team, computing the spectra of nuclei up to nickel [743]. We show as an example their result for the nickel spectrum in Figure 9.18.

For elements beyond iron, the production mechanism by stellar nucleosynthesis is radically different as discussed in Chapter 4. The neighbouring nickel nucleus is the most abundant, however suppressed by a factor of 20 with respect to iron. The abundance of elements above nickel is even further suppressed, therefore the nickel flux spectrum is relatively easy to measure since the background from fragmentation of heavier nuclei in the detector is low.

Pioneering experiments [190, 197, 199, 201, 235, 247, 494] have measured the nickel spectrum only below 80 GeV/n, however measurements above 35 GeV/n suffered from uncertainties larger than 100%. Very recently the CALET collaboration has published a measurement of the nickel spectrum as function of kinetic energy per nucleon in the range 8.8 GeV/n to 240 GeV/n (corresponding to a rigidity range from 20 to 500 GV) [839] with uncertainties ranging from 13% to 20% below 140 GeV/n and 31% above. We show the published nickel measurements in Figure 9.18 together with the latest GALPROP-HelMod calculation [743], tuned on nuclei data from AMS-02 for $Z \leq 14$ (silicon) and HEAO3-C2 for heavier nuclei. The GALPROP-HelMod calculation and the CALET measurement roughly agree above 13 GeV/n.

The CALET collaboration has found that the ratio Fe/Ni is constant in the overall energy range of their measurement within the experimental accuracy (see Figure 9.16), suggesting that below 500 GV nickel nuclei share the same spectral shape as iron and light primaries – He, C and O. If confirmed with larger statistics and extended energy range data this hints at a common origin as well as acceleration and propagation history of cosmic-ray iron and nickel nuclei.

The AMS-02 collaboration has presented a preliminary measurement as function of rigidity from 3 GV up to 1 TV at recent conferences,[2] which is substantially in agreement with the CALET measurement above 13 GeV/n. These preliminary AMS data also agree that the Fe/Ni ratio is constant over the entire rigidity range within the experimental uncertainty, which above 140 GeV/n is about the same order of magnitude as the CALET measurement.

---

[1]See for instance ICHEP2022 https://www.ichep2022.it/, or COSPAR2022 https://www.cosparathens2022.org, where a special session was dedicated to the AMS-02 results with 10-years data.

[2]Q. Yan at ICHEP2022 https://www.ichep2022.it/, or V. Choutko at COSPAR2022 https://www.cosparathens2022.org and ECRS2022 https://indico.nikhef.nl/event/2110/.

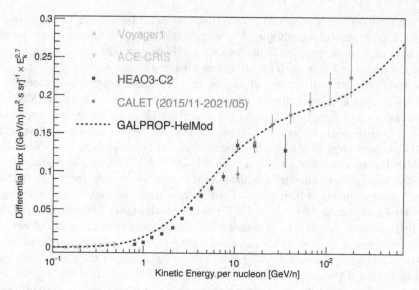

Figure 9.18 Differential flux of nickel nuclei as function of kinetic energy per nucleon measured by HEAO3-C2 [235], ACE-CRIS [494], Voyager 1 [575] and CALET [839]. The dashed line shows the prediction of the nickel spectrum near Earth from the latest GALPROP-HelMod [743], tuned on measurements of the local interstellar spectrum from Voyager 1, the low-energy inner heliosphere spectrum from ACE-CRIS, and high-energy near-Earth spectrum from HEAO3-C2. It corresponds to the solar modulation calculated for the CALET data taking period (retrieved from the HelMod website https://www.helmod.org/).

Measurements of nuclei beyond nickel are interesting to understand the origin of heavy elements, i.e their sources and the nucleosynthesis processes occurring therein, and also to gain insight on cosmic ray acceleration sites. Currently, there are no measurements of fluxes beyond nickel because of their extremely low abundance. However, integrated relative abundances have been measured by several balloon experiments, for instance by TIGER for iron to selenium ($26 \leq Z \leq 34$) at GeV/n energies [415], by SuperTIGER for Fe up to Zr ($26 \leq Z \leq 40$) [585] of energies above 700 GeV/n, and by LDEF for the ultra-heavy elements up to the actinides ($60 \leq Z \leq 90$) of energy above 1.5 GeV/n (corresponding to a median energy of 4 GeV/n) [273], and the satellite experiment HEAO3-HNE for ($33 \leq Z \leq 70$) of energy above 1.5 GeV/n [232]. These measurements, together with those of lighter nuclei, have allowed to study cosmic ray source abundances on an extended portion of the table of elements. The calculation of source abundances generally requires the use of a propagation model.

The TIGER team has carried out such calculations finding that the relative source abundances of cosmic-ray nuclei seem to indicate OB star associations[3] as prominent sites of cosmic ray acceleration, where material ejected from massive stars mixed with interstellar material could give the right mixture of elements to explain the data. Then cosmic rays would be accelerated in superbubbles inflated by supernovae explosions occurring occasionally inside these star clusters rather than in plain interstellar medium. They have also observed that refractory elements, which aggregate in dust grains, are more preferentially accelerated than volatile elements, found in the interstellar gas. This would partially explain why refractory elements are generally more abundant than volatile elements in cosmic rays

---

[3]O and B stars are massive stars often found in groups called OB association. O stars are short lived dying in a supernova explosion after only a million years.

while the opposite holds for the solar system [415, 585]. The TIGER team is now developing a space-born experiment, TIGERISS, to extend their measurements up to Pb ($Z = 82$) on the International Space Station [835].

9.3.4 ISOTOPES

Going to a deeper level of detail, the measurement of mass composition for each cosmic ray species and ideally the isotopic fluxes would give access to a plethora of new information on the origin and propagation of cosmic rays. Indeed, for most species different isotopes have different origin, primary or secondary, or might even be produced by different types of sources. This is the case for instance for neon, for which the observed overabundance of ^{22}Ne in cosmic rays with respect to the solar wind by about a factor of five suggests that this isoptope originates from material ejected by massive stars at advanced stages of their evolution, the so-called Wolf Rayet stars [281, 315]. Whether the sites of production of ^{22}Ne coincide with OB star associations considered a prominent cosmic ray acceleration sites is still under debate [677, 810]. Another example is iron, for which the detection of ^{60}Fe of cosmic ray origin in the Antarctica deep ice hints to nearby past supernova activity [699, 803]. The origin of lithium, where the underabundance with respect to big bang nucleosynthesis is still not fully understood (see Section 4.1), might become clearer if a differential flux of ^{7}Li is measured and the existence of a primary component from novae is ascertained.

In this section, however, we focus on isotopes which have a secondary origin, giving access to additional secondary-to-primary ratios, and on unstable secondaries, the so-called radioactive clocks, which can bring complementary information on the cosmic ray residence time in the galaxy, or the galactic halo size.

In Focus Box 9.2, we have outlined how the cosmic ray grammage $X(R) \propto H/D(R)$, where H is the halo size and D is the diffusion coefficient, is extracted from secondary-to-primary spectrum ratios measured near Earth. However, knowing the grammage leaves unconstrained the cosmic ray confinement time in the galaxy $\tau_{esc}(R) = H^2/D(R)$ because the galactic halo size H is not known a priori. But Nature is generous and has provided radioactive secondary nuclei, such as ^{10}Be, ^{26}Al, ^{36}Cl and ^{54}Mn, which undergo beta decay to the stable secondaries ^{10}B, ^{26}Mg, ^{36}Ar and ^{54}Fe respectively, with a rest-frame lifetime of $\mathcal{O}(\text{Myr})$ similar to the estimated cosmic ray residence time in the galaxy. While stable secondary cosmic ray spectra, like ^{9}Be, are sensitive to both the diffusion coefficient and the galactic halo size, radioactive secondary cosmic rays, as ^{10}Be, are only sensitive to the diffusion coefficient because they decay before reaching the galaxy boundary. Therefore one can extract the galactic halo size from the ratio of stable to unstable secondary cosmic rays, such as ^{9}Be/^{10}Be. We sketch how this is done in Focus Box 9.11.

A complete review of existing isotope ratios measurements is beyond the scope of this book. We rather discribe the detection techniques used to identify isotopes in cosmic ray direct detection experiments, and then discuss a few results of recent isotopic fluxes measurements.

Measuring cosmic-ray isotopic composition in space-born or balloon experiments is not an easy task because of the restrictions in size and weight. The nuclear mass is measured by combining a velocity measurement with an energy or a momentum measurement. Velocity measurements are only meaningful at rather low energies. At high energy, the particle velocity comes too close to the speed of light and the difference between two isotopes quickly becomes smaller than the resolution of any velocity measurement technique. For current detection technologies, this limit translates to a kinetic energy per nucleon of about $10\,\text{GeV/n}$. The energy is measured with the stopping range technique, which can only be employed at very low energies (hundreds of MeV/n) since it requires the nucleus to stop in the material

Following the formalism used in [446], the spectrum of surviving radioactive secondary isotopes of type r, $d\Phi_r/dR$, observed near Earth can be expressed as:

$$\frac{d\Phi_r}{dR} = f_s \frac{d\Phi_s}{dR}$$

where $d\Phi_s/dR$ is defined in Focus Box 9.2 and describes the flux resulting from the net production rate of secondary nuclei of the same type, and $f_s < 1$ is a particle survival factor that takes into account that some of these unstable isotopes will decay before arriving near Earth. The survival factor f_s is a function of the cosmic ray residence time $\tau_{esc}(R)$ and of the decay time τ_s of the radioactive secondary cosmic ray isotope. The exact relation depends on the specific diffusion model [820]. Therefore, measuring the suppression factor allows to determine the cosmic ray residence time and, using the grammage derived from secondary-to-primary ratios, obtain the galactic halo size.

The determination of the survival factor f_s requires the measurement of the spectrum of the radioactive secondary isotope, e.g. ^{10}Be, and to compute its expected spectrum at Earth, $d\Phi_{^{10}\text{Be}}/dR$, if its lifetime were infinite. The calculation involves the knowledge of the spallation cross sections for each isotope of the primary nuclei that contribute to the ^{10}Be production, which are not well known as discussed in Focus Box 9.3. In addition, there are no currently published measurement of separate fluxes of beryllium isotopes; a measurement from AMS-02 up to ~ 10 GeV/n has been presented at recent conferences [836] and is expected to be published soon. Meanwhile, the galactic halo size has been obtained from the Be/B ratio using the B and Be nuclei spectra measured by AMS-02 up to TV rigidities using the following relation:

$$\frac{d\Phi_{\text{Be}}}{dR} \Big/ \frac{d\Phi_{\text{B}}}{dR} = \frac{\frac{d\Phi_{^7\text{Be}}}{dR} + \frac{d\Phi_{^9\text{Be}}}{dR} + f_{^{10}\text{Be}}\frac{d\Phi_{^{10}\text{Be}}}{dR}}{\frac{d\Phi_{^{10}\text{B}}}{dR} + \frac{d\Phi_{^{11}\text{B}}}{dR} + \xi(1 - f_{^{10}\text{Be}})\frac{d\Phi_{^{10}\text{Be}}}{dR}}$$

It links the measured Be/B spectrum ratio to the spectra of each of their isotopes and to the ^{10}Be flux suppression factor $(1 - f_{^{10}\text{Be}})$. The isotopic fluxes in the right term are computed from the measured spectra and isotopic composition of primary nuclei producing the boron and beryllium isotopes by spallation and the spallation cross sections. The ξ factor takes into account that in the beta decay ^{10}Be $\rightarrow$ ^{10}B the daughter nucleus inherits the kinetic energy per nucleon of the parent nucleus. This method allows to determine the cosmic ray confinement time up to TV rigidities, but the calculation of the right-hand side term is affected by large uncertainties on the spallation cross sections which are of the order of tens of percent [820]. This limits the constraining power of the Be/B ratio on the galactic halo size [736]. Therefore, accurate measurements of the beryllium isotopes spectra from AMS-02 or the future HELIX balloon experiment [804] are eagerly awaited.

In principle the same method can be applied to derive the galactic halo size from the Al/Mg, Ar/Cl and Fe/Mn ratios, but in practice most of the needed spallation cross sections are currently unknown [820].

Focus Box 9.11: Cosmic ray radioactive clocks

after having deposited all its energy (see Focus Box 3.4). Momentum measurements can be performed also for more energetic particles. Earlier experiments measured the momentum with the multiple scattering technique in nuclear emulsions or the geomagnetic cutoff but these techniques cannot reach above few GV. Above, a magnetic spectrometer is needed (see Focus Box 3.2). Magnetic spectrometers measures rigidities up to multi-TV, here however we are interested in the region below a few tens of GV where multiple scattering in the tracking detector may affect the momentum resolution. It is then essential to have as little as possible material and a high spatial resolution employing a strong magnetic field. Several detection techniques are used to measure the velocity: indirectly, from the stopping range in a stack of passive material layers or from the specific energy loss dE/dx (see Focus Box 1.1), or directly employing time-of-flight or Cherenkov detectors. The time-of-flight technique requires a sufficiently large distance between the two time measuring layers, and a good time resolution. The most accurate technique to measure the velocity is the Cherenkov technique. As explained in Focus Box 2.11, the aperture of the light cone is proportional to the particle velocity and the number of emitted photons is proportional to the particle charge squared. Cherenkov detectors can reach velocity resolution of per mille. But they are sensitive only above a certain threshold depending on the refractive index of the radiator material. To extend the velocity range of the measurement often a dual material radiator is used, as in the AMS-02 detector. Often several detection techniques are combined together since they are sensitive at different velocity ranges.

Measurements of nuclei isotopic composition were performed already in early experiments at low energies. Some of them measured also isotopic fluxes, however at few energies and with large uncertainties. Complete datasets and references can be found in the CRDB database [532].

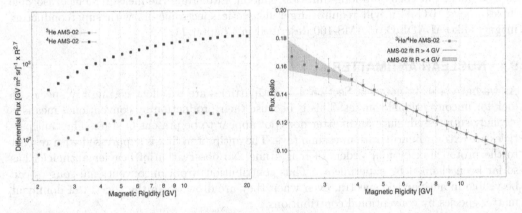

Figure 9.19 Isotopic fluxes for ^{3}He and ^{4}He (left) and their ratio (right) as function of rigidity as measured by AMS-02 [685].

Currently, isotopic fluxes measurements over extended energy ranges exist only for hydrogen and helium nuclei. For heavier nuclei up to silicon only one or two measurements exist at a few hundred MeV/n with large uncertainties since the isotopic fluxes were indirectly derived from the average nuclei flux and the relative isotopic composition without an assessment of systematic uncertainties, detector acceptance and efficiencies for each isotope. Cosmic ray hydrogen and helium nuclei both have two isotopes. Deuterons ^{2}H and ^{3}He isotopes are secondaries, while the more abundant protons and ^{4}He nuclei are primaries. Both ^{2}H and ^{3}He are produced by spallation of ^{4}He, while heavy secondaries are produced by multiple fragmentation channels. The measurement of fluxes and flux ratios for light isotopes would give complementary information on cosmic ray propagation as their interaction mean free path is larger than their escape mean free path from the galaxy, contrary to heavier secondaries [468]. The deuteron and helium isotope fluxes have been measured from 10 MeV/n up to 1 GeV/n by several space-born and balloon experiments (PAMELA, AMS01, BESS, CAPRICE, etc.), mostly employing a magnetic spectrometer and a time-of-flight or a Cherenkov detector. The CAPRICE98 experiment has extended the measurement of the deuteron flux to 22 GeV/n [347]. The deuteron flux is found to be only 2% of the total hydrogen flux. The ^{2}H/^{4}He data are well described by common diffusion models combined to solar modulation models and agree with calculations where diffusion parameters are tuned on B/C within spallation cross sections uncertainties [490, 727].

AMS-02 has provided helium isotopic fluxes up to 10 GeV/n (Figure 9.19) together with their time dependence [685]. The ^{3}He/^{4}He flux ratio is time-independent above 4 GV, as seen in Figure 8.19 and follows a single power law with a spectral index value in good agreement with the spectral index of B/C and B/O ratios at TV rigidities. Analysis of these data together with heavier secondary-to-primary ratios has shown that helium isotopic ratios give better constraints on propagation models with re-acceleration [745]. Preliminary results on the deuteron flux extending from 0.4 to 10 GeV/n presented by AMS-02 at recent conferences[4] seem to indicate that the very light secondary-to-primary ratios ^{2}H/^{4}He and ^{3}He/^{4}He follow different power laws.

The AMS-02 collaboration has presented at recent conferences preliminary results on the lithium and the beryllium isotope fluxes up to about 10 GeV/n.[5] These data extend into a previously unexplored region and are expected to be published soon. The beryllium isotopic fluxes are eagerly awaited by the community because ^{10}Be is a radioactive clock. The future HELIX detector has been specifically designed to perform this measurement and is expected to be launched in 2023 on a long-duration balloon. Extending the measurement of isotopic fluxes above 10 GeV/n will require larger magnetic spectrometers with superconducting magnets like ALADInO or AMS-100 described in Chapter 11.

9.4 NUCLEAR ANTIMATTER

As we have already noted in Section 4.4, antiparticles are excellent cosmic ray species to look for unconventional sources. This is because their production by conventional means is strongly suppressed, since antimatter does not appear to be present in stars. Thus antiparticles are extraordinarily rare in cosmic rays. The antiproton flux is suppressed with respect to the proton flux by four orders of magnitude. No observation of heavier antinuclei has so far been claimed by experiments. Thus contributions from unconventional sources can be visible in antiparticle spectra, even when they are drowned in the spectra of dominant matter species by conventional contributions.

Prominent among the potential non-conventional sources of antinuclei (and positrons, as we will see in Section 9.6) is the annihilation or decay of dark matter particles. As we have already seen in Section 4.4, the existence of dark matter has been put in firm evidence by the observation of galactic rotation curves and by gravitational lensing through galaxy clusters, which both show the presence of a significant amount of non-luminous matter. Despite the fact that dark matter makes up about 85% of the total matter in the Universe, its nature remains still unknown. Among possible dark matter candidates are weakly interacting massive particles, which we will concentrate on here. If they exist, their masses may range from GeV to TeV.

The matter-antimatter asymmetry is also a long-standing issue in the understanding of our Universe. It is thought that at the time of the Big Bang matter particles (quarks and leptons) and antimatter particles (anti-quarks and anti-leptons) were produced in equal amounts. However, the world we observe today is made by matter particles, i.e. atoms with positively charged nuclei surrounded by electrons. Anti-hydrogen atoms were synthesised at CERN for the first time more than 30 years ago [278]. Today, they are used to systematically

---

[4]C. Delgado at ICHEP2022 https://www.ichep2022.it/, or P. Zuccon at COSPAR2022 https://www.cosparathens2022.org.

[5]J. Wei at ICHEP2022 https://www.ichep2022.it/, or L. Derome at COSPAR2022 https://www.cosparathens2022.org

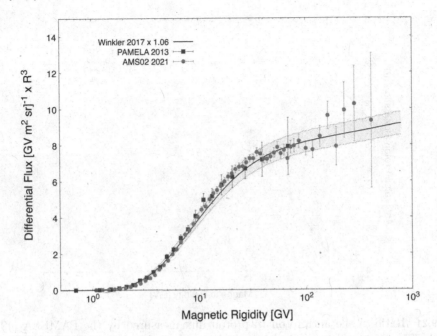

Figure 9.20 Differential flux of antiprotons ($\times R^3$) measured by PAMELA [509] and AMS-02 [775], compared to a calculation based on secondary production only [636]. The shaded band corresponds to the uncertainty from the cosmic ray propagation model and the antiproton production cross sections.

study possible differences between matter and antimatter [732]. Another experimental approach to address the issue of the apparent matter-antimatter asymmetry in nature is to search for leftovers of primordial anti-nuclei – relics from the Big Bang – in the cosmic rays.

Accurate measurements of antiproton spectra require a magnetic spectrometer which has the capability of distinguishing the sign of the particle charge and a high rejection power against the more abundant protons and electrons. As we pointed out in Focus Box 9.4, the AMS-02 detector indeed has the required capabilities. Confusion between positive and negative charges is at the level of 10^{-4} for rigidities up to $100\,\mathrm{GV}$ and less than 8% up to $1\,\mathrm{TV}$ [775, Sec. 1.2.4]. And antiprotons are distinguished from the like-sign, but much lighter electrons at a level of one in 10^4. At lower energies, the BESS Polar II balloon mission (see Figure 2.16) and the PAMELA space spectrometer (see Figure 2.17) pioneered the study of cosmic ray antiprotons.

Figure 9.20 shows the antiproton spectra measured by the space observatories PAMELA and AMS-02, compared to a recent calculation [636] assuming a purely secondary origin of antiprotons, as covered in Section 4.5.3. The agreement is satisfactory, given the uncertainties from propagation parameters and antiproton production cross sections represented by the error band of the calculation result. It is also clear that the data leave little room for contributions to the flux other than conventional processes. Interaction of matter particles with the interstellar medium appears to produce the observed antiprotons in the energy range currently studied.

In principle one can expect less of a systematic uncertainty when considering the ratio of the antiproton to proton flux as a function of rigidity, $(d\Phi_{\bar{p}}/dR)/(d\Phi_{p}/dR)$. Experimental results from AMS-02 are shown in Figure 9.21, compared to a recent calculation [697], which allows for a contribution of dark matter annihilation as discussed in Section 4.4. If any, this

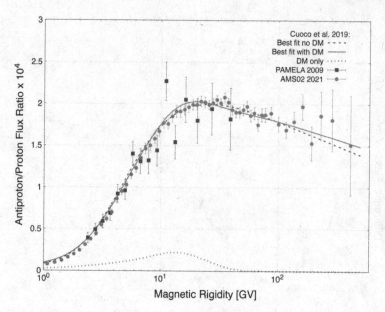

Figure 9.21 Ratio of the antiproton to proton flux measured by the PAMELA [421] and AMS-02 [775], compared to a calculation without (dashed line) and with (solid line) a potential contribution from dark matter annihilation [697] fitted to earlier AMS-02 data [569]. The small discrepancy at rigidities less than 10 GV is due to solar modulation. The dotted curve shows the best fit contribution by dark matter alone.

contribution is marginal, and comparable to the uncertainty due to propagation effects, as e.g. discussed in [775, Sec. 5] using numerous examples. The conclusion that suspected dark matter signals are insignificant in this channel is also reached in other studies [790, 822, 846]. Rather stringent upper limits on the dark matter annihilation rate $\langle \sigma v \rangle$ are obtained for masses of dark matter particles between ten and a few hundred GeV.

The search for heavier antinuclei, like antideuterium or antihelium, is particularly fascinating. They are extremely difficult to produce through interactions between matter particles. As discussed in Section 4.5, the probability to form them decreases by orders of magnitude with each additional antinucleon [754]. There are, however, a good handful of antihelium candidates in the AMS-02 data among 200 billion cosmic particles. Sam Ting presented some of them in a 2018 CERN seminar.[6] The rate is roughly equivalent to one $\overline{\text{He}}$ per year, or one per 100 million He nuclei. On the one hand, this is a very small rate; the systematic significance is difficult to assess because, in particular, the expected background has to be quantified. The fact that this has not happened so far, and that there is no publication, is probably also due to the tiny rate. To ensure that there is no detector malfunction at the level of 1 in 100 million, one needs to know more about the properties of these rare events. Thus the AMS-02 collaboration does not (yet?) claim to have discovered complex cosmic antinuclei.

On the other hand, if at least some of these candidates are taken seriously, there are too many to explain them by conventional nuclear astrophysics. Thus there should be a small, ready-made supply of antimatter somewhere in the galaxy which might have been left over from the Big Bang. That motivated a new search for anti-stars in the Milky Way, i.e. stars made of antimatter. In the catalog of the Fermi satellite there are 14 objects whose photon spectra are compatible with antimatter [785], about 2.5 candidates per million normal stars

---

[6]See https://cds.cern.ch/record/2320166

at a distance between a few 10^{14} and 10^{16} km. They could be considered as sources of $\overline{\text{He}}$ if one knew the release and acceleration mechanism. According to the authors, however, it is more likely that they are normal γ-ray emitters such as pulsars or black holes.

9.5 STRANGELETS

Three generations of quarks and leptons have been observed at collider experiments, but only the first generation is present in the ordinary forms of matter observed in nature: the protons and neutrons which make the atomic nuclei are made by quarks of the first generation (up and down), the electron belongs to the first generation of leptons. The standard model of particle physics, however, does not exclude the existence of astrophysical objects made also by quarks of the second or third generation, such as stars made of compounds of up, down and strange quarks under extreme conditions [211]. Such stars would produce so-called strangelets: exotic bound states made by up, down and strange quarks in roughly equal proportions. Nuclei made by the first generation of quarks have a characteristic charge-to-mass ratio of about $1/2$, while strangelets would have a charge-to-mass ratio less than $1/10$. Such states of matter have been searched for in collider experiments [291, 314] and in lunar soil [419], without success.

The existence of such exotic forms of matter might be revealed by the observation of strangelets in cosmic rays. In the absence of an observation, the best limits on strangelets previously came from the PAMELA satellite experiment [562], which found upper limits on the flux of a few thousand per $(\text{m}^2\,\text{yr}\,\text{sr})$ for strangelets with $Z = 1$ or 2, and flux limits between about 10^4 and 10^7 per $(\text{m}^2\,\text{yr}\,\text{sr})$ for higher charges. The AMS-02 experiment also found no candidates with the required characteristics of a low charge-to-mass ratio in about 80 million helium and eight million heavier nuclei with charge between 3 and 8. Their upper limits on the strangelet flux thus comes out much more stringent than previous limits [775, Sec. 14]. For all charges less than 8 the upper limit is less than about 100 strangelets per $(\text{m}^2\,\text{yr}\,\text{sr})$, for charges 6, 7 and 8 even less than 60 strangelets per $(\text{m}^2\,\text{yr}\,\text{sr})$, for baryon numbers ranging from about 15 to 150. These limits are below the expectation from recent model calculations [366]. Such exotic states of matter have thus not been seen in cosmic rays.

9.6 ELECTRONS AND POSITRONS

Electrons and positrons are the lightest cosmic ray species. Electrons are components of ordinary matter and easily liberated by ionisation. The pair production of electrons and positrons has a low "cost" in energy, thus electromagnetic processes can more easily generate these than other matter-antimatter pairs. In fact, electrons and especially positrons have long been suspected to partially come from unidentified sources, already since the results of the PAMELA space spectrometer [508]. On the other hand, electrons and positrons suffer stronger energy losses during propagation through interstellar matter, as shown in Focus Box 2.7. At high energy they do not travel very far, astrophysical and unconventional sources – if any – must thus be close to the solar system, no farther than a kpc or so.

The flux of the sum of electrons and positrons, $d\Phi_{e^\pm}/dE$, can be measured by spectrometers and calorimeters alike. Figure 9.22 shows recent results from the AMS-02 spectrometer [775] and the DAMPE [601] and CALET calorimeters [644]. The Cherenkov telescope H.E.S.S. can also observe atmospheric showers from high-energy $e^\pm$ and measure their energy [402, 424, 622, 664], as described in Focus Box 7.5. The differential flux is multiplied by the kinetic energy to the third power, E^3, since the spectrum is supposed to be softer than for nuclei due to the higher energy loss. It is obvious in this representation that – taken at face value – the results of the experiments do not agree very well at high energies.

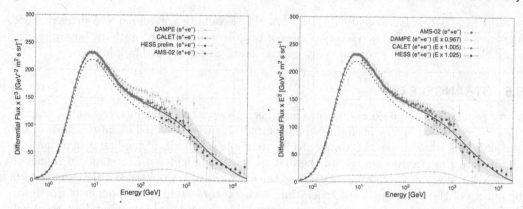

Figure 9.22 The differential spectrum of the sum of electrons and positrons, $d\Phi_{e^\pm}/dE$, scaled by E^3. Results of the AMS-02 spectrometer [775] are shown with those of the calorimeters DAMPE [601] and CALET [644], as well as preliminary results from the Cherenkov Telescope H.E.S.S. [622], for which the shaded area indicates the range of systematic errors. The left graph shows the data as published, the right graph applies small energy scale adjustments to improve the agreement among experiments. The solid curve is the sum of fits to the separate electron (dashed curve) and positron spectra (dotted curve) from AMS, as described in Focus Box 9.12.

The reason can be sought in unidentified background from more abundant species or in uncertainties concerning the energy scale. Background is certainly not an issue for AMS-02, since the TRD signal, the comparison between calorimeter energy and spectrometer rigidity as well as the shower shape in the calorimeter give three handles for a very high background rejection power (see Focus Box 9.4). The deeper calorimeters of CALET and DAMPE likewise provide excellent background rejection by shower shape alone. The energy scale of the space experiments, on the other hand, has an uncertainty of two to three percent [571, 598, 648, 775], and 15% for H.E.S.S.. Using this margin, one can improve the agreement, as also shown in Figure 9.22.

The spectrum clearly has more than one component, as evident from the shoulder for energies beyond about thirty GeV. Before going into details about the properties of this joint spectrum, we look into the separate contributions of electrons and positron. Obviously only spectrometers are equipped to measure these separately.[7]

Figure 9.23 shows the differential spectrum of electrons measured by AMS-02 during its first seven years of exposure on the ISS [775]. There is a clear change in spectral index beyond an energy of about 30 GeV. The total spectrum in the GeV to TeV energy range is well descried by a two-component fit described in Focus Box 9.12. It consists of two power law components, a soft one dominant at low energies and a harder one apparent at high energies. Again, neither spectral index is equal to −3, as one would naively expect from Fermi acceleration and diffusion alone. And numerical codes like GALPROP [458, 470] do not match the spectral behaviour either. The origin can again be sought in different astrophysical mechanisms. One can e.g. invoke separate sources for the two components, a break in the injection spectrum, a break in the diffusion coefficient, or a re-acceleration phenomenon.

The spectrum of positrons measured by AMS in Figure 9.24 shows even more astonishing deviations from prejudice. The two components of the flux are even more apparent than

---

[7]An exception is an effort by the Fermi-LAT collaboration to use the Earth's magnetic field as a spectrometer magnet [619].

Cosmic ray fluxes often suggest that more than one component contributes. Multiple contributions can be due to different particle sources – e.g. near or far ones – or different astrophysical conditions at the sources. They can also arise from different acceleration mechanisms at work or different diffusive conditions met during propagation.

In the simplest case, such contributions just add. One can then describe the joint differential flux by the superposition of the components, e.g. for electrons:

$$\frac{d\Phi_{e^-}}{dE} = c_a \frac{d\Phi_a}{dE} + c_b \frac{d\Phi_b}{dE}$$

$$= c_a \left(\frac{E}{E_a}\right)^{\gamma_a} + c_b \left(\frac{E}{E_b}\right)^{\gamma_b}$$

where the second equation applies in case of power law spectra. For particles with electric charge different from 1, energy can of course be substituted by magnetic rigidity R where appropriate. The constants E_i have no physical significance, but can be chosen to minimise the correlation between the normalisations c_i and the spectral indices γ_i. The estimation of solar modulation (see Section 8.6), relevant at low energies, is done here in the force field approximation. This introduces an estimator $\hat{E} = E + \phi$ of the energy before particles enter the heliosphere, with the so-called solar potential ϕ dependent on the solar cycle and particle type. The total spectrum then reads:

$$\frac{d\Phi_{e^-}}{dE} = \left(\frac{E}{\hat{E}}\right)^2 \left[c_a \left(\frac{\hat{E}}{E_a}\right)^{\gamma_a} + c_b \left(\frac{\hat{E}}{E_b}\right)^{\gamma_b} \right]$$

This function is fitted to the observed electron spectrum from AMS-02 and shown in Figure 9.23. The fitted parameters for $E_a = 20\,\text{GeV}$ and $E_b = 300\,\text{GeV}$ are given in the table below, with normalisation constants in $[\text{m}^{-2}\text{sr}^{-1}\text{s}^{-1}\text{GeV}^{-1}]$:

c_a	γ_a	c_b	γ_b	$\phi\,[\text{GeV}]$
$(1.87 \pm 0.02) \times 10^{-2}$	-4.63 ± 0.03	$(4.05 \pm 0.02) \times 10^{-6}$	-3.17 ± 0.09	1.97 ± 0.02

For positrons a similar decomposition of the flux applies:

$$\frac{d\Phi_{e^+}}{dE} = \left(\frac{E}{\hat{E}}\right)^2 \left[c_d \left(\frac{\hat{E}}{E_d}\right)^{\gamma_d} + c_s \left(\frac{\hat{E}}{E_s}\right)^{\gamma_s} e^{-\hat{E}/E_c} \right]$$

with an additional exponential cut-off at energy E_c which applies to the second component. The indices suggestively indicate a "diffuse" and a "source" term. The curve in Figure 9.24 corresponds to this function, with parameters given below, for $E_d = 7\,\text{GeV}$ and $E_s = 60\,\text{GeV}$:

c_d	γ_d	c_s	γ_s	$\phi\,[\text{GeV}]$
$(6.51 \pm 0.14) \times 10^{-2}$	-4.07 ± 0.06	$(6.80 \pm 0.15) \times 10^{-5}$	-2.58 ± 0.05	1.10 ± 0.03

The cut-off energy comes out as $1/E_c = (1.23 \pm 0.34)/\text{TeV}$ or $E_c = 810^{+310}_{-180}\,\text{GeV}$.

The AMS electron spectrum does not require a cut-off at high energies, in fact a cut-off below an energy of 1.9 TeV is excluded by the data [775]. The sum of electron and positron spectra of Figure 9.22 is well described by the sum of the two functions quoted above, but it does require a cut-off also for the electron spectrum. This cut-off energy is $(8 \pm 2)\,\text{TeV}$, well above the limit from AMS-02 data.

Focus Box 9.12: Two-component electron and positron fluxes

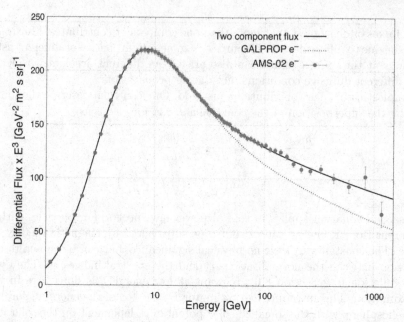

Figure 9.23 AMS-02 measurement of the differential flux of electrons as a function of the energy E, scaled with E^3 [775]. Error bars represent the total statistical and systematic errors. The solid curve is the result of a fit with two flux components as described in Focus Box 9.12, the dotted curve the result of GALPROP [458, 470] with the normalisation adjusted to fit the peak.

in the electron spectrum, and they have very different characteristics as shown in Focus Box 9.12. At low energies, a contribution similar to what one would expect from secondary production of positrons dominates the differential flux. This is shown in Figure 9.24 by the AMS fit result (light grey curve) and the GALPROP result [458] (dotted curve), which show a qualitatively similar behaviour. The AMS-02 collaboration calls this contribution a "diffuse term" in their publications. At energies beyond about 20 GeV, a component with a much harder spectrum becomes apparent, suggestively called a "source term" by AMS-02. It appears to have a cut-off at an energy of the order of 1 TeV, which the electron spectrum does not show. The AMS electron data do, however, allow the addition of the positron source term, without spoiling the agreement with the measured shape. Thus, the source term can contribute symmetrically to electron and positron spectra.

Since positrons are not present in normal matter, one would expect them to be of secondary nature, produced by electromagnetic phenomena. Not only interactions between ordinary cosmic rays and interstellar matter, but also environments with exceptionally high electromagnetic fields like pulsars can be a source of positrons without invoking any unconventional physics. We have discussed these and their characteristics in Section 4.5.1.

To test hypotheses about the additional source of positrons, one often uses the so-called positron fraction, the ratio $(d\Phi_{e^+}/dE)/(d\Phi_{e^\pm}/dE)$. Not only does a good portion of experimental systematics cancel in forming this ratio, the theoretical uncertainty coming from propagation parameters is also reduced, since both suffer similar energy loss (see Focus Box 2.7). An astrophysical example for electromagnetic production of electron-positron pairs is the rapidly changing magnetic fields of pulsars. These are rotating neutron stars whose rotational axis does not coincide with the direction of the magnetic axis. This leads to strong fields in some regions, which can be suitable for generating high-energy electron-positron

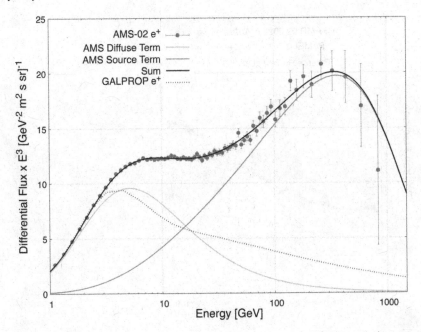

Figure 9.24 AMS-02 measurement of the differential flux of positrons as a function of the energy E, scaled with E^3 [775]. Error bars represent the total statistical and systematic errors. The black curve is the result of a fit with two contributions as described in Focus Box 9.12: the diffusion term (light grey) represents a secondary production of positrons through the interaction of cosmic rays with interstellar matter; the source term (dark grey) corresponds to primary production by a new, not yet identified source. The dotted curve is the result of GALPROP [458] for secondary positrons, similar to the fitted diffusion term.

pairs. As we have seen in Section 4.5.1, single pulsars tend to produce a spectrum which is more peaked than the AMS measurement and the fit results. There is also no reason why a single pulsar should be the source of additional positrons. Since positrons easily disappear in interaction with interstellar matter, point sources should not be located too far from the solar system. Positrons of 100 GeV can e.g. only come from sources some ten thousand years old and located no farther than about one kiloparsec ($\simeq 3 \times 10^{16}$ km) from us [705]. This limits the choice of pulsar source candidates to a few known ones and potentially also some undetected ones [838].

As an example, Figure 9.25 shows a calculation of the positron fraction caused by a collection of near-by pulsars [607]. The candidate pulsar sources are selected not only on the basis of their known characteristics like distance and age, but also using constraints from an observation of their high-energy photon spectra. The latter allows to also construct a sub-dominant contribution from pulsars which have not yet been observed. The agreement of the calculation with the AMS data is satisfactory, given the uncertainties coming from propagation parameters.

If point sources were the origin of the observed positron excess, one could expect a slight anisotropy in their directions of incidence. This is, however, not observed. Since anisotropies are expected to be more important at very high energies, we defer a more detailed discussion to the next chapter in Section 10.5. Here we just quote upper limits on a dipole asymmetry, defined in Focus Box 10.7. AMS finds the distribution of positron arrival directions for $E > 16$ GeV compatible with isotropy and sets an upper limit on the amplitude of a possible dipole anisotropy of $\delta < 0.019$ at 95% confidence level. For the

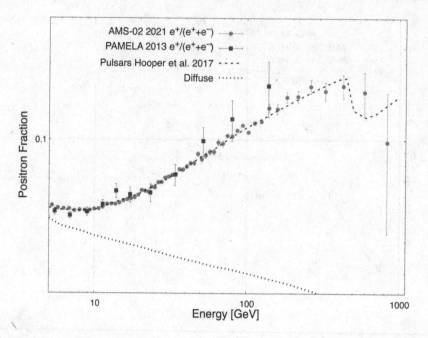

Figure 9.25 The positron fraction, i.e. the ratio of the positron flux to the sum of electrons and positrons, as a function of energy. Data are shown from PAMELA [508] and AMS-02 [775]. The dashed curve is the result of a calculation [607] for the potential contributions of the pulsars Geminga and B0656+14 (Monogem) and other as yet unidentified pulsars in the Milky Way, constraint by the observation of TeV photon emission by HAWK [838]. The dotted curve corresponds to the diffuse contribution.

sum of electrons and positrons and lower limits on energy between 60 GeV and 480 GeV, the Fermi-LAT collaboration finds limits on a dipole amplitude between $\delta < 0.005$ and $\delta < 0.1$ [437]. Because of its uneven exposure map, the H.E.S.S. telescope cannot give such a global upper limit. However, an analysis of their data in the specific directions of the Vela, Geminga and Monogem pulsars [664] finds upper limits on a dipole anisotropy from these sources of $\delta < 0.06$ to 0.07.

While astrophysical point sources are so far not identified, the way is open to consider more diffuse unconventional sources. The shape and energy scale of the source term invites interpretations also in terms of dark matter annihilation, which we have discussed in Section 4.4. There is a large body of literature dealing with a possible dark matter origin of the excess; without an attempt for completeness we quote [492, 495, 520, 524, 559, 570, 581, 641, 659, 808]. If annihilation between two dark matter particles or their decays cause the high-energy feature in the positron spectrum, their origin follows the density profile of dark matter itself. There are several models for this, including the Navarro-Frank-White profile [279, 433] used here and others listed in [455]. For dark matter annihilation, the reaction rate is given by the product of cross section σ and velocity v, $\langle\sigma v\rangle$. For cross sections of typical weak interactions and a thermal velocity distribution, the rates are of the order of $\langle\sigma v\rangle \simeq 10^{-26}\,\mathrm{cm}^3/\mathrm{s}$. To saturate the source term observed by AMS-02, much higher rates are usually required. An example is shown in Figure 9.26, where we have overlayed the expected spectrum of positrons from the reactions $\chi\chi \to \mu^+\mu^-$, $\tau^+\tau^-$, $b\bar{b}$ and $t\bar{t}$, using PPPC 4 DM ID results [455], for an annihilation rate about 70 times larger than the thermal one. Definite conclusions cannot be drawn, given the important uncertainties from cosmic ray propagation.

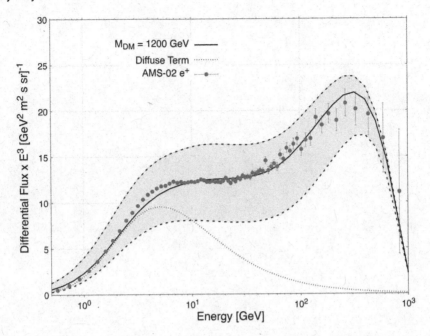

Figure 9.26 AMS-02 measurement of the differential flux of positrons as a function of the energy E, scaled by E^3 [775], compared to the sum of the diffuse term and the expectation from dark matter annihilation calculated using PPPC 4 DM ID results [455]. Dark matter is assumed to have a Navarro-Frank-White profile [279] in the Milky Way. The mass is $m_\chi = 1200\,\text{GeV}$ and annihilation is supposed to produce heavy leptons and quarks. The assumed annihilation rate is $\langle \sigma v \rangle = 7.1 \times 10^{-25}\text{cm}^3/\text{s}$. Uncertainties due to the diffusion through the galaxy are indicated by the shaded band.

FURTHER READING

J.N. Bahcall and J.P. Ostriker, *Unsolved Problems in Astrophysics*, Princeton Univ. Press 2018

A. Bykov, E. Amato, J. Arons, M. Falanga, M. Lemoine, L. Stella and R. von Steiger (Edts.), *Jets and Winds in Pulsar Wind Nebulae, Gamma-Ray Bursts and Blazars*, Springer 2019

P. Fisher, *What is Dark Matter?*, Princeton Univ. Press 2022

T. Stanev, *High Energy Cosmic Rays*, 3rd Edition, Springer 2022

10 In the Cosmos

Above a few hundred TeV, the total cosmic ray flux is too low to be captured by current space borne spectrometers or calorimeters. The very detailed studies of the composition as a function of energy that we have encountered in previous chapters are thus no longer feasible. A decade above this limit, at the so-called knees in the cosmic rays spectrum, between a few 10^6 and 10^8 GeV, the slope of the power law weakens in a notable fashion as we have seen in Chapter 3. At the so-called ankle, around 10^9 GeV, it hardens again. In between, the spectrum does not appear to be entirely featureless either. Somewhere in this region, one can reasonably assume that the spectrum of galactic cosmic rays dies out and extragalactic sources contribute in a noticeable way. The argument has to do with the so-called Hillas criterion for the containment of cosmic rays by a magnetic structure. Containment is achieved as long as the Larmor radius of a particle (see Focus Box 5.1) in the ambient magnetic field is smaller than the size of the region where the particle can be accelerated and contained. This simple argument yields a maximum energy $E_{max} = ZBR$ for a particle of charge Z in a magnetic field B extending to radius R. For the Milky Way, with its large scale field of a few μG and a size of 5×10^{17} m, we get an order of magnitude estimate of a few 10^7 GeV for a unit charge particle. Beyond that energy, cosmic protons start to strongly leak out of the galaxy in which they have been produced. Heavier nuclei are contained up to an energy which is Z times larger. If acceleration is kinetic rather than magnetic, as in the cannonball model outlined in Section 5.4, the spacing would rather be proportional to the nuclear mass A.

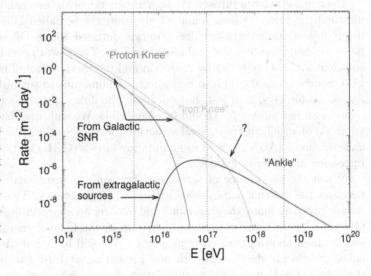

Figure 10.1 Schematic illustration of the rate of cosmic rays as a function of energy and possible contributions to the observed spectral breaks. The question mark indicates the transition between galactic and extragalactic contributions. (Credit: Andrew Taylor-Castillo, reproduced by permission.)

The limits are signalled by exponential cut-offs in energy as schematically shown in Figure 10.1, and are thus observed as spectral features. This could e.g. be the case for the knees in the energy spectrum, which can be attributed to the end of the spectra for light and heavy nuclei from galactic sources. Higher energy particles would then gradually be of extragalactic origin. However, at this point in time, nobody knows exactly where and how the transition between galactic and extragalactic sources of cosmic rays takes place, and

DOI: 10.1201/9781003181385-10

227

how this feeds into the spectrum and composition observed at or near Earth. We therefore take the liberty of treating all cosmic rays beyond the reach of space borne instruments in this chapter. After discussing experimental techniques, we start to review results at an energy of 1 PeV (10^6 GeV) and cover the potential end of the spectrum from galactic sources in Section 10.3. Ultrahigh energies beyond EeV (10^9 GeV) – likely to be of extragalactic origin – are discussed in Section 10.4. Indications for an anisotropy at the high end of the cosmic ray spectrum are discussed in Section 10.5.

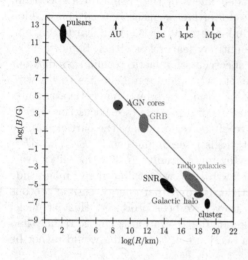

Figure 10.2 The so-called Hillas diagram [404] classifying potential sources for high-energy cosmic rays according to their size R and magnetic field B. Sources which can accelerate protons to $E > 10^{21}$ eV should lie above the upper line, sources above the lower line can accelerate iron nuclei up to 10^{20} eV. (Credit: M. Kachelriess)

Extragalactic structures with high magnetic fields are candidates for acceleration to ultrahigh energies, as seen qualitatively from the so-called Hillas diagram shown in Figure 10.2 [404], based on the Hillas criterion outlined above. Of course more criteria have to be considered than just size and magnetic field. The power provided by the source has to be sufficient and energy losses by radiation and interaction should not exceed the acceleration gain. Source types should also be sufficiently numerous to sustain the observed high-energy flux of cosmic rays. And last, but not least, the fluxes of pointing messengers from a source – discussed in Chapter 7 – should be compatible. We will come back to evidence for the contribution of specific astrophysical structures later in Section 10.6, taking as examples active galactic nuclei (AGN) as well as gamma ray bursts (GRB) and other energetic accelerators imbedded in starburst galaxies.

When the acceptance of satellite instruments is insufficient, the instrument of choice becomes the Earth's atmosphere acting as a calorimeter. This allows to be sensitive to cosmic rays over many thousand km^2 and to cover an appreciable fraction of 2π sr solid angle. One can thus measure cosmic ray fluxes up to the highest energies, but with considerably less detailed knowledge of their properties. We will limit our discussion to the current so-called hybrid air shower experiments, presenting at least two out of the three detection types shown in Figure 10.3: ground arrays, fluorescence telescopes and underground muon detectors. We thus cover results from the Pierre Auger Observatory (Auger for short) and the IceTop/IceCube observatory (see Section 7.2) in the southern hemisphere, as well as the Telescope Array (TA for short) experiment in the northern one. We will go into quite some detail about the experimental methodology of extensive air shower observatories in Section 10.2, so that the reader can appreciate the systematics associated with these difficult measurements. For a review of previous experiments see e.g. [308].

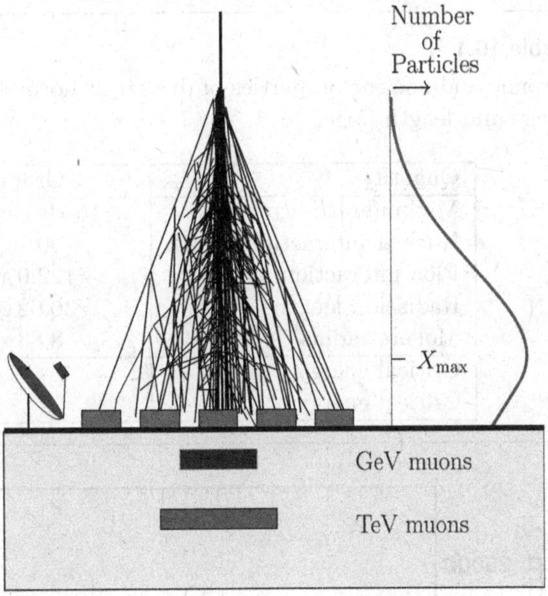

Figure 10.3 Schematic of the development of an extensive air shower [344] and its observation by air shower arrays on the ground, fluorescence telescopes and underground muon counters. The curve on the right sketches the Gaisser-Hillas function describing the evolution of the shower population.

10.1 EXTENSIVE AIR SHOWERS

Let us start by recalling some basic properties of the Earth's atmosphere. In approximately isothermal condition, the air density ρ diminishes exponentially with altitude z:

$$\rho(z) = \rho_0 e^{-z/z_s}$$

with $\rho_0 \simeq 1.225 \times 10^{-3}\,\text{g/cm}^3$ the density at sea level and $z_s \simeq 8.4\,\text{km}$ the so-called scale height. The grammage X_V traversed by a particle penetrating vertically to altitude z (in km) is then:

$$X_V(z) = \int_z^\infty \rho(h)\,dh = \rho_0 z_s e^{-z/z_s} \simeq 1029 \cdot e^{-z/8.4}\,\text{g/cm}^2$$

In air shower physics slang, this is called the vertical depth of the atmosphere. If the Earth were flat, the grammage would increase with zenith angle θ like $X(z) = X_V(z)/\cos\theta$. This quantity is called the slant depth. Taking into account the curvature of Earth and its atmosphere, the result is slightly smaller as shown in Figure 10.4 [344]. Some relevant atomic and nuclear properties of air under normal conditions are listed in Table 10.1.

The total vertical depth of the atmosphere at sea level is thus equivalent to about 26 radiation lengths X_0 (see Focus Box 2.8) or about 15 nuclear interaction lengths λ_I (see Focus Box 2.9). Strikingly enough, these numbers are about the same as the properties of calorimeters for modern collider detectors. For example, the calorimeters of the CMS detector at the CERN LHC are about $25X_0$ and $11\lambda_I$ thick at normal incidence. The Earth atmosphere would thus make an excellent calorimeter if we were able to read out an electrical signal proportional to the energy deposited per layer by incident particles. As we will see, this is only feasible in a rather indirect way.

One can operationally distinguish two phases of the development of hadronic air showers. The first one is the primary interaction of the incoming cosmic ray with an air molecule. The second phase concerns what happens to the final state particles from this primary interaction as they travel through the atmosphere. Interacting with air molecules they enrich the shower with particles and widen it, as schematically shown in Figure 10.3.

Table 10.1

Atomic and nuclear properties of dry air at normal atmospheric pressure in terms of grammage and length [519].

Quantity	Grammage	Length
Minimum dE/dx	$1.815 \,\mathrm{MeV\,cm^2/g}$	$2.187 \times 10^{-3} \,\mathrm{MeV/cm}$
Nuclear interaction length	$90.1 \,\mathrm{g/cm^2}$	$7.477 \times 10^4 \,\mathrm{cm}$
Pion interaction length	$122.0 \,\mathrm{g/cm^2}$	$1.013 \times 10^5 \,\mathrm{cm}$
Radiation length	$36.62 \,\mathrm{g/cm^2}$	$3.039 \times 10^4 \,\mathrm{cm}$
Molière radius	$8.83 \,\mathrm{g/cm^2}$	$7.330 \times 10^3 \,\mathrm{cm}$
Critical energy e^-	$87.92 \,\mathrm{MeV}$	
Critical energy e^+	$85.96 \,\mathrm{MeV}$	

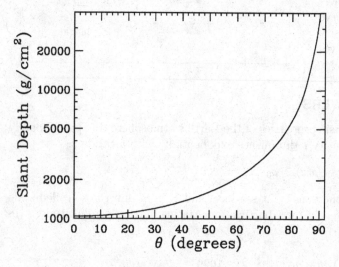

Figure 10.4 Slant depth of the Earth atmosphere as a function of the zenith angle, taking into account the curvature of the Earth [344].

Since we are talking about very high energies, binding energies inside nuclei are negligible. The primary and all subsequent interactions will be incoherent ones between an individual incoming hadron and a nucleon bound in an air molecule. Moreover, since the difference between the interactions of light quarks is also negligible, we can treat all nucleons as protons, all secondaries as pions. The total cross section for proton-proton collisions has been measured at colliders up to LHC energies and estimated from cosmic ray data beyond these energies [765]. The result is shown in Figure 10.5, together with the cross section for pion-proton interactions. Both share the same energy dependence, a modest logarithmic growth with energy. Their ratio is $\sigma_{\pi p}/\sigma_{pp} = 0.614 \pm 0.002$, roughly constant with energy and of order 1. To qualitatively understand the gross features of hadronic showers in a simple model – analogous to the one by Heitler, Rossi and Greisen for electromagnetic showers introduced in Focus Box 2.8 – we are thus entitled to assume that all hadronic processes are characterised by a single length, the nuclear interaction length λ_I. Expressed by grammage in units of $\mathrm{g/cm^2}$, it is given by:

$$\lambda_I = \frac{A}{N_A \sigma}$$

Here A is the atomic weight in gram, N_A is Avogadro's constant and σ is the relevant cross section. The characteristic length is thus of the order of $100 \,\mathrm{g/cm^2}$ for air. Also shown in

Figure 10.5 are data on the multiplicity of charged particles in the final state, mainly light mesons. Again, a modest logarithmic rise with energy is observed, which we can ignore. To get a qualitative insight, we can thus assume that the final states of hadronic interactions have a constant charged multiplicity of order 10.

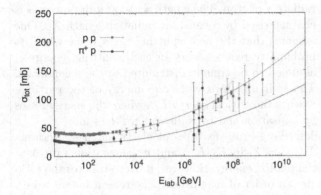 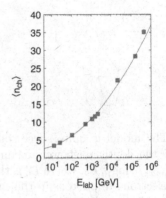

Figure 10.5 Left: Total cross section for proton-proton and pion-proton interactions as a function of the projectile energy [765]. Right: Mean charged multiplicity in proton-proton interactions as a function of the projectile energy [187, 192, 209, 217]. The curves are fits to the data with a function $a + b \ln E + c \ln^2 E$ to guide the eye.

A short reminder might be required here. In Focus Box 2.8, we introduced the simple model of electromagnetic showers due to Heitler, Rossi and Greisen. The shower is characterised by a single length, the radiation length X_0, which applies to electron and positron as well as photon interactions and is assumed to be constant. Energy loss by ionisation is neglected. The shower develops as a sequence of bremsstrahlung reactions from electrons and positrons, and pair conversions of photons. Each such reaction multiplies the number of particles by two and roughly halves the energy per particle. The shower develops exponentially until the electrons and positrons reach the critical energy, where non-radiative energy loss takes over.

Hadronic showers are more complex, since the reactions of hadrons have a larger diversity and involve more massive particles. One can, however, decide to ignore this for the moment and concentrate on a few basic features. An important gross feature already introduced in Focus Box 2.9 is the splitting into an electromagnetic part of the shower, mainly coming from the basically instantaneous $\pi^0 \to \gamma\gamma$ decay, and the hadronic part which comprises all other hadrons. At each stage, the split is such that roughly 1/3 of the incoming energy is transferred to the electromagnetic cascade. You can think of one π^0 being produced for every two $\pi^\pm$. It is thus easy to see that after a few stages the vast majority of the energy deposited by a hadronic shower, about 90%, will turn up in electrons and photons. One can construct a simple model of hadronic showers [367] with the simplifying assumptions outlined above. The model is explained in Focus Box 10.1. Like in the ansatz of Heitler, Rossi and Greisen, the charged particle shower is characterised by a single length, the nuclear interaction length λ_I, which is assumed constant and the same for all hadrons. The neutral part of the shower follows the Heitler-Rossi-Greisen scheme. The splitting of energy proceeds between 1/3 neutral and 2/3 charged pions at each step with a constant multiplicity and constant energy share per particle. The splitting ceases when the charged pions reach an energy where their decay length, $\gamma c \tau_{\pi^\pm} \simeq \gamma \cdot 7.8 \,\mathrm{m}$, is shorter than λ_I. At this point, all of them decay into muons and neutrinos. Despite its simplicity, the model is quite successful in explaining the gross features of high-energy hadronic air showers.

Matthews [367] has constructed an instructive analytical model of hadronic air showers along the lines of the model by Heitler, Rossi and Greisen for **electromagnetic showers** (see Focus Box 2.8).

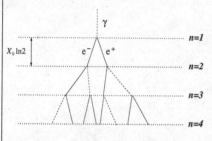

In the electromagnetic model, an initial photon will split into an electron-positron pair, each of which will radiate a photon. The length scale of both processes is characterised by a constant radiation length X_0. One assumes that the next splitting takes place when its probability reaches 50%. In each splitting, energy is assumed to be equally distributed to the secondaries. The splitting stops when the energy of the particles reaches the critical energy E_C, where the energy losses by ionisation and by radiation are the same.

The model is quite successful in describing gross features of electromagnetic showers like the depth at shower maximum $X_{max} = X_0 \ln(E_0/E_C)$ and its elongation rate $\Lambda = dX_{max}/d\log_{10} E_0$, where E_0 is the initial photon energy. However, it overestimates the ratio of electrons to photons in the shower by an order of magnitude. The reason is that an $e^\pm$ can radiate multiple photons and it looses energy by ionisation also before reaching critical energy. The shower size N_{max}, i.e. the number of charged particles at shower maximum, is thus overestimated by an order of magnitude.

In a **hadronic shower**, the initial interaction of the primary cosmic ray creates N_{ch} charged particles (mostly $\pi^\pm$) and about $N_{ch}/2$ neutral ones (mostly π^0). It is assumed that they all get an equal share of the initial energy. The π^0 almost instantly decay to two photons and thus feed an electromagnetic shower as described by the Heitler-Rossi-Greisen model. The shower initiated by the $\pi^\pm$ is characterised by the nuclear interaction length λ_I, assumed to be constant with energy and the same for all hadrons. This is justified by the constant ratio of interaction cross sections for pions and protons (see Figure 10.5) and their logarithmic energy dependence.

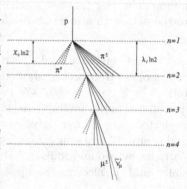

Likewise, it is assumed that charged and neutral multiplicities are constant with energy. The splitting continues as sketched in the scheme above, at each stage multiplying the total multiplicity by $3N_{ch}/2$. The energy per pion thus decreases as $E_\pi = \frac{2}{3}E_0/N_{ch}$. When the energy and the relativistic factor $\gamma = E_\pi/m_\pi$ decrease such that the decay length $\lambda = \gamma c\tau_\pi \simeq \gamma \cdot 7.8\,\text{m}$ is less than the distance λ_I to the next interaction, the splitting ceases.

All pions are then assumed to decay via $\pi^\pm \to \mu^\pm \overset{(-)}{\nu_\mu}$. Neutrinos escape, muons continue down to ground level. For energies of interest here, this happens already after a few λ_I. It is thus assumed that the shower multiplicity is dominated by the electromagnetic showers generated by the interaction of the primary cosmic ray. The shower maximum is reached for $X_{max} \simeq \lambda_I \ln 2 + \ln[E_0/(3N_{ch}E_C)]$. This falls short of the X_{max} predicted by numerical simulation by about $100\,\text{g/cm}^2$. The constant shift is probably due to the fact that in reality the energy is not at all shared equally by the secondaries, but a large part is taken away by a so-called leading particle [662]. The model is, however, successful to qualitatively reproduce gross shower characteristics and their dependence on energy. It can be extended to take into account the energy dependence of hadronic cross sections and multiplicities [533]. This extension also follows through all stages n for both the electromagnetic and hadronic part of the shower, but does not really improve the agreement with numerical simulation.

Focus Box 10.1: Extensive air showers

This simple model qualitatively describes the evolution of a hadronic shower along its axis, which is given by the direction of the primary cosmic ray. The multiplicity $N(X)$ of the shower as a function of the depth X can be described by the semi-empirical Gaisser-Hillas parametrisation [193]:

$$N(X) = N_{max} \left(\frac{X - a}{X_{max} - a} \right)^{(X_{max} - a)/\lambda} e^{(X_{max} - X)/\lambda}$$

where a denotes the depth of the primary interaction, N_{max} is the multiplicity at shower maximum and λ is a length parameter in the spirit of X_0 and λ_I. The functional form is sketched in Figure 10.3. The multiplicity of electrons at the shower maximum and the number of muons from the last stage allow to construct an estimator of the primary energy. The relation between charged and neutral pion multiplicity allows to reduce this estimator to rely solely on the number of electrons at shower maximum. All these considerations of course only concern the average behaviour of air showers. The model comes close to what is actually done in experiments, as we will see in the next section. However, simple models do not do justice to the rich physics presented by strong interactions. This physics and the fluctuations from one shower to the next are best handled by using Monte Carlo methods.

Let us try to get more than a qualitative insight by basing our considerations on the quantum theory of strong interactions, Quantum Chromodynamics (QCD). The momentum distribution of quarks and gluons inside the nucleon are described by structure functions as sketched in Focus Box 10.2. The total cross section of the primary interaction is dominated by soft and semi-soft interactions between the constituents of the two nuclei, quarks and gluons. Tree-level QCD has both a soft and a collinear divergence for the emission and absorption of gluons from a quark. The strong coupling constant α_s, which characterises both the quark-gluon and the gluon-gluon interaction, strongly rises at low momentum transfer. Therefore, the soft part of what happens inside a nucleon is not amenable to perturbative calculations within QCD. Soft interactions have to be treated by models inspired by QCD, but not based on first principles. Semi-hard processes, which give rise to the so-called mini-jet phenomenon, can in principle be treated by perturbative QCD. And so can the hard processes giving rise to the high transverse momentum jets observed in LHC experiments. However, the initial state is dominated by quarks and gluons with a very low fraction of the nucleon momentum, $x = p_{q,g}/p_{\mathcal{N}}$. We give a short introduction to the distribution of quark and gluon momenta inside the nucleon in Focus Box 10.2. These x distributions have been measured down to about 10^{-6} by electron-nucleon scattering experiments [750]. Dominant contributions for primary cosmic ray interactions come from fractions which are several orders of magnitude lower.

The x distributions thus have to be extrapolated to tiny x values, again using QCD inspired models for which there is no unique choice. The one closest to high-energy experimental data is contained in the code EPOS LHC [553, 708], constructed to describe soft processes (so-called "minimum bias" events) observed by LHC experiments. Codes more inspired by theory are QGSjet [463, 506] and Sibyll [615, 740]. They differ mostly in their way of extrapolating parton distribution functions to very low x values.

These successful models of primary hadronic interactions come in the form of so-called event generators: final states of single reactions are simulated by computer programs implementing a model of choice. The output is a numerical description of single events with their complete final state and its kinematics, which can be fed into programs to follow these particles through the atmosphere and simulate their further interactions as well as energy loss and decay processes.

In the primary interaction the cosmic ray energy is already broken down by at least an order of magnitude, such that secondaries will quickly enter into an energy range covered

In QCD, nucleons are described as bound states of quarks, held together by gluons, the force carriers of strong interactions. Both are collectively called partons. The internal structure of hadrons is dynamic, thanks to the coupling constant α_s characterising the coupling between quarks and gluons as well as among gluons, which is large and grows with decreasing momentum transfer. Scattering experiments are sensitive to the structure via so-called structure functions describing the fraction x of the nucleon momentum p_N carried by a quark q or gluon g, $x = p_{q,g}/p_N$. They modify the cross section as a function of energy and momentum transfer. The most important one is the structure function F_2, which is asymptotically proportional to the parton momentum distribution $f_i(x) = dp_i/dx$, $F_2 = \sum_i e_i^2 x f_i(x)$, where i denotes the quark flavour. The relevant charge e_i depends on the particle which probes the structure; for electromagnetic probes it is the electric charge, 2/3 for up-type quarks, 1/3 for down-type quarks and 0 for gluons. The deep inelastic cross section thus can be pictured as the incoherent sum of elastic interactions with individual quarks, which behave as almost free particles when the momentum transfer is much larger than their binding energy.

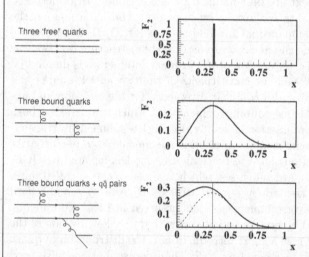

At high momentum transfer, the proton mainly looks like three almost free quark. The structure functions centre around their mean value, but are widened by gluon exchange as sketched on the left. In addition, gluons can pair-produce quarks and thus enrich the distribution at low x. In any case, quarks only carry about half of the nucleon's momentum, the rest being carried by gluons. For up quarks in a proton, the momentum distribution $x f_u(x)$ thus peaks at a little less than 1/3, for down quarks $x f_d(x)$ has a maximum at half that value.

Quark-antiquark pairs created by gluon splitting $g \to q\bar{q}$ populate a "sea" at low x with $x f_s(x)$. The gluon distribution $x f_g(x)$ becomes dominant at low x. Parton distribution functions have been extracted from experimental data down to $x \simeq 10^{-6}$ [301].

An example of the resulting momentum distributions is shown on the right. The development of hadronic showers is dominated by the population at extremely low x, which cause the soft and semi-soft processes responsible for the majority of shower particles. The measured functions are hard to extrapolate towards the soft edge because the strong coupling constant quickly becomes too large to allow perturbative calculations.

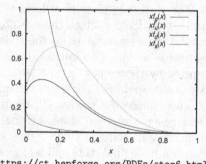

https://ct.hepforge.org/PDFs/cteq6.html

Focus Box 10.2: Parton distribution function

by collider experiments. One can thus describe the rest of the shower development based on collider data on cross sections and final state properties. This is done in Monte Carlo codes like EGS for electromagnetic showers[1], Fluka[2] and Gheisha [213] for hadronic showers and GEANT[3] for both. These Monte Carlo codes can be embedded in air shower simulation systems like e.g. CORSIKA.[4] Properties of simulated events are then statistically compared to those of observed air showers. Parameters of the models can thus be optimised. And physical properties of observed events like energy and direction of the primary cosmic ray particle can be extracted from the observed data as described in the next section.

10.2 AIR SHOWER OBSERVATORIES

One can imagine an air shower as a thin curved shell of relativistic particles, starting at the primary interaction and moving along the direction of the initial cosmic ray at the speed of light. It is populated by all kinds of hadrons, leptons and photons at the beginning. When it hits the ground, basically only electrons, positrons, photons, muons and neutrinos are left.

There are two ways to measure the geometry of air showers and extract physical properties, as sketched in Figure 10.3. The first method is to observe the shower from the side, detecting fluorescence light emitted by the air when excited by ionising particles (see Focus Box 2.6). The detectors of choice are photosensitive devices in the focal plane of telescope mirrors. They allow to picture the shower development basically from beginning to end, measuring its direction and energy deposition as a function of slant depth. This is the most direct way of analysing air showers. However, one can observe the very faint fluorescence light only during moonless nights with clear skies. The duty cycle of such observatories is thus of the order of 10%.

The second method is to observe the particles impacting the ground with large arrays of particle detectors on and/or below the surface. Detectors of choice are plastic or liquid scintillators and Cherenkov counters. They count the number of minimum ionising particles hitting the ground, mainly electrons, and penetrating below ground, mainly muons. From the arrival times and lateral distribution of particles, direction and energy of the shower can be reconstructed. This methodology is more complex and suffers from larger systematics. But the duty cycle of ground arrays approaches 100%, with important statistical benefits when searching for the rare events at the end of the cosmic rays spectrum.

Modern observatories for ultra-high-energy cosmic rays combine both methodologies in order to cross-calibrate their findings and thus reduce systematics. The most important examples are the Pierre Auger Observatory in the southern hemisphere and the Telescope Array observatory in the northern one.

The Auger ground array [540] covers about $3000 \, \text{km}^2$ in the Pampas Amarilla located in western Argentina, about 30 times the surface of the city of Paris. The layout of the observatory is shown in Figure 10.6. The ground array consists of more than 1600 water Cherenkov stations arranged on an isometric grid with about 1.5 km spacing in a roughly hexagonal shape. Their elevation ranges from 1300 m to 1600 m above sea level. Each station [516] is a cylindrical water tank covering $10 \, \text{m}^2$. The container is filled to a depth of 1.2 m with highly purified water. Its walls are covered by a reflective liner. The water volume is viewed from above by three 9" photomultiplier tubes (see Focus Box 2.10), which detect the Cherenkov light emitted by the passage of ionising particles. The timing and pulse height measured by the tubes is recorded.

---

[1] See http://rcwww.kek.jp/research/egs/kek/.
[2] See http://www.fluka.org/fluka.php.
[3] See https://geant4.web.cern.ch.
[4] See https://www.iap.kit.edu/corsika/70.php.

The ground array is overlooked by 24 air fluorescence telescopes [441], with fields of view indicated in Figure 10.6. Each telescope covers an elevation up to $30°$ above the horizon and can thus follow the complete evolution of high-energy showers. The hybrid array is sensitive in the energy range from a few 10^9 GeV to a few 10^{11} GeV, where the cosmic ray spectrum ends. A subarray in one of the corners fills a small area of $23.5\,\mathrm{km}^2$ surface with a tighter $0.75\,\mathrm{km}$ spacing between stations. Three tilted telescope in addition allow to observe fluorescence light from the higher elevations of low-energy air showers [447]. This subarray lowers the threshold for cosmic ray observations to 0.25×10^9 GeV.

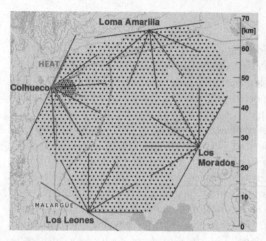

Figure 10.6 Left: The geographic location of water Cherenkov tanks (black dots) and the fields of view of air fluorescence telescopes (delimited by long lines) in the Pierre Auger Cosmic Ray Observatory [540]. The tanks are separated by a distance of about $1.5\,\mathrm{km}$ and cover a surface of about $3000\,\mathrm{km}^2$. The 24 florescence telescopes overlook the same surface. A small sub-array of $23.5\,\mathrm{km}^2$ surface with $0.75\,\mathrm{km}$ spacing (grey dots) is overlooked by three additional fluorescence telescopes (HEAT, short lines) covering high elevations. Right: Photo of a fluorescence telescope site on a hill top, and a water Cherenkov tank in the plane.

As an example, Figure 10.7 shows a shower reconstructed with the Auger ground array, with a reconstructed zenith angle of $\theta = 55.2°$ and an energy of $38.7\,\mathrm{EeV}$. The dots indicate stations with a significant signal. The size is proportional to the logarithm of the number of minimum ionising particles crossing the station. The grey scale indicates the time of arrival of the shower at each station, with early hits dark, late hits lighter. The projection of the reconstructed shower axis onto the ground is shown by the line, ending at the impact point of the shower centre indicated by the square dot. An upgrade program is ongoing, called AugerPrime, deploying additional detector elements on and below the existing water Cherenkov tanks of the ground array, as well as antennae for the detection of radio emission from air showers. We will come back to this upgrade program in Section 11.2.

The Telescope Array experiment is situated at an altitude of $1400\,\mathrm{m}$ above sea level in the desert area of Millard County, Utah, south of Salt Lake City [411]. It has been built by pioneers of fluorescence telescope observatories, formerly operating the High Resolution Fly's Eye (HiRes), and the Akeno Giant Air Shower Array (AGASA) in Japan. Its ground array [488] uses plastic scintillators in more than 500 detector units arranged in a square grid of $1.2\,\mathrm{km}$ spacing. It covers a total area of about $700\,\mathrm{km}^2$ in a roughly triangular shape. Each surface detector has two layers of plastic scintillator with $3\,\mathrm{m}^2$ surface and $1.2\,\mathrm{cm}$ thickness each, and a $1\,\mathrm{mm}$ thick steel plate in between. The layers are cut into two halves, each viewed by photomultiplier tubes from both ends. They measure the arrival time and pulse

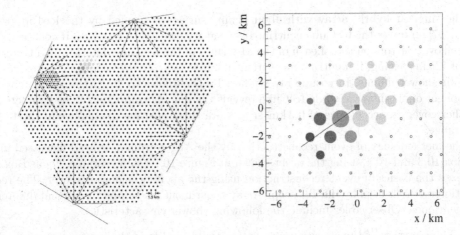

Figure 10.7 A cosmic ray registered by the ground array of the Pierre Auger Cosmic Ray Observatory. [768]. The left plot shows the complete array with its water Cherenkov stations (dots). The right plot zooms in on the ground stations participating in the reconstruction. The filled dots indicate stations with a significant signal. The size is proportional to the logarithm of the number of minimum ionising particles crossing the station. The grey scale indicates the time of arrival of the shower at each station, with early hits dark, late hits lighter. The projection of the reconstructed shower axis onto the ground is shown by the line, ending at the impact point of the shower centre indicated by the square dot.

height of scintillation light emitted in the doped plastic material. The array is overlooked by three fluorescence telescope stations [487], with a total of 36 mirror-camera assemblies, covering elevations from 3° to 31° above ground. One of the stations is shown in Figure 10.8. At the northern fluorescence telescope station, a low-energy extension, called TALE, has been added [676]. It consists of ten additional fluorescence telescopes covering high elevation angles up to 59° and 103 additional shower counters in a spacing progressing from 400 m to 600 m and 1200 m, i.e. towards the spacing of the main array. The combination of TALE and TA allows a measurement of the all-particle cosmic rays spectrum over nearly five decades in energy, a remarkable achievement indeed. Recently, TA has been upgraded to what is called TA×4, as yet not quite four times as large as TA [829]. Results presented here are, however, based on the configuration described above. More information about the upgrade will be provided in Section 11.2.

Figure 10.8 One of the fluorescence detector stations of the Telescope Array experiment, with the doors open [487]. In each opening, four telescopes watch the sky above the surface array. The upper ones cover elevations between 3° and 18.5°, the lower ones between 17.5° to 33°. (Credit: J.N. Matthews, Telescope Array)

The smallest hybrid array with about $1\,km^2$ surface is formed by the IceTop ground array [497] on top of the IceCube neutrino observatory (see Focus Box 7.8). It consists of two Cherenkov detectors on top of each of the 81 strings of the IceCube detector, in a triangular grid of 125 m spacing. In neutrino measurements, it serves as a veto array shielding IceCube from air shower muons. In cosmic ray studies, IceTop allows to study the electromagnetic component of air showers, while IceCube reveals the muonic component. The two detectors together cover an acceptance of $0.3\,km^2sr$ and are thus sensitive to energies in the PeV to EeV range.

The methodology of event reconstruction by the Auger ground array is discussed in Focus Box 10.3 and 10.4, using the event shown in Figure 10.7 as an example. Focus Box 10.5 discusses the basics of event reconstruction using the air fluorescence detector. The reconstruction methods used by the Telescope Array experiment are similar, important differences are mentioned. Observables include the following shower characteristics:

- The direction of the shower, representing the direction of the incoming particle.

- The shower size N_e, defined as the total number of charged particles contained in the shower.

- The depth X_{max} at which the shower reaches its maximum.

- The muon component N_μ of the shower.

The first three features are measured by ground arrays and fluorescence telescopes. The last one additionally requires detectors buried underground, so that the electron component of the shower is absorbed. This methodology and its use to look into the elemental composition of high-energy cosmic rays has been pioneered by the KASCADE Grande array in Germany [815] operated until 2013. The IceTop ground array is situated above the IceCube neutrino observatory deeply buried in the antarctic ice (see Focus Box 7.8), which is used to identify and count muons. The Pierre Auger Observatory is in the process of installing underground muon detectors [825], with completion foreseen in 2022 (see Section 11.2).

Reconstruction of the arrival direction and impact point relies on ground array timing information as discussed in Focus Box 10.3 [768]. The results are the coordinates $\vec{x}_c$ of the center as the shower hits the ground, and the zenith and azimuthal angles, θ and ϕ, of the shower axis. A proxy of the shower size, called $S(r_{opt})$ in the Auger analysis, is derived from ground array pulse height measurements as explained in Focus Box 10.4. It expresses the particle number in units of "vertical equivalent muons" (VEM) recorded by the detector at a fixed distance of $r_{opt} = 1000\,m$ for Auger, $r_{opt} = 800\,m$ for TA and $r_{opt} = 125\,m$ for IceTop. The optimum distances are selected for each observatory to minimise shower fluctuations. Differences are due to size, spacing and detector technology.

The determination of the shower axis and energy from the fluorescence data is more direct [441], since the whole shower development is accessible. It is covered in Focus Box 10.5. Figure 10.9 shows an example event from the Auger fluorescence telescopes. The arrival time for pixels hit by fluorescence photons is indicated by the colour of the dots. The associated stations of the ground array are marked by squares on the bottom. The full line corresponds to the fitted shower-detector-plane as explained in Focus Box 10.5. The location of the shower axis inside this plane is found by analysing the arrival time of photons in the camera pixels, the position of which corresponds to the elevation and azimuthal angles of emission sites along the shower axis. The light intensity then allows to reconstruct the total dE/dx of the shower particles as a function of slant depth. Integrating it yields an estimator of the shower energy, as sketched in Focus Box 10.5, with an accuracy of about 7%. It is used to calibrate the energy proxy of the ground array using events where both detector types yield a measurement.

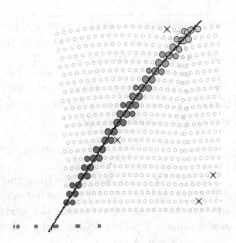

Figure 10.9 Display of an example event from the Auger fluorescence telescopes [441]. The camera pixels are indicated by circles, grey scale represents arrival time from early (lighter) to late (darker). Random hits eliminated from the analysis are indicated by crosses. The line marks the fitted position of the shower-detector-plane (see Focus Box 10.5), projected onto the camera focal plane. Little squares mark the position of associated ground array stations in the same projection.

When converting the shower size proxy $S(r_{opt})$ into an energy, a number of corrections are required [767]. A small correction of a few percent corrects for atmospheric conditions and the curvature of charged particles in the geomagnetic field. More importantly, the relation of $S(r_{opt})$ – defined on the ground – to shower size depends on the zenith angle. The reason is that more inclined showers start at a larger distance from the ground array and are thus more attenuated beyond their maximum. A correction is derived by requiring that the zenith angle distribution be flat [767], corresponding to the hypothesis of an isotropic primary flux. All shower sizes are then corrected to S_{38}, a shower size proxy corresponding to a standard zenith angle of 38°. One finally uses so-called hybrid events to calibrate the resulting size proxy and determine a shower energy E. Hybrid events are those for which both the fluorescence telescopes and the ground array have recorded the shower parameters. Figure 10.10 shows the correlation between the florescence energy E_{FD} – defined according to the procedure described in Focus Box 10.5 – and S_{38}. The line describing the average behaviour follows a power law, such that $E_{FD} = AS_{38}^B$. The best fit parameters are $A = (1.86 \pm 0.03) \times 10^{17}\,\mathrm{eV}$ and $B = (1.031 \pm 0.004)/\mathrm{VEM}$ independent of zenith angle. The correction for the energy lost in non-ionising neutrinos and muons which are not stopping before reaching Earth is contained in the calibration procedure as described in Focus Box 10.5. The calibration procedure used by the Telescope Array experiment is again similar [690]. Since IceTop does not have fluorescence telescopes, the conversion of $S(r_{opt})$ to energy is done using simulated showers [501, 716] with an assumed elemental composition [478]. The systematics incurred by this procedure are obviously larger.

Figure 10.10 Correlation [767] between the ground array shower size proxy S_{38} and the energy E_{FD} determined from the fluorescence telescopes, for over 3000 events where both measurements were present in the Pierre Auger Observatory. The solid line is the best fit for the power law dependence as described in the text.

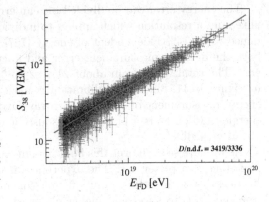

One can imagine an air shower as a thin curved shell of relativistic particles, starting at the primary interaction and moving along the direction of the initial cosmic ray at the speed of light. When the shell hits the Earth's surface inside the ground array, each station records the arrival time of the first particles. Reconstructed observables are the zenith angle θ and the azimuth ϕ, which define a unit vector $\hat{a}$. A first simple estimator is derived assuming that the shower front is a flat surface orthogonal to $\hat{a}$. The arrival time of the shower front at a given location $\vec{x}$ is then:

$$ct(\vec{x}) = ct_b - \hat{a}(\vec{x} - \vec{x}_b)$$

The parameters t_b and $\vec{x}_b$ are time and location where the shower centre hits the ground. A first approximation of $\hat{a}$ can be derived using a seed triangle of stations. In a second step, one can then take into account the curvature of the shower front. There are two possible model shapes, a parabolic one and a spherical one. For the parabolic approximation, arrival times are given by:

$$ct(\vec{x}) = ct_c - \hat{a}(\vec{x} - \vec{x}_c) + \frac{R_0}{2} \left[|\hat{a} \times \vec{x}|(\vec{x} - \vec{x}_c) \right]^2$$

where t_c and $\vec{x}_c$ are new estimators of time and location where the shower centre hits the ground, and R_0 is a curvature parameter. The spherical approximation requires an estimate of the shower starting point $\vec{x}_0$:

$$ct(\vec{x}) = ct_0 - \hat{a}(\vec{x} - \vec{x}_0)$$

with $\hat{a} = (\vec{x}_0 - \vec{x}_c)/|\vec{x}_0 - \vec{x}_c|$. Summing over all arrival times t_i of stations i participating in the reconstruction, the direction $\hat{a}$ and the parameters of the shower impact point are then obtained by minimising the total χ^2:

$$\chi^2 = \sum_i \frac{(t_i - t(\vec{x}_i))^2}{\sigma_{t_i}^2}$$

where σ_{t_i} is an estimator of the measurement error of the time measurement, typically of the order of picoseconds. The angular resolution is of the order of $1°$ for vertical showers. It decreases with zenith angle and reaches $0.5°$ at $60°$. It also depends logarithmically on the total shower energy.

Focus Box 10.3: Shower geometry from ground array timing information [768]

The energy error can be divided into an error on the energy scale applying to all events alike, and a resolution which applies to individual measurements. The scale error is determined to be 14% independent of energy [767]. The resolution contains contributions from both the detector measurement error and fluctuations from shower to shower at fixed energy. The total varies from about 21% at 10^{18} eV to 7% at the highest energies, as shown in Figure 10.11. Since the calibration is most accurate for showers of large size, the final energy measurement of the ground array favours fluctuations towards high size at lower energies and presents a bias, which is also quantified in Figure 10.11. At high energies the bias is negligible.

As we have seen all over this book, cosmic ray spectra fall rapidly with energy, in general according to a power law. The experimental resolution has the tendency to diminish the steepness. A symmetric resolution function like a Gaussian will transport events with a true energy E to lower and higher observed energies in equal proportion. However, there

The energy of the primary cosmic ray is estimated using the pulse height of signals in the water Cherenkov or scintillation counters. These are calibrated to represent multiples of a standard pulse height, which corresponds to the signal of a single vertical particle traversing the counter with minimum ionisation. The pulse height units are suggestively called Vertical Equivalent Muon (VEM). The proxy for the shower size called $S(r_{opt})$ is the proportionality factor between the observed pulse height $S(r)$ (in VEM) at lateral distance r from the shower centre (in meters) and an assumed lateral distribution function $f_L(r)$:

$$S(r) = S(r_{opt}) f_L(r)$$

$S(r_{opt})$ is thus the average pulse height at a fixed lateral distance r_{opt}. The lateral distribution function is normalised such that $f_L(r_{opt}) = 1$. Its functional form is empirically chosen. The Auger Collaboration uses two different forms [768]. The first one is a log-log parabola:

$$\ln f_L(r) = \beta \ln\left(\frac{r}{r_{opt}}\right) + \gamma \ln^2\left(\frac{r}{r_{opt}}\right)$$

For small lateral distances of less than 300 m, a linearisation of the function is used. The second lateral distribution function is a power law inspired by the lateral distribution of particles in a purely electromagnetic shower, the so-called Nishimura-Kamata-Greisen (NKG) function [149, 152, 157]:

$$f_L(r) = \left(\frac{r}{r_{opt}}\right)^{\beta} \left(\frac{r + r_s}{r_{opt} + r_s}\right)^{\beta + \gamma}$$

with a fixed parameter $r_s = 700$ m. In both cases, average values of the slope parameters β and γ are determined for a sub-sample of events providing adequate data, and parametrised as a function of zenith angle and $S(r_{opt})$. The models assume that the lateral distribution of the detected signal is azimuthally symmetric around the shower axis. This is not exactly the case for geometric reasons. A small correction of $S(r)$ depending on azimuth is thus applied where necessary.

The parameters $S(r_{opt})$ and the shower impact point are determined for each event using a maximum likelihood method. The plot on the right [768] gives the lateral distribution of signals from the event shown in Figure 10.7 reconstructed by the two methods sketched above. The results for $S(r_{opt})$ (dashed lines) are 112 VEM and 107 VEM, respectively. The resolution for the shower size varies between 14% for small showers and 6% for large ones, almost independent of the zenith angle.

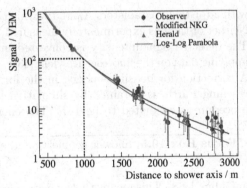

Focus Box 10.4: Shower size from ground array pulse heights [768]

are less events with true energy larger that E. The migration to higher energies is thus proportionally more important. As a consequence, the folding of the true spectrum with the resolution function artificially reduces the spectrum steepness. The experimental spectra are corrected to determine the true spectral index. In the Auger analysis this is done by forward

One first determines the shower-detector plane, i.e. the plane which contains both the shower axis and the telescope camera, as shown in the sketch on the right [441]. It is determined by the best fit to the light track projected onto the camera focal plane as shown in the example of Figure 10.9. Because of the telescope optics it is slightly curved. The accuracy of the plane orientation is of the order of a few 0.1°. The shower axis position within the shower-detector plane is then determined by the arrival time t_i of the fluorescence light at each pixel i:

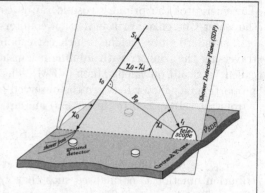

(Credit: American Physical Society)

$$t_i = t_0 + \frac{R_P}{c} \tan \frac{\chi_0 - \chi_i}{2}$$

with the offset t_0 determined by timing from the ground array. The angle χ_0 is the elevation of the shower axis, χ_i is the elevation of the line of sight for pixel i. R_P is the orthogonal distance of the shower axis from the camera. The procedure is calibrated by measuring fluorescence light excited by laser beams at fixed positions. The direction and position accuracies are 0.6° and 50 m, respectively.

Light in a camera pixel mainly comes from isotropic fluorescence emission, with small contributions from direct and scattered Cherenkov light [401]. Taking into account attenuation in the atmosphere, the pulse height in each pixel can thus be converted into a particle intensity along the shower axis. This is as close as one can get to a direct measurement of the signal from a man-made calorimeter. The density can then be fitted to the semi-empirical Gaisser-Hillas function [193] as a function of slant depth X:

$$N(X) = N_{max} \left(\frac{X - X_0}{X_{max} - X_0} \right)^{\frac{X - X_0}{\lambda}} e^{(X_{max} - X)/\lambda}$$

with adjustable parameters X_0 and λ which depend on the type and energy of the primary. Neither of these is experimentally measured, values are fixed using simulated showers [300]. The detected light intensity is converted into an energy loss per unit depth, dE/dX, by applying data on the fluorescence yield [538]. An energy estimator is obtained by integration. A correction for invisible energy, in the form of neutrinos and muons not stopped before reaching Earth, is obtained by simulation [343]. The total scale error on energy by this procedure is estimated to be 14%. The energy resolution for individual showers is about 7%.

Focus Box 10.5: Shower geometry and energy from fluorescence measurements [441]

unfolding [388]. This means that one assumes a trial spectral shape that one then folds with the known resolution function. The result is compared to the raw observed spectrum and the hypothesis adjusted to iteratively reveal the true spectral shape. Alternatively, one can fold the experiment's exposure as a function of energy, i.e. its (acceptance × observation time) with the known resolution function [266]. This method is used by the Telescope Array analysis. These methodological differences make a direct comparison of results difficult.

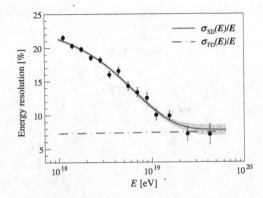 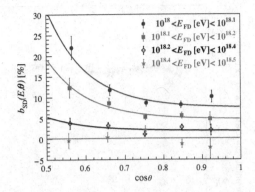

Figure 10.11 Left: Resolution of the Auger ground array energy estimator E_{SD} as a function of energy [767]. The constant resolution of the fluorescence telescope E_{FD} is shown for comparison. Right: Relative bias of the Auger ground array energy estimator E_{SD} as a function of energy and zenith angle [767]. A positive bias has to be subtracted from the uncorrected result.

10.3 KNEES

Figure 10.12 shows a summary of the spectra of all particles lumped together, as measured by the hybrid experiments briefly described above. In this section, we discuss the spectrum from PeV to EeV energies, i.e. the left three decades of energy in Figure 10.12. Statistical errors are almost invisible, since the observed rates are large. The total systematic errors of the TALE measurement in this energy range are indicated by the left light band.

The dominant systematic error of all air shower experiments comes from the energy scale as discussed in Section 10.2. The uncertainty concerns the logarithm of the energy and thus corresponds to an error on a multiplicative factor. Its dominance is obviously even stronger when the flux is scaled by a power of the energy as in Figure 10.12, since the error then concerns the ordinate as well as the abscissa of the plot. The systematics in general and the energy scale error in particular have been studied by TA and Auger in a joint working group [694, 824]. A difference of about 9% in the absolute energy scale of the two experiments has been identified, due to differences in the calibration by the fluorescence measurements. When different assumptions about the fluorescence yield and missing energy are removed, the difference reduces substantially. The working group recommends to correct both energy scales by half the difference, upwards for Auger, downwards for TA. Since the IceTop energy measurement is calibrated by simulation only, we shift its energy scale down by 8.9% with respect to TALE [676].

The scaling by the third power of energy makes the two spectral features known as knees apparent. The first one is situated at about 3 PeV, where the spectral index changes so that the spectrum softens. This is often called the first knee, or more suggestively the proton knee, as in Figure 10.1. After a slow recovery, the index softens again at about 100 PeV, forming what is called the second or iron knee. The left curves in Figure 10.12 try to convince you that the shape of the spectrum can be described by a two component toy model. Its two components each follow a power law with exponential cut-off:

$$\Phi(E) = a_l E^{-\gamma_l} e^{-E/E_c} + a_h E^{-\gamma_h} e^{-E/(fE_c)}$$

In this model, parameters a_l and γ_l characterise the spectrum of a light component, a_h and γ_h a heavy one. The maximum energies of the two components are E_c and fE_c, respectively. It turns out that fixing $f = 26$ gives about the right relation between the two cut-offs

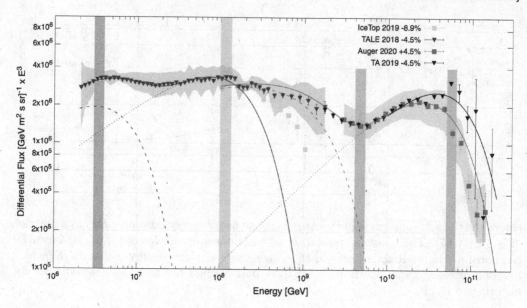

Figure 10.12 Differential fluxes for the sum of all particles, scaled by E^3, measured by current hybrid observatories: TALE [676], IceTop/IceCube [716], Telescope Array [691] and Pierre Auger Observatory [767]. The measured energies of the experiments are scaled by the percentages indicated, following the recommendations of the Auger-TA energy spectrum working group [824]. Error bars show statistical errors, the left shaded area indicates the systematic errors of TALE, the right one that of Auger. The left curves are simple models for contributions from light (dashed), heavy (dotted) and total nuclei (solid) of galactic origin. The right curves show the same for potential extragalactic contributions when using Auger data (lower solid) or TA data (upper solid) at the highest energies. Vertical shaded bars indicate the position of four spectral features, from left to right: first and second knee, ankle and GZK cut-off.

associated to first and second knee. In the framework of Fermi acceleration, the maximum energies are given by the Hillas criterion, refined by including an efficiency factor η and the shock velocity β:

$$E_c = \eta \beta Z B R$$

The factor would thus be $f = Z_h/Z_l$ and the components could be identified as dominated by H/He and the iron group, respectively. If one adopts a cannonball model [723], the maximum energy would be:

$$E_c = 2\gamma^2 M$$

where γ is the initial relativistic factor of the cannonball and M is the mass of the accelerated particle. In that case $f = A_h/A_l$ and the heavy component would be dominated by the CNO group.

A simplistic analysis of just the energy spectrum with two components does not do justice to the capabilities of hybrid experiments. In TALE [854], the distribution of the depth at shower maximum, X_{max}, as observed by the fluorescence telescopes in bins of the total energy, gives an additional handle on the elemental composition. Since the nuclear binding energy is negligible, a nucleus of mass A at high energy E basically acts as A nucleons of energy E/A each. As a consequence, an iron shower has an X_{max} about 15% lower than a proton shower of the same primary energy. To extract composition from the X_{max} distribution in bins of energy, the moments of the distribution are compared to simulated showers caused by different elements. The results obtained by TALE show that, as in our naive model, the composition shifts towards heavier components as the energy rises from 10^6 to 10^8 GeV. Detailed TALE composition data indicate that the fraction of light elements starts of at around 90% at 2×10^6 GeV and reduces to about 40% at 2×10^9 GeV. The iron component starts out low and reaches about 50% at the high end of the range. Beyond, the heavier component appears to reduce, as expected.

The IceTop/IceCube hybrid analysis of the composition [716] uses the energy loss by muons in IceCube as an additional handle on the composition, since the multiplicity of charged mesons is typically larger in interactions of heavy nuclei. Size, zenith angle and muon data are fed into an artificial neural network trained on simulated data with light (p/He), medium (O) and heavy (Fe) primaries. The results confirm that the fractions of light nuclei decrease, the fraction of iron increases from about 20% to about 50% as the energy grows from 3×10^6 GeV to 10^9 GeV. The IceTop/IceCube results thus confirm the findings of the TALE analysis. And our toy model is not too far off, either.

10.4 ANKLE

In this section, we discuss the spectrum from EeV energies to the end at a few times 10^{11} GeV, i.e. the right two and a half decades of energy in Figure 10.12. Statistical errors become important at the high end of the energy range, since rates fall below the level of an event per square kilometre and year. The total systematic errors of the Auger measurement in this energy range are indicated by the shaded band. Again, energies have been scaled up by 4.5% for Auger and down by the same amount for TA, as their joint working group recommends [824]. The spectra then agree well below 3×10^{10} GeV. Above that energy, both show a cut-off which can potentially be identified with the Greisen-Zatsepin-Kuzmin (GZK) effect [167, 171] shortly discussed at the end of Chapter 6. This effect is due to the large photoproduction cross section for the Δ resonance, an excited state of the nucleon. At an energy of about 5×10^{10} GeV, interaction of a proton with a photon from the cosmic microwave background will excite the resonance and severely degrade the proton energy. For protons in excess of this energy, the GZK mechanism limits their distance of travel to about 50 Mpc. For nuclei of mass A, this is true for a total energy A times larger. However, Auger and TA spectral measurements differ systematically on the position and shape of the cut-off, as can be seen in the rightmost part of Figure 10.12.

Using Auger data in our simplistic two-component model, we find a cut-off for the "light" component in this range of $E_C = 1.5 \times 10^9$ GeV, as shown by the lower right curve in Figure 10.12. This is an order of magnitude lower than the GZK prediction. In a much more serious analysis of the spectrum, Unger and collaborators find a proton cut-off a factor of two higher [568]. The cut-off for the "heavy" component is then about 4×10^{10} GeV for us, our magic factor of 26 larger. The composition analysis of Auger in this energy range [632, 826] appears to confirm our simple picture by pointing out that above a few times 10^{10} GeV, the composition is dominated by medium ($5 \leq A \leq 22$) and heavy nuclei ($23 \leq A \leq 38$).

If on the other hand we use the highest energy TA data in our fit, the cut-off position for the high-energy component comes out at $E_C = 5 \times 10^{10}$ GeV, as shown by the upper right curve in Figure 10.12. This is closer to the nominal GZK prediction for protons. On the other hand, the TA composition analysis [689] finds that all the way from 2×10^9 GeV to the highest energies, the composition of cosmic rays is dominated by the light component, with no more than 10% contributions from medium and heavy elements at the highest energies. Auger and TA thus disagree on both the spectrum and the composition of cosmic rays close to the GZK cut-off energy. In an analysis of the elongation rate, i.e. the rate of change of X_{max} per decade of energy, A.A. Watson [860] finds indication in all published data that the composition does change above 3×10^9 GeV towards a heavier average mass, but the statistical and systematic significance of the effect is not overwhelming.

It is obvious – as we noted in the search for antinuclei with AMS in Chapter 9 – that more detailed studies of systematics also need a decent statistics of observed events, so as to better characterise their properties. Given the fact that it took the Pierre Auger Observatory 13.7 years to collect 270 events above 5×10^{10} GeV [767], we cannot expect rapid progress in resolving this issue.

In fact we do not know for sure that the cut-off beyond 5×10^{11} GeV is really due to the GZK mechanism, i.e. to a limitation in cosmic ray propagation. Attempts to describe the energy dependence of the flux and the composition at the end of the cosmic rays spectrum by a model incorporating a GZK-type propagation cut do not appear to be very success-ful [704, Sec. 3.2.5], regardless of whether Auger of TA data are used. On the other hand, an acceleration limited in magnetic rigidity fits at least the Auger results [632, 671], like it does in our toy model. If confirmed, this would mean that the cut-off is rather due to limitations in the highest energies reachable by astrophysical sources, i.e. in the accelera-tion mechanisms. This would be compatible with the source characteristics shown in the Hillas plot of Figure 10.2. We will shortly review the properties of a few plausible candidate accelerators near the maximum energy in Section 10.6.

10.5 ANISOTROPIES

We have so far argued that the flux of cosmic rays is isotropic because of the diffusing effect of turbulent magnetic fields in the galaxy. This argument finds its limits in the existence of an ordered component of the galactic magnetic field and for sufficiently strong and near-by sources. The latter cause may become more significant with energy, since the curvature of cosmic ray trajectories in the ambient magnetic fields diminishes.

Anisotropies can be analysed in terms of equatorial or galactic coordinates. The relevant coordinate systems are explained in Focus Box 10.6. Due to the Earth's rotation, cosmic ray observatories with long exposure times have an almost uniform exposure in right ascension, i.e. in one of the equatorial coordinates. The first harmonic modulation in right ascension then provides a direct measurement of the projection of a dipolar anisotropy onto the equatorial plane. The principles of the measurement are explained in Focus Box 10.7.

An irreducible anisotropy comes from the motion of the Solar system with respect to a coordinate system where the diffusion is anchored. This tiny anisotropy is called the Compton-Getting effect [82, 175]. When the detector moves through a homogeneous and isotropic flux in a given direction, the flux from the opposite direction is increased due to the increase of relative velocity.[5] The size of this effect is quite uncertain but in any case very small. The resulting relative anisotropy is of the order of 10^{-4} and negligible compared to other sources of anisotropy [683].

As we have seen in Chapter 9, galactic cosmic rays present a small anisotropy, if any. It takes very large statistics and well controlled systematics to see an effect at all. The anisotropy is usually expressed in terms of a multipole expansion of the intensity as a function of right ascension, i.e. in equatorial coordinates. At the high end of the galactic spectrum, i.e. towards the two knees in the PeV energy range, IceTop [502] and IceCube [589, 814] observed a significant dipole amplitude in the distribution of arrival directions with more than 300 billion events, KASCADE-Grande [718] found a meaningful upper limit with 20 billion events. At higher energies, starting from 30 PeV and up to 8 EeV, Auger and its

---

[5]The reader is reminded that the flux of a homogeneous particle density ρ impinging on an orthogonal surfaced with velocity v is simply ρv.

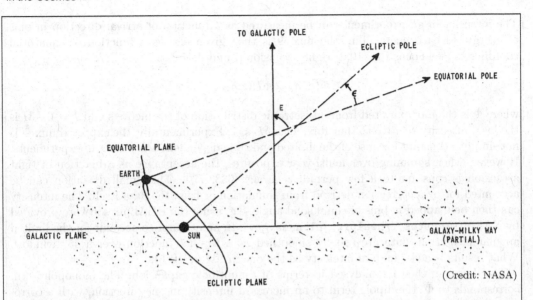

(Credit: NASA)

The scheme above [176] shows the position of the Solar system and Earth with respect to the galactic plane, with angles largely exaggerated to make the relations visible. **Equatorial coordinates** use a geocentric system, which does not co-rotate. The equatorial plane is given by the Earth's equator projected onto the celestial sphere. The principle axis on this plane points in the direction of the vernal equinox, such that spherical coordinates called right ascension (RA, α) and declination (DEC, δ) correspond to latitude and longitude on the celestial sphere.

The **galactic coordinate** system has the Sun as its centre. Its fundamental plane is parallel to an approximation of the galactic plane but offset to its north. Its primary direction is aligned with the approximate centre of the Milky Way. The coordinates of an object are then given as galactic longitude l and latitude b. The transformation between the two systems is given by [599, p. 900–901]:

$$\sin\delta = \sin b \sin\delta_{\mathrm{NGP}} + \cos b \cos\delta_{\mathrm{NGP}} \cos(l_{\mathrm{NCP}} - l)$$
$$\cos\delta \cos(\alpha - \alpha_{\mathrm{NGP}}) = \sin b \cos\delta_{\mathrm{NGP}} - \cos b \sin\delta_{\mathrm{NGP}} \cos(l_{\mathrm{NCP}} - l)$$
$$\cos\delta \sin(\alpha - \alpha_{\mathrm{NGP}}) = \cos b \sin(l_{\mathrm{NCP}} - l)$$

The index NGP refers to the coordinate values of the north galactic pole and NCP to those of the north celestial pole.

Focus Box 10.6: Equatorial and galactic coordinates

low-energy extension reported upper limits of the order of 10^{-2} for the amplitude of a dipole anisotropy [766]. The measurements are reported in Figure 10.13. It is interesting to note that while the anisotropies at energies below a few EeV do not differ significantly from zero, the observed dipole phase points in the direction of the galactic centre as also shown in Figure 10.13. This indicates that a small anisotropy might be caused by an outflow of cosmic rays from the centre of the Milky Way.

Above a few EeV, the amplitude of a dipole anisotropy increases, becoming significant above 8 EeV [633, 670, 766]. Figure 10.14 [670] shows the intensity map measured by Auger in equatorial coordinates for the energy bin from 4 to 8 EeV and above 8 EeV. For the highest energies, an overpopulation around a right ascension of $\simeq 100°$ (and a corresponding underpopulation opposite to it) is clearly visible. The amplitude and phase of the dipole

The intensity in an experiment can be measured as a function of arrival direction in bins of energy. In each energy bin, the flux Φ is then given e.g. as a function of equatorial coordinates (see Focus Box 10.6) right ascension α and declination δ:

$$\Phi(\alpha, \delta) = \bar{\Phi} I(\alpha, \delta)$$

where $\bar{\Phi}$ is the flux expected from an isotropic distribution of cosmic rays and $I = 1 + \delta I$ is the local intensity relative to that flux, with $\delta I \ll 1$. Experimentally, the expected flux $\bar{\Phi}$ is not simply a constant because of the limited and time-dependent acceptance of experiments. However, when summing over multi-year exposures, the acceptance as a function of right ascension is constant to a few permill accuracy [633, 770]. This small deviation can be accounted for by a correction derived from simulation or from the data itself. The intensity can then be mapped in bins of equal solid angle, e.g. using the HEALpix software provided by NASA (see https://healpix.jpl.nasa.gov). In an attempt to account for the limited angular resolution, data are usually smoothed by a top-hat function over a 45° window. What results is a sky map of intensity as the one in Figure 10.14.

The data can then be analysed in terms of a multipole expansion. The monopole term corresponds to $\bar{\Phi}$, the dipole term to an increased intensity in one direction, with a corresponding decrease in the opposite direction, and so on. No experiment currently has enough sensitivity to find significant multipoles beyond the dipole. A standard tool is to perform a harmonic analysis in right ascension α by summing over δ (and vice-versa). The first-harmonic Fourier components of the distribution in right ascension are given by:

$$a_\alpha = \frac{2}{\mathcal{N}} \sum_{i=1}^{N} w_i \cos \alpha_i \quad ; \quad b_\alpha = \frac{2}{\mathcal{N}} \sum_{i=1}^{N} w_i \sin \alpha_i$$

where w_i accounts for exposure differences, $\mathcal{N} = \sum_{i=1}^{N} w_i$ and N is the total number of events. The two components define a vector in the equatorial plane which points towards the over-intensity region. A measure of the anisotropy is the amplitude $r_\alpha = \sqrt{a_\alpha^2 + b_\alpha^2}$ and the phase angle $\tan \phi_\alpha = b_\alpha / a_\alpha$. An analogous quantity b_ϕ can be defined for the distribution of the geographic azimuth ϕ measured counterclockwise from the east, summing over right ascension. Using the average declination angles of the sample, $\langle \cos \delta \rangle$ and $\langle \sin \delta \rangle$, the dipole components in equatorial coordinates, α_d and δ_d are then given by [543]:

$$\alpha_d = \phi_\alpha \quad ; \quad \tan \delta_d = \frac{d_z}{d_\perp}$$

$$d_\perp \simeq \frac{r_\alpha}{\langle \cos \delta \rangle} \quad ; \quad d_z \simeq \frac{b_\phi}{\cos l_{\text{obs}} \langle \sin \delta \rangle}$$

where l_{obs} is the average latitude of the observatory. The components $d_\perp$ and d_z refer to the dipole amplitudes parallel and orthogonal to the galactic plane.

Focus Box 10.7: Dipole anisotropy

term is again summarised in Figure 10.13. The amplitude grows with energy to reach of the order of 10^{-1} in the highest energy bin above $32 \, \text{EeV}$. The direction also changes, moving away from the galactic centre. This corroborates our conclusion that cosmic rays above the ankle predominantly come from outside the Milky Way. Anisotropy measurements from the Telescope Array confirm these findings with somewhat less significance [770, 832].

A large scale anisotropy cannot arise from an isotropic extragalactic flux because of Liouville's theorem, which states that the phase space density of particles is constant along

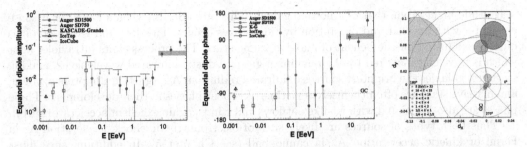

Figure 10.13 Reconstructed equatorial dipole amplitude (left) and phase (middle) [766]. When the measured amplitude is not statistically significant, an upper limit at 99% confidence level is also shown. The grey bands show the results for the wide energy bin above 8 EeV. Results are shown from Auger [766], IceCube [502, 589] and KASCADE-Grande[718]. The right plot maps the x- and y-components of the dipole anisotropy onto the equatorial plane for energy bins above 0.25 EeV. The radius of the circle is proportional to the one standard deviation error in $|\vec{d}|$. The galactic centre direction (GC) is indicated by the grey line.

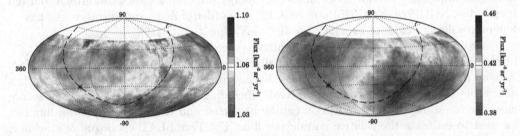

Figure 10.14 Skymap [670] in equatorial coordinates showing the cosmic ray flux observed by Auger in a given direction (smoothed in 45° windows). The left map shows data in the energy bin between 4 and 8 EeV, the right map for energies above 8 EeV.

the trajectory. Thus diffusion cannot create an anisotropy where there is none to begin with. From the dipolar anisotropy observed beyond the ankle we thus conclude that an anisotropy must be present in extragalactic cosmic rays before they are scrambled by galactic magnetic fields. Such an anisotropy can take the form of an inhomogeneous distribution of sources at large, a few strong sources near the Milky way, or both. Indeed, the direction of the dipole observed by Auger appears to be compatible [633] with the distribution of galaxies around ours, which has a significant dipolar component [378]. We take a look at a few plausible source types in the following Section.

10.6 EXTRAGALACTIC SOURCES

The Hillas plot of Figure 10.2 summarises source classes which can have the required size and magnetic field to accelerate particles to energies which cannot be contained in their host galaxies. We already discussed electromagnetic acceleration near pulsars in Chapter 5. Another class which sticks out [614] are Active Galactic Nuclei (AGN) in general, radio galaxies and their relativistic jets in particular. These are obvious candidates for acceleration towards the highest energies by their size and magnetic fields (see Chapter 5).

In addition, Gamma Ray Bursts and other energetic phenomena like hypernovae or magnetars are particularly frequent in so-called starburst galaxies. These are galaxies in a phase of their development where the star formation rate is particularly high, orders of magnitude higher than that of a normal galaxy. This often occurs when a galaxy is in close encounter with another and disruption by tidal waves occurs. The dust and gas reservoir is quickly exhausted for the benefit of massive young stars. The process emits ultraviolet light which is wavelength-shifted by the remaining gas and dust to infrared wavelengths, making starburst galaxies the brightest sources of infrared radiation. As argued in Chapters 4, young and massive stars are often themselves victims of violent phases in their development. These phases are also associated with jet-like outflows and winds, again excellent accelerators [614].

Thus both types of sources are likely sites of electromagnetic shock acceleration Ã la Fermi or kinetic acceleration Ã la cannonball (see Chapter 5). In addition, sites must provide the necessary energy production to sustain the flux of highest energy cosmic rays. The power density of cosmic rays above 3×10^8 GeV has been estimated by Unger and collaborators to be about 6×10^{47} GeV Mpc^{-3}yr^{-1} and must be matched by sources, taking into account the efficiency of converting their energy yield into cosmic ray energy. Moreover, the distance to candidate sources is limited by the horizon defined by the Greisen-Zatsepin-Kuzmin effect (see Section 10.4). This leaves a rather limited catalogue of sources [430], identified from spectra measured by the Fermi Gamma-Ray Space Telescope (see Figure 7.4 and Focus Box 7.4). As far as AGN are concerned, mostly near-by Faranov-Riley class 1 radio galaxies qualify as sources for ultra-high energies. Examples are Centaurus A, our old friend M87, NGC1275, NGC6251 and NGC1218, in order of increasing distance. All have a luminosity density close to or above 5×10^{48} GeV Mpc^{-3}yr^{-1} [430].

Near-by starburst galaxies with a luminosity density above 1×10^{48} GeV Mpc^{-3}yr^{-1} include NGC253, M82 and NGC4945 [430], all within the distance of Centaurus A, 3.7 Mpc. A more recent catalogue from the Fermi-LAT instrument lists many more sources with a detectable photon flux above $E_\gamma > 50$ GeV, including the ones listed above. These are close enough so that γ-ray absorption can be neglected and the observed photon flux can be used to estimate the relative cosmic ray flux. The Fermi-LAT catalogue of starburst galaxies [476] lists 23 objects within the quoted limits, again including the ones named above. The observed flux at photon energies or radio wavelengths can be used to estimate the relative cosmic ray flux. Since these sources are close, expected energy losses and bend angles at high energies are manageable. It is thus tempting to attempt a top-down model trying to associate a given source to a subset of observed ultra-high-energy cosmic rays.

Using the above and similar source catalogues, the Auger collaboration [669] built a sky model as the sum of an isotropic component and a component coming from a given source type. The model takes into account attenuation and uses a hard emission index $\gamma = 1$ for all sources. It neglects deflections in magnetic fields inside and outside the galaxy. This map was then compared to the one of observed cosmic rays above $E > 39$ EeV. Taking into account the exposure of the array as a function of direction, one can determine a likelihood for a fraction of the observed events to come from a given source type, and an r.m.s. angular deviation between source and cosmic ray incidence. The hypothesis of about 10% of the observed events coming from near-by starburst galaxies best fits the data, which also admit a smaller contribution by AGN. The contribution of starburst sources has a statistical significance of four standard deviations when tested against the null hypothesis of an isotropic flux. The resulting r.m.s angular deviation of 13° is compatible with the expected bend angles in this energy range. The data from the Telescope Array collected up to 2018 did not have the required sensitivity to distinguish between the Auger sky model and isotropy of the flux [675].

The Auger analysis only takes into account the arrival direction of the cosmic rays. A very interesting next step in the development of a complete top-down model from source to terrestrial observation has been taken by Capel and Mortlock [687]. They include the measured energy of each observed cosmic ray event, in addition to the arrival direction. They construct a model starting from the location, emission spectral index and luminosity of a source. Propagation is accounted for by calculating a likelihood for energy loss and magnetic deflection, checked against CRPropa simulations [584]. The likelihood for detection is constructed taking into account the exposure of a terrestrial observatory as well as uncertainties in angle and energy measurement. To make the problem manageable simplifications are necessary. The most important ones are that the luminosity is assumed the same for all sources, only protons are admitted in the composition and the experimental energy scale uncertainty is ignored. The result confirms that the association to starburst galaxies is stronger than that to active galactic nuclei. The resulting fractions of events associated to a source class are about twice as large as those based on directional information alone. This emphasises the importance of energy information in trying to associate ultra-high-energy cosmic rays to plausible extragalactic sources.

One can only hope that this top-down program is completed one day, taking into account all experimental observables and all properties of plausible sources. When comprehensive descriptions of composition, acceleration, propagation and detection are included, the dream to finally understand the functioning of the cosmic laboratory may be within reach.

FURTHER READING

Luis A. Anchordoqui, *Ultra-high-energy cosmic rays*, Phys. Rept. 801 (2019) 1–93

Raphael Alves Batista *et al.*, *Open Questions in Cosmic Ray Research at Ultrahigh Energies*, Frontiers Astron. Space Sci. 6 (2019) 23

11 The Next Revolution

In this chapter, we give an outlook on where we go from here. We present projects for future cosmic ray detectors, both space borne and terrestrial ones. There are more advanced ones – like HERD, TA×4 or AugerPrime – which are either on the way or rather likely to be realised. There are also more ambitious projects – like ALADInO, AMS-100 or POEMMA – which will probably take more time and effort to see the first cosmic ray passing. Their common goal is to widen the energy range covered by cosmic ray measurements, increase the capabilities of identifying their nature and search for rare components like complex antinuclei.

11.1 DIRECT DETECTION

Direct detection of cosmic rays in space will see future projects following the two basic approaches: calorimetric measurements and magnetic spectroscopy. Probably the next observatory in line will be the High Energy Radiation Detector (HERD) [523, 841] of the Chinese Space Agency – with design and construction led by WiZARD and AMS members. As shown in Figure 11.1, the detector will be accommodated on the Chinese Space Station CSS, which is currently under construction. The core of the detector is a novel calorimeter made of small cubic elements with three-dimensional read-out, inspired by the CaloCube project [703]. It will allow to not only accept cosmic rays entering through the zenith face but also through the lateral faces, thus greatly increasing the accepted solid angle. Despite its compact size of less than a m^3 and a weight of less than 4 t, the acceptance of HERD will reach several m^2 sr. The fine segmentation, highly sensitive material and unprecedented depth of the calorimeter will allow to reliably separate e$^\pm$ from nuclei up to 100 TeV and measure their energy with percent resolution. Large acceptance and long term exposure on the space station will enable identification and energy measurement of nuclei up to PeV energies. The calorimeter is embedded in a series of detectors for particle identification. Details of the detector concept and the expected performance are found in Focus Box 11.1. It is currently foreseen to install the observatory on the CSS around 2027. It will orbit the Earth at an altitude of around 400 km with an orbit inclination angle of 42°.

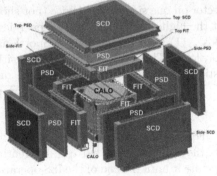

Figure 11.1 Left: Accommodation of the High Energy Radiation Detector (HERD) on the Chinese Space Station CSS (Credit: IHEP HERD Team, http//herd.ihep.ac.cn). Right: Exploded view of the HERD detector [841] showing its components described in Focus Box 11.1.

DOI: 10.1201/9781003181385-11

The components of the High Energy Radiation Detector are shown in the exploded view of Figure 11.1. The heart of the detector is a calorimeter made of 7500 LYSO crystals. The signal from each crystal is individually read out in a two-fold system. A wavelength shifting fibre couples to an image intensified CMOS camera. A photodiode glued onto each crystal is connected to front-end electronics through a flat PCB cable. The crystals are made of lutetium-yttrium oxyorthosilicate, a high performance scintillator material, and arranged in an octagonal barrel. The total depth of the calorimeter is 55 radiation length and about 3 nuclear interaction lengths. The fine segmentation allows to measure the lateral and longitudinal shower shape in great detail, such that the expected misidentification probability of protons as electrons is of the order of 10^{-6}. The calorimeter will measure $e^{\pm}$ with energies between 10 GeV and 100 TeV and an energy resolution of order 1%. For nuclei, the energy reach is larger because of their higher rate, from 30 GeV to 3 PeV, and the energy resolution of the order of 20%.

Five sides of the calorimeter are covered by the tracking device FIT made of scintillating fibres. Multiple planes of fibres are oriented in orthogonal directions so as to give a three-dimensional measurement of the cosmic ray trajectory impinging on the calorimeter. The readout is based on silicon photomultipliers. Beam tests have established the excellent detection efficiency and spatial resolution of such a device.

On the outside of the FIT follows a layer of plastic scintillator, PSD, to measure the absolute charge of incident particles and discriminate between cosmic rays and photons. Two configurations are considered. Bar shaped scintillators arranged in two orthogonal layers, or tiles individually read out will ensure the required spatial granularity of the device.

The outermost layer of the detector is again a charge measuring tracking device, the SCD. It is based on silicon microstrip detectors supported by carbon fibre honeycomb planes. The eight layers are arranged as shown in the sketch on the right [841], such that multiple scattering and nuclear interactions are minimised and angular resolution is maximised.

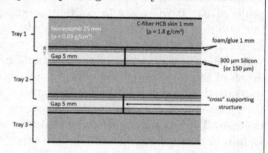

Since beams at accelerators do not reach PeV energies, the calibration of the detector cannot be ensured by pre-flight exposure to particles. Therefore, a transition radiation detector on one side of the device (not shown in Figure 11.1) will provide in-flight calibration at the highest energies. At TeV energies, also protons emit transition radiation (see Focus Box 2.11) with an intensity proportional to the logarithm of the particle relativistic factor, $\log\gamma$. That way, the HERD transition radiation detector will provide an independent estimator of the proton energy in the range $10^3 < \gamma < 10^4$.

Focus Box 11.1: The High Energy Radiation Detector (HERD)

For the foreseeable future, AMS-02 will be the only magnetic spectrometer in space, i.e. the only detector that can differentiate between matter and antimatter. The plan is to operate it until the end of the ISS operation, currently foreseen for January 2031. To improve the performance of AMS-02 further, the collaboration intends to add two large layers of tracking devices at the very top of the current set-up shown in Figure 9.1. The two layers will cover the whole of the TRD surface with two layers of silicon microstrip detectors, about $4\,m^2$ each, while in the existing configuration only about a third is covered. The detectors will be similar to the DAMPE sensors (see Figure 9.4), with single-sided read-out. One layer

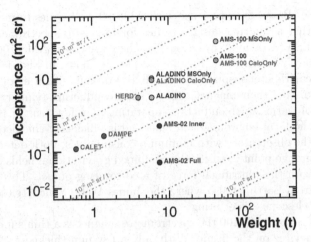

Figure 11.2 Acceptance as a function of weight for operating (full dots) and proposed cosmic ray observatories (open dots) [851]. Note that the acceptance of a device depends on the analysis, thus the point positions represent only benchmark values. The diagonal lines correspond to a fixed acceptance per ton.

will measure the bending direction, while the second will be rotated by 45° tagging each particle entry with a space point. This addition will increase the detector acceptance by about a factor of three.

If one intends to extend the energy reach of a device beyond existing cosmic ray observatories, the aim must be to increase the detector acceptance by orders of magnitude. At the same time, one must take into account the transport capabilities of rockets as far as volume and weight is concerned. Figure 11.2 shows an interesting study [851] of the acceptance (in m^2sr) of calorimeters and spectrometers as a function of their weight, both for existing devices and for future projects. It shows that spectrometers can in fact compete with calorimeters for these basic figures of merit.

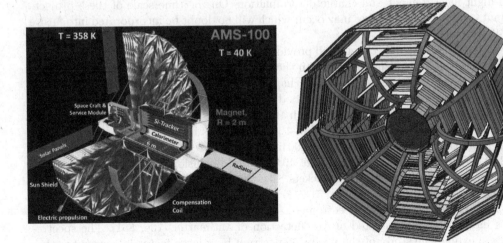

Figure 11.3 Left: The AMS-100 design study [706, 844] showing the large sun shield on the left. The solenoidal magnet surrounds a tracker and central 3D calorimeter. To avoid torque, the field is compensated through the large outside coil. Right: The ALADInO design study of a central 3D calorimeter surrounded by a toroidal spectrometer instrumented with layers of tracking detectors.

Based on experience with AMS-02, colleagues from Germany [706] and Italy [805] have thus submitted design studies for follow-up spectrometer projects in connection with the ESA program "Voyage 2050". These are called AMS-100 [844] and ALADInO [851], respectively. Their basic design concepts are shown in Figure 11.3. Both intend to surround a 3D calorimeter inspired by the HERD design with a magnetic spectrometer. To avoid having to cool their magnet coils to superconducting temperatures, both designs foresee the use of warm superconductors, operating at a few tens of Kelvin, in thermal equilibrium with the local environment if sunshine is efficiently shielded. To ensure long term exposure of the observatory with minimum consumables, both detectors are intended to operate at Lagrange point 2, where the combined gravitational fields of Sun and Earth are in equilibrium with the centrifugal force of a co-rotating object. This position, about 1.5×10^9 km away from Earth, is also where the James Webb Space Telescope – the successor of the Hubble telescope – is orbiting.

For AMS-100 the spectrometer magnet is a thin superconducting solenoid with tracking devices on the inside. With only a few mm thickness, the magnet will generate a homogeneous field of about 1 T inside its cylindrical volume of 75 m^3. Since the solenoid cannot have a return yoke due to weight limitations, its stray field must be compensated on the outside. Otherwise the dipole magnetic field of the solenoid would interact with the ambient magnetic field of a few nT and create a torque aligning the two fields. Since the ambient field is not constant, the torque cannot be compensated by gyroscopes.

For ALADInO, the magnetic field is configured as an open toroid filled with tracking layers. Since the field outside the toroid is negligible, no compensation is necessary. The ten coils will create a doughnut shape field of about 0.8 T.

The basic design and depth of the calorimeter is similar for both studies, reaching an average thickness of the order of $70X_0$ or $4\lambda_I$, reached by stacking individually read-out LYSO crystals into the required cylindrical shape. Tracking devices are based on silicon microstrip detectors, as inherited from AMS-02. However, the required surfaces are about an order of magnitude larger. Both detectors are covered by scintillators for triggering, time-of-flight measurement and charge determination. On the time scale of these projects important technological advances may occur which will no doubt be incorporated into future design iterations.

All in all, the ALADInO concept will provide an acceptance in excess of 10 m^{2}sr, roughly one or two orders of magnitude larger than the current AMS-02. Its spectrometer will have a maximum determined rigidity of more than 20 TV, commensurate with the increased acceptance. Its weight of less than 7 t and its dimensions allow transport to Lagrange point 2 by e.g. the Ariane heavy lift launcher which already successfully deployed the James Webb Space Telescope. The more ambitious AMS-100 design aims for an acceptance yet another order of magnitude larger, of the order of 100 m^{2}sr, and a spectrometer measuring rigidities up to 100 TV. It will weigh of the order of 40 t and its dimensions will require delivery by the next generation of heavy weight rockets, like the SLS Block 2 of NASA or the Long March 9 of CAST.

As appropriate for magnetic spectrometers, studies of the physics potential of the two projects have so far concentrated on the detection of antimatter [706, 851]. The spectrometers will extend the study of the positron spectrum by a large factor in energy to reach tens of TeV. In this way, the exact shape of the drop in positron flux will be measured, potentially distinguishing the abrupt monotonic drop expected from a dark matter source from the stepwise cut-off expected from the contribution of multiple pulsars (see Section 9.6). Light nuclear antimatter like $\overline{D}$, $^3\overline{He}$ and $^4\overline{He}$ will be searched for with unprecedented sensitivity. In particular, even if only secondary anti-deuterium is detected, its spectrum will be measured and potential spectral features identified in the GeV energy region. In addition, the two projects will be able to extend the AMS-02 catalogue of nuclear spectra by about an order of magnitude in rigidity, search for spectral features and make contact with composition studies by air shower arrays.

11.2 ULTRA-HIGH ENERGIES

We already mentioned the ongoing upgrade projects to the major air shower detectors in Chapter 10. Here we go into somewhat more detail about TA×4 [829, 831] and Auger-Prime [842].

Figure 11.4 shows a map of the planned extension to the Telescope Array. The narrowly spaced hexagons in the centre west indicate the locations of the existing TA and TALE surface arrays described in Section 10.2. The corresponding fluorescence cameras are shown in Figure 10.8. The wider spaced hexagons indicate the locations of the stations forming the northern and southern extensions which together will increase the surface of the array by about a factor of four, hence the acronym TA×4 [854]. As of 2019, about half of the 500 additional scintillator stations were operational. Since the aim of the upgrade is to become sensitive to very high energies, in excess of 57 EeV, the spacing of the extensions is about 2 km, almost twice the TA spacing of 1.2 km. Two additional fluorescence detector stations, with four cameras in the north and in the south, were installed in 2018 and 2019, respectively [829].

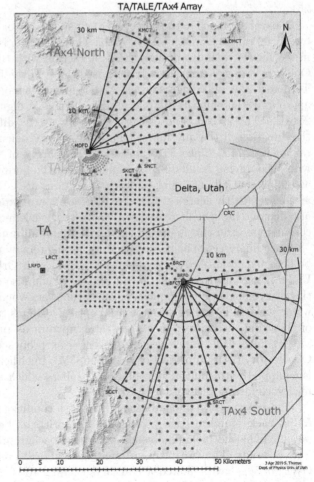

Figure 11.4 Map of the Telescope Array (TA) site [831]. Wider spaced hexagons in the northeast and southeast correspond to the planned location of a TA×4 surface detector (SD) with a spacing of 2.08 km. Dark hexagons in the centre west show the location of a TA SD with 1.2 km spacing, lighter ones the location of the low-energy TALE SD (see Chapter 10). The two fan shapes drawn with black lines are the fields of view of the TA×4 fluorescence detectors (FD) [829]. Four telescopes have been built overlooking the northern, eight telescopes in the southern area.

AugerPrime, on the other hand, does not increase the surface of the Auger air shower array described in Section 10.2. The project instead augments the instrumentation of the existing array with the aim of being more sensitive to the composition of ultra-high-energy cosmic rays [733]. In addition, AugerPrime will collect more information about hadronic interactions at extreme energies inaccessible to man-made accelerators. The water Cherenkov tanks of Auger will be equipped with additional smaller photomultipliers, so that energy deposits which saturate the existing large ones, will become measurable. These large signals occur close to the shower core. All stations of the surface array will be equipped with plastic scintillators, of dimensions $3.8 \times 1.3 \, \text{m}^2$, on top of the existing water Cherenkov tanks. Data transmission will be triggered by the larger Cherenkov detector below. All have been produced and many already deployed [733].

Cherenkov detectors (see Focus Box 2.11) and plastic scintillators (see Focus Box 2.6) have a different response to the electromagnetic and muonic components of air showers when they hit the ground. While the sensitivity to muons is about the same, the response of the scintillation counters to the electromagnetic component of the shower is lower. The two detection techniques will thus allow to disentangle the two components. In addition, underground muon detectors, called AMIGA [574], will be deployed close to each surface station. Segmented plastic scintillator modules with a total surface of about $30 \, \text{m}^2$ will be buried about 2.5 m underground, so that the electromagnetic component of the shower does not reach them. When the deployment is completed, 61 AMIGA stations will cover about $25 \, \text{km}^2$.

A novel approach to the detection of air showers is based on radiation emitted in the radio frequency range, i.e. with tens of MHz. This radiation is emitted by electrons and positrons in the shower, mainly through their interaction with the Earth's magnetic field [168, 715]. One expects a coherent linearly polarised pulse in the frequency range from 30 to 80 MHz. To detect the radio signal, each Auger surface array station will be equipped with an appropriate antenna [828]. The antenna signal is digitised with a sampling frequency of 25 MHz. The arrival time of the wave front at the different antennae allows to estimate the direction of the air shower, the radio energy density gives an estimate of the cosmic ray energy [715]. Since 2019, an engineering array of ten antennas has been operational, with encouraging results [828]. Simulations of the performance of the complete array [827], assuming that all 1661 surface detector stations are equipped, show that several thousand air showers with energies exceeding $10^{10} \, \text{GeV}$ are expected in a 10 year exposure. They also demonstrate the increased sensitivity to the chemical composition of ultra-high-energy cosmic rays. And this is indeed the punchline of all important Auger upgrades.

Since fluorescence light is emitted by air showers in all directions, it can in principle also be detected from above, by cameras in Low Earth Orbit pointing in the nadir direction. Under favourable conditions, such a camera would also capture Cherenkov light emitted by the electromagnetic component of an air shower, when it is reflected e.g. by a terrestrial water surface. There are two potential optical systems capable of focalising ultraviolet light of this kind, Schmidt-type concentrator mirrors as proposed for the Russian KLYPVE project [592, 613], and Fresnel-type lenses as proposed for the Extreme Universe Space Observatory (EUSO) project of ESA [556, 557]. In either case, the large curved focal surface is equipped with multi-anode photomultipliers which sample the field of view every few microseconds in of the order of 10^5 pixels.

The experimental difficulties and the harsh conditions in orbit complicate these projects. Major backgrounds are the passages of meteorites and other transient phenomena, anthropogenic light and bioluminescence. But these backgrounds constitute themselves valid subjects of study. The two projects mentioned above – as well as the analogous US project OWL [503] – have followed a long and winding road. First ideas for an air shower space mission have been published in the 1980s [198]. An overview of the steps followed by the EUSO project is shown in Figure 11.5. A first small prototype for a camera, called EUSO-TA, has been installed on the ground at the TA site in 2013 [734]. Balloon flights at heights of 33 and 40 km followed in 2014 and 2017 [843]. A further flight, EUSO-SPB2, is foreseen for 2023 [609]. Space borne prototypes have been flown on a Russian satellite in 2016 (TUS) [613] and installed on the ISS in 2019 (Mini-EUSO) [807]. A few candidate air shower events have indeed been identified, encouraging the proponents to pursue the project further.

EUSO [782] was originally an ESA project, which passed the mandatory Phase A review but was put on hold after the Columbia accident (see Section 2.6). Subsequently, the Japanese space agency JAXA stepped in and proposed to accommodate EUSO on the Japanese Experimental Module JEM; the project thus became JEM-EUSO. Due to budgetary constraints, JAXA ceased to be the leading agency in 2013. The Russian KLYPVE team joined the project and it became K-EUSO [853], with Roscosmos as the leading agency. The goal is to install a camera with about 10^5 pixels on one of the Russian modules of the ISS after 2025. It will cover a field of view of about 0.3 sr, i.e. close to 50000 km^2 viewed from the ISS orbit, with uniform night time exposure in the northern and southern hemispheres. The aperture will be rectangular, using refractive Fresnel-type optics, with an entrance window of about 3 m^2. Once deployed, the exposure will thus be about three times larger than that of Auger or TA×4 above an energy of 100 EeV. However, due to sanctions imposed on the Russia Federation in 2022 and the Russian decision to pull out of the ISS in 2024, the fate of the project is uncertain.

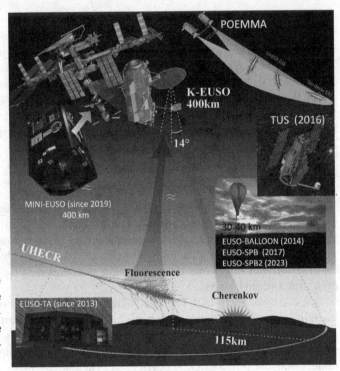

Figure 11.5 Evolution of the EUSO project with ground based, balloon and space borne prototypes mentioned in the text. (Credit: JEM-EUSO Collaboration, courtesy of Etienne Parizot, reproduced by permission)

An even more ambitious project is POEMMA [823], a pair of satellites with nadir pointing UV cameras making stereo images of air shower fluorescence. It is in some sense a successor to the OWL project. The exposures of these future projects are compared to past

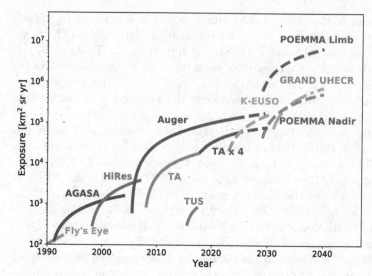

Figure 11.6 Evolution of the exposure of past, present and upcoming observatories (solid lines) for ultra-high-energy cosmic rays [704]. The dashed lines show the expected exposures for proposed new experiments. (Credit: F. Oikonomou, reproduced by permission.)

and present observatories in Figure 11.6 [704]. The exposure $\mathcal{E}$ is defined here as:

$$\mathcal{E}(t) = \int_{t_0}^{t} A\,dt$$

with the acceptance A in $[\text{km}^2\,\text{sr}]$ and the start time t_0 of the experiment. Exposure for ultra-high energies is thus measured in $[\text{km}^2\,\text{sr}\,\text{yr}]$. When pointing in the nadir direction, POEMMA will match the exposure of K-EUSO. When pointing towards the limb of the atmosphere, it will surpass existing ground based observatories by a large margin.

In fact, space observatories of air showers will also be sensitive to the Cherenkov light from showers caused by upward going neutrinos [606]. This brings us to our last subject, the synergy between the observation of different cosmic messengers from the same sources or source classes.

11.3 MULTIPLE MESSENGERS

If our dream of a complete top-down model of cosmic ray generation, acceleration and transport is to advance, we will need more than their study at all energies. We will also need to know more about their potential sources, acceleration mechanisms and propagation properties of the interstellar and intergalactic media, one by one, class by class. And that information can only come from pointing messengers, photons, neutrinos and eventually even gravitational waves.

Especially at ultra-high energies, where we probably know the least about the necessary ingredients, we cannot be content with measuring the spectrum and identifying the chemical composition of cosmic rays themselves. We need to identify and understand the acceleration sites capable of reaching energies unreachable for men-made accelerators. And we need to understand the physics of the particles on their way to us.

As far as high-energy photons are concerned, progress in constructing world class observatories is on its way. The most prominent example is the imaging Cherenkov Telescope Array (CTA) mentioned in Focus Box 7.5. It will have two sites, a larger one in the southern hemisphere, where the galactic center is visible, and a smaller one in the northern hemisphere. The telescopes come in three sizes. The largest ones, named LST, will have parabolic mirrors of 23 m diameter; their focal plane will contain 1855 pixel PMT cameras. Their role is to capture the lowest energy photon showers, down to about 20 GeV and up to 150 GeV, which are abundant but emit the least Cherenkov light. This low-energy threshold will allow to make contact with the direct detection e.g. by the FERMI-LAT (see Focus Box 7.4). The smallest size telescopes SST will allow to sparsely fill large areas with relatively low cost devices, 4.3 m diameter Schwarzschild-Couder type telescopes with 2048 pixel SiPM cameras. They will collectively be sensitive to the rare ultra-high-energy photons, with energies from a few to several hundred TeV. The intermediate size telescopes MST will be the work horses of the CTAO, Davis-Cotton type with 12 m diameter dishes and comparable to the current H.E.S.S. and MAGIC devices. Two different cameras have been proposed for this telescope type.

Figure 11.7 shows the currently planned layout of the two telescope sites [783]. The northern one on La Palma island at about 2200 m altitude will have four LST surrounded by nine MST for a total area of $0.25\,\text{km}^2$. Its main scientific target will be the observation of the extragalactic sky, with its transient phenomena from strongly absorbed sources at cosmological distances. The southern array will initially have 14 MST and 37 SST in a configuration shown in the right map of Figure 11.7 covering in total about $3\,\text{km}^2$ at 2100 m altitude. Four locations are reserved for a later installation of LST. The southern location in the Atacama desert – in the vicinity of the ESO Paranal observatory – allows the observation of galactic targets and especially the search for photon emission from galactic ultra-high-energy accelerators. As a rule of thumb one might say that when charged particles are accelerated to PeV energies, photons of a few hundred TeV, a tenth of the maximum energy, should be emitted. Thus the southern observatory will initially privilege such targets by covering a large area. The two sites will later evolve according to science needs.

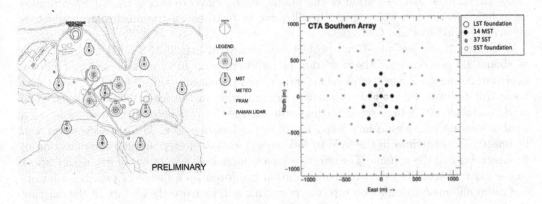

Figure 11.7 Preliminary layouts of the two CTAO arrays with Large (LST), Medium (MST) and Small Size Telescopes (SST). Left: Northern Array on the Canary island of La Palma (Spain). Right: Southern Array at Paranal (Chile). [783].

In fact, the CTA Observatory also marks a transition in organisation, away from experiments operated solely by the collaborations who built them and towards an observatory open to the scientific community. After an initial proprietary period, CTA will thus publish announcements of opportunity to submit scientific proposals which will compete for observation time. All data will be calibrated and eventually made publicly available in standard formats to increase their usefulness. In line with these organisational requirements, CTA will soon become a European Research Infrastructure Consortium (ERIC), a legal entity according to the rules of the European Community. Most of the observation time will thus be reserved to scientists belonging to ERIC member institutions, a small fraction attributed worldwide.

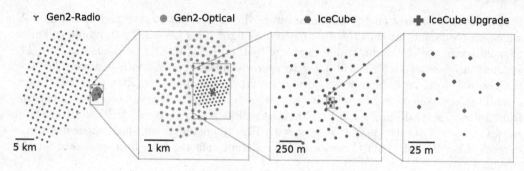

Figure 11.8 A sketch of the planned IceCube-Gen2 extension viewed from above [781]. A giant array of radio antennae will equip the surface, covering an extended array of optical detectors embedded in the ice. The two rightmost sketches show the locations of existing strings as well as the ongoing upgrade.

As mentioned in Section 7.2, an upgrade of the neutrino observatory IceCube is underway. As a first step, seven new columns of photosensors will be deployed near the bottom centre of the existing array [717]. They will provide enhanced neutrino detection capabilities due to a high density of high efficiency photosensors. In addition, they will provide new calibration abilities by including fast LED and CCD cameras as well as a system of acoustic emitters and receivers [793]. All of these devices serve to better understand the properties of the ice and the exact location of the photosensors. Novel optical sensors [787] will also be included in the new strings, which will serve as test-bed for the much more ambitious extension, called IceCube-Gen2 [753].

The idea of the IceCube-Gen2 project is sketched in Figure 11.8 [781]. A giant array of about 200 antennae for the detection of radio emission from particle showers will be deployed, covering a surface of close to $500\,\mathrm{km}^2$ and potentially sensitive to EeV neutrinos. Some will be installed on the ice surface, some buried in the ice at shallow depth. The newly installed 120 optical sensor strings will enlarge the volume of instrumented ice to a total of about $8\,\mathrm{km}^3$ with a larger spacing than the IceCube core. The complete array will be sensitive to neutrinos in the TeV to PeV energy range, mostly through the detection of muons traversing the volume. The effective area is increased by a factor of five with respect to the existing IceCube, the angular resolution improved by a factor of two. Scintillators and radio antennae installed on top of every string will increase the surface of the existing air shower array by an order of magnitude. It will reach sensitivity to PeV energies of cosmic rays and substantially improve the veto against atmospheric neutrinos. IceCube-Gen2 will thus have an unprecedented sensitivity to both transient and steady sources of neutrinos and investigate hadronic processes in cosmic ray physics.

Many next generation observatories will come into operation in the next decade, also including improved detectors for gravitational waves. The prospects for shedding new light on the origin, acceleration and propagation of cosmic rays are thus bright. The next decades of research in the Cosmic Laboratory will remain exciting.

Bibliography

[1] I. Newton. *Opticks: or, A Treatise of the Reflexions, Refractions, Inflexions and Colours of Light*. Sam. Smith and Benj. Walford, London, 1704.

[2] C.-A. de Coulomb. Troisième mémoire sur l'électricité et le magnétisme. *Mem. de l'Acad. royale*, 88:612–638, 1785.

[3] W. Crookes. On electrical insulation in high vacuum. *Proc. R. Soc. Lond.* 28:347–352, 1879.

[4] H. Hertz. Über einen Einfluss des ultravioletten Lichtes auf die electrische Entladung. *Ann. Phys.* 267:983–1000, 1887.

[5] W.C. Röntgen. Über eine neue Art von Strahlen, I. Mittheilung. *Sitzungsberichte der phys.-med. Ges. zu Würzburg*:132, 1895.

[6] H. Becquerel. Sur les propriétés différentes des radiations invisibles émises par les sels d'Uranium, et du rayonnement de la paroi anticathodique d'un tube de Crookes. *Compt. Rend. Hebd. Seances Acad. Sci.* 122:689, 1896.

[7] W.C. Röntgen. Über eine neue Art von Strahlen, II. Mittheilung. *Sitzungsberichte der phys.-med. Ges. zu Würzburg*:110, 1896.

[8] J.J. Thomson. Cathode rays. *Proceedings of the Royal Institution*, 15:419, 1897.

[9] M. Sklodowska Curie. Rayons émis par les composés de l'uranium et du thorium. *Compt. Rend. Hebd. Seances Acad. Sci.* 126:1101, 1898.

[10] P. Curie, M. Curie and M.G. Bémont. Sur une nouvelle substance fortement radio-active, contenue dans la pechblende. *Compt. Rend. Hebd. Seances Acad. Sci.* 127:1215–1217, 1898.

[11] P. Curie, M. Curie and M.G. Bémont. Sur une substance nouvelle radio-active contenue dans la pechblende. *Compt. Rend. Hebd. Seances Acad. Sci.* 127:175, 1898.

[12] J. Elster and H. Geitel. Über die Existenz electrischer Ionen in der Atmosphere. *Terrestrial Magnetism and Atmospheric Electricity*, 4:213–234, 1899.

[13] J.J. Thomson. On the masses of the ions in gases at low pressures. *Phil. Mag. 48*:547, 1899.

[14] F.E. Dorn. Die von radioactiven Substanzen ausgesandte Emanation. *Abh. der Naturforschenden Ges. zu Halle*, 23:1–15, 1900.

[15] E. Rutherford. A radioactive substance emitted from thorium compounds. *Phil. Mag.* 49:1–14, 1900.

[16] J.J. Thomson. The genesis of ions in the discharge of electricity through gases. *Phil. Mag.* 50:278–283, 1900.

[17] H. Geitel. Über die durch atmosphärische Luft induzierte Radioaktivität. *Phys. Z.* 3:76–79, 1901.

[18] C.T.R. Wilson. On the ionisation of atmospheric air. *Proc. Roy. Soc. London*, 68:151–161, 1901.

[19] J.J. Thomson. On the charge of electricity carried by a gaseous ion. *Phil. Mag.* 5:346–355, 1903.

[20] H.A. Wilson. A determination of the charge on the ions produced in air by Röntgen rays. *Phil. Mag.* 5:429, 1903.

[21] F. Linke. Luftelektrische Messungen bei zwölf Ballonfahrten. *Abh. der Ges. der Wiss. in Göttingen, Math.-Phys. Klasse*, 3:1–90, 1904.

[22] A. Einstein. Über einen die Erzeugung und Verwandlung des Lichtes betreffenden heuristischen Gesichtspunkt. *Ann. Phys.* 17:132–148, 1905.

[23] E. Rutherford and H. Geiger. An electrical method of counting the number of α particles from radioactive substances. *Proc. Royal Soc. A*, 81:141–161, 1908.

[24] Th. Wulf. Über die in der Atmosphäre vorhandene Strahlung von hoher Durchdringungsfähigkeit. *Phys. Z.* 10:152–157, 1909.

[25] K. Bergwitz. Die Gammastrahlung des Erdkörpers und ihr Anteil an der spontanen Ionisierung der Atmosphäre. *Jahresbericht des Braunschweiger Vereins für Naturwissenschaft*, 10:196–239, 1910.

[26] A. Gockel. Luftelektrische Beobachtungen bei einer Ballonfahrt. *Phys. Z.* 11:280, 1910.

[27] S. Kinoshita. The photographic action of the α-particles emitted from radioactive substances. *Proc. Roy. Soc A*, 83:432–453, 1910.

[28] Th. Wulf. Beobachtungen über Strahlung hoher Durchdringungsfähigkeit auf dem Eiffelturm. *Phys. Z.* 11:811–813, 1910.

[29] M. Reinganum. Streuung und photographische Wirkung der α-Strahlen. *Phys. Z.* 12:1076–1077, 1911.

[30] C.T.R. Wilson. On a method of making visible the paths of ionising particles through a gas. *Proc. Roy. Soc.* A85:285–288, 1911.

[31] V.F. Hess. Über Beobachtungen der durchdringenden Strahlung bei sieben Freiballonfahrten. *Phys. Z.* 13:1084–1091, 1912.

[32] D. Pacini. Penetrating radiation at the surface of and in water. *Nuovo Cim.* 6:93, 1912.

[33] C.T.R. Wilson. On an expansion apparatus for making visible the tracks of ionising particles in gases and some results obtained by its use. *Proc. Roy. Soc. A*, 87:277–292, 1912.

[34] W. Kolhörster. Messungen der durchdringenden Strahlung im Freiballon in grösseren Höhen. *Phys. Z.* 14:1153–1155, 1913.

[35] W. Kolhörster. Über eine Neukonstruktion des Apparates zur Messung der durchdringenden Strahlung nach Wulf und die damit bisher gewonnenen Ergebnisse. *Phys. Z.* 14:1066–1069, 1913.

[36] W. Kolhörster. Messungen der durchdringenden Strahlungen bis in Höhen von 9300m. *Verh. der DPG*, 16:719–721, 1914.

[37] S. Kinoshita and H. Ikeuti. The tracks of alpha particles in sensitive photographic films. *Phil. Mag.* 29:420–425, 1915.

[38] H.D. Curtis. Descriptions of 762 nebulae and clusters photographed with the Crossley reflector. *Publ. of the Lick Observatory*, 13:9–42, 1918.

[39] M. Blau. Die photographische Wirkung von H-Strahlen aus Paraffin und Aluminium. *Z. Phys.* 34:285–295, 1925.

[40] Editorial. Millikan rays. *Science*, 62:461–462, 1925.

[41]   V.F. Hess. Über den Ursprung der Höhenstrahlung. *Phys. Z.* 27:159–164, 1926.

[42]   W. Kolhörster. Bemerkungen zu der Arbeit von R.A. Millikan: Kurzwellige Strahlen kosmischen Ursprungs. *Ann. Phys.* 14:621–628, 1926.

[43]   R.A. Millikan and I.S. Brown. High frequency rays of cosmic origin: I. Sounding balloon observations at extreme altitudes. *Phys. Rev.* 27:353–363, 1926.

[44]   R.A. Millikan and G.H. Cameron. High frequency rays of cosmic origin: III. Measurements in snow-fed lakes at high altitudes. *Phys. Rev.* 28:851–869, 1926.

[45]   R.A. Millikan and R.M. Otis. High frequency rays of cosmic origin: II. Mountain peak and airplane observations. *Phys. Rev.* 27:645–659, 1926.

[46]   D.V. Skobeltsyn. Die Intensitätsverteilung in dem Spektrum der γ-Strahlen von Ra C. *Z. Phys.* 43:354–378, 1927.

[47]   H. Geiger and W. Müller. Elektronenzählrohr zur Messung schwächster Aktivitäten. *Die Naturwissenschaften*, 6:617–618, 1928.

[48]   R.A. Millikan and G.H. Cameron. The origin of cosmic rays. *Phys. Rev.* 32:533–557, 1928.

[49]   W. Bothe and W. Kolhörster. Das Wesen der Höhenstrahlung. *Z. Phys.* 56:751–777, 1929.

[50]   E. Regener. Messungen über das kurzwellige Ende der durchdringenden Höhenstrahlung. *Naturwiss.* 17:183–185, 1929.

[51]   D.V. Skobeltsyn. Über eine neue Art sehr schneller β-Strahlen. *Z. Phys.* 54:686–702, 1929.

[52]   W. Bothe. Zur Vereinfachung von Koinzidenzzählungen. *Z. f. Phys.* 59:1–5, 1930.

[53]   B. Rossi. Method of registering multiple simultaneous impulses of several geiger counters. *Nature*, 125:636, 1930.

[54]   E. Hubble and M.L. Humason. The velocity-distance relation among extra-galactic nebulae. *Astrophys. J.* 74:43–80, 1931.

[55]   B. Rossi. Nachweis einer Sekundärstrahlung. *Nuovo Cim.* 8:49–70, 1931.

[56]   B. Rossi. Über den Ursprung der durchdringenden Korpuskularstrahlung der Atmosphäre. *Z. Phys.* 68:64–84, 1931.

[57]   A. Sommerfeld. Über die Beugung und Bremsung der Elektronen. *Ann. Phys.* 403:257–330, 1931.

[58]   C. Störmer. Ein Fundamentalproblem der Bewegung elektrisch geladener Korpuskel im kosmischen Raume: I. Teil. *Z. Astrophys.* 3:31–52, 1931.

[59]   C. Störmer. Ein Fundamentalproblem der Bewegung elektrisch geladener Korpuskel im kosmischen Raume: II. Teil. *Z. Astrophys.* 3:227–252, 1931.

[60]   A. Piccard. Douze heures dans notre cabine ronde. *L'Illustration*, 4669:525–528, 1932.

[61]   A. Piccard and M. Cosyns. Etude du rayonnement cosmique en grande altitude. *Compt. Rend. Acad. Sci.* 195:604–606, 1932.

[62]   C.D. Anderson. The apparent existence of easily deflectable positives. *Science*, 76:238–239, 1932.

[63]   M. Blau and H. Wambacher. Über Versuche, durch Neutronen ausgelöste Protonen photographisch nachzuweisen. *Anz. Akad. Wiss. Wien*, 69:180–181, 1932.

[64]   M. Blau and H. Wambacher. Über Versuche, durch Neutronen ausgelöste Protonen photographisch nachzuweisen II. *Sitz. Ber. Akad. Wiss. Wien*, 141:617–620, 1932.

[65] J. Martin. A la conquète de la stratosphère: La deuxième ascension. *L'Illustration*, 4669:523–525, 1932.

[66] B. Rossi. Nachweis einer Sekundärstrahlung der durchdringenden Korpuskelstrahlung. *Phys. Z*, 33:304–305, 1932.

[67] C. Störmer. Ein Fundamentalproblem der Bewegung elektrisch geladener Korpuskel im kosmischen Raume: III. Teil. *Z. Astrophys.* 4:290–318, 1932.

[68] W. Regener. Intensity of cosmic radiation in the high atmosphere. *Nature*, 130:364, 1932.

[69] W. Regener. Über das Spektrum der Ultrastrahlung. *Z. Phys.* 74:433–454, 1932.

[70] L. Alvarez and A.H. Compton. A positively charged component of cosmic rays. *Phys. Rev.* 43:835–836, 1933.

[71] C.D. Anderson. The positive electron. *Phys. Rev.* 43:491–494, 1933.

[72] P.M.S. Blackett and G.P.S. Occhialini. Some photographs of the tracks of penetrating radiation. *Proc. Roy. Soc. A*, 139:699, 1933.

[73] T.H. Johnson and E.C. Stevenson. The asymmetry of the cosmic radiation at Swarthmore. *Phys. Rev.* 44:125–126, 1933.

[74] B. Rossi. Über die Eigenschaften der durchdringenden Korpuskularstrahlung im Meeresniveau. *Z. Phys.* 82:151–178, 1933.

[75] F. Zwicky. Die Rotverschiebung von extragalaktischen Nebeln. *Helv. Phys. Acta*, 6:110–127, 1933.

[76] W. Baade and F. Zwicky. Cosmic rays from super-novae. *Proc. Natl. Acad. Sci. USA*, 20:259–263, 1934.

[77] W. Baade and F. Zwicky. On super-novae. *Proc. Natl. Acad. Sci. USA*, 20:254–259, 1934.

[78] E. Hubble. The distribution of extra-galactic nebulae. *Astrophys. J.* 79:8–76, 1934.

[79] P. A. Cherenkov. Visible luminescence of pure liquids under the influence of γ-radiation. *Dokl. Akad. Nauk SSSR*, 2:451–454, 1934.

[80] W. Regener. Intensity of the cosmic ultra-radiation in the stratosphere with a tube-counter. *Nature*, 134:325, 1934.

[81] W. Regener and G. Pfotzer. Messung der Ultrastrahlung in der oberen Atmosphäre mit dem Zählrohr. *Phys. Z.* 35:779–784, 1934.

[82] A.H. Compton and I.A. Getting. An apparent effect of galactic rotation on the intensity of cosmic rays. *Phy. Rev.* 47:817–821, 1935.

[83] E. Regener and G. Pfotzer. Vertical intensities of cosmic rays by threefold coincidences in the stratosphere. *Nature*, 136:718–719, 1935.

[84] H. Yukawa. On the interaction of elementary particles I. *Proc. Phys. Math. Soc. Japan*, 17:48–57, 1935.

[85] C.D. Anderson and S.H. Neddermeyer. Cloud chamber observations of cosmic rays at 4300 meters elevation and near sea-level. *Phys. Rev.* 50:263–271, 1936.

[86] V.F. Hess. *Unsolved Problems in Physics: Tasks for the Immediate Future in Cosmic Ray Studies*. Nobel Lecture. 1936. URL: https://www.nobelprize.org/prizes/physics/1936/hess/lecture/.

[87] G. Pfotzer. Dreifachkoinzidenzen der Utrastrahlung aus vertikaler Richtung in der Stratosphäre: I. Messmethode und Ergebnisse. *Z. Phys.* 102:23–40, 1936.

[88]   G. Pfotzer. Dreifachkoinzidenzen der Utrastrahlung aus vertikaler Richtung in der Stratosphäre: II. Analyse der gemessenen Kurve. *Z. Phys.* 102:41–58, 1936.

[89]   S. Smith. The mass of the Virgo cluster. *Astrophys. J.* 83:23–30, 1936.

[90]   C.D. Anderson and S.H. Neddermeyer. Note on the nature of cosmic ray particles. *Phys. Rev.* 51:884–886, 1937.

[91]   H.J. Bhabha and W. Heitler. The passage of fast electrons and the theory of cosmic showers. *Proc. Roy. Soc.* 259:432–458, 1937.

[92]   M. Blau. Über die schweren Teilchen in der Ultrastrahlung. *Verh. der DPG*, 18:123, 1937.

[93]   M. Blau and H. Wambacher. Disintegration processes by cosmic rays with the simultaneous emission of several heavy particles. *Nature*, 140:585, 1937.

[94]   M. Blau and H. Wambacher. II. Mitteilung über photographische Untersuchungen der schweren Teilchen in der kosmischen Strahlung. *Sitz. Ber. Akad. Wiss. Wien*, 146:623–641, 1937.

[95]   J.F. Carlson and J.R. Oppenheimer. On multiplicative showers. *Phys. Rev.* 51:220–231, 1937.

[96]   A.H. Compton and R.N. Turner. Cosmic rays on the pacific ocean. *Phys. Rev.* 52:799–814, 1937.

[97]   S.E. Forbush. On the effects in cosmic-ray intensity observed during the recent magnetic storm. *Phys.Rev.* 51:1108–1109, 1937.

[98]   J.C. Street and E.C. Stevenson. New evidence for the existence of a particle of mass intermediate between the proton and electron. *Phys. Rev.* 52:1003–1004, 1937.

[99]   C.F. von Weizsäcker. Über Elementumwandlungen im Innern der Sterne I. *Phys. Z.* 38:176–191, 1937.

[100] F. Zwicky. On the masses of nebulae and of clusters of nebulae. *Astrophys. J.* 86:217–246, 1937.

[101] M. Blau. Photographic tracks from cosmic rays. *Nature*, 142:613, 1938.

[102] E. Miehlnickel. *Höhenstrahlung (Ultrastrahlung)*. Theodor Steinkopff, Dresden und Leipzig, 1938.

[103] K. Schmeiser and W. Bothe. Die harten Ultraschauer. *Ann. Phys.* 32:161–177, 1938.

[104] W. Kolhörster, I. Matthes and E. Weber. Gekoppelte Höhenstrahlen. *Naturwissenschaften*, 26:576, 1938.

[105] C.F. von Weizsäcker. Über Elementumwandlungen im Innern der Sterne II. *Phys. Z.* 39:633–646, 1938.

[106] H.A. Bethe. Energy production in stars. *Phys. Rev.* 55:434–456, 1939.

[107] P. Auger *et al.* Extensive cosmic-ray showers. *Rev. Mod. Phys.* 11:288–291, 1939.

[108] A.C.B. Lovell and J.G. Wilson. Investigation of cosmic ray showers of atmospheric origin, using two cloud chambers. *Nature*, 144:863–863, 1939.

[109] R. Maze. Etude d'un appareil à grand pouvoir de résolution pour rayons cosmiques. *J. Phys. et Le Radium*, 9:162–168, 1939.

[110] J.R. Oppenheimer and G.M. Volkoff. On massive neutron cores. *Phys. Rev.* 55:374–381, 1939.

[111] P. Auger, R. Maze and A.F. Roblet. Extension et pouvoir pénétrant des grandes gerbes de rayons cosmiques. *Comptes rendus*, 208:1641–1643, 1939.

[112] R.C. Tolman. Static solutions of Einstein's field equations for spheres of fluid. *Phys. Rev.* 55:364–373, 1939.

[113] F. Zwicky. Types of novae. *Rev. Mod. Phys.* 12:66–86, 1940.

[114] F. Rasetti. Disintegration of slow mesotrons. *Phys. Rev.* 60:198, 1941.

[115] B. Rossi and K. Greisen. Cosmic-ray theory. *Rev. Mod. Phys.* 13:240–309, 1941.

[116] M. Schein, W.P. Jesse, and E.O. Wollan. The nature of the primary cosmic radiation and the origin of the mesotron. *Phys. Rev.* 59:615–615, 1941.

[117] H. Alfvén. Existence of electromagnetic-hydrodynamic waves. *Nature*, 150:405–406, 1942.

[118] M. Conversi, E. Pancini and O. Piccioni. On the decay process of positive and negative mesons. *Phys. Rev.* 68:232, 1945.

[119] F.R. Elder *et al.* Radiation from electrons in a synchrotron. *Phys. Rev.* 71:829–830, 1947.

[120] M. Conversi, E. Pancini and O. Piccioni. On the disintegration of negative mesons. *Phys. Rev.* 71:209–210, 1947.

[121] R.E. Marshak and H.A. Bethe. On the two-meson hypothesis. *Phys. Rev.* 72:506–509, 1947.

[122] G.D. Rochester and C.C. Butler. Evidence for the existence of new unstable elementary particles. *Nature*, 160:855, 1947.

[123] J.A. Van Allen and H.E. Tatel. The cosmic-ray counting rate of a single Geiger counter from ground level to 161 kilometers altitude. *Phys. Rev.* 73:245–251, 1948.

[124] Ph. Freier *et al.* The heavy component of primary cosmic rays. *Phys. Rev.* 74:1818–1827, 1948.

[125] R.L. Hulsizer and B. Rossi. Search for electrons in the primary cosmic radiation. *Phys. Rev.* 73:1402–1403, 1948.

[126] G. Molière. Theorie der Streuung schneller geladener Teilchen II: Mehrfach- und Vielfachstreuung. *Z. Naturforschg. A*, 3:78–97, 1948.

[127] H. Alfvén, R.D. Richtmyer, and E. Teller. On the origin of the cosmic radiation. *Phys. Rev.* 75:892–893, 1949.

[128] E. Fermi. On the origin of the cosmic radiation. *Phys. Rev.* 75:1169–1174, 1949.

[129] W.B. Fretter. In *Proceedings of the Echo Lake Cosmic Ray Symposium, Office of Naval Research*, 37. 1949.

[130] A.H. Millikan. The present status of the evidence for the atom-annihilation hypothesis. *Rev. Mod. Phys.* 21:1–13, 1949.

[131] M.A. Pomerantz and F.L. Hereford. The detection of heavy particles in the primary cosmic radiation. *Phys. Rev.* 76:997–998, 1949.

[132] V.F. Hess, J. Eugster and P. Scherrer. *Cosmic radiation and its biological effects.* Fordham University Press, 1949.

[133] A.G. Carlson, J.E. Hooper and D.T. King. LXIII. Nuclear transmutations produced by cosmic-ray particles of great energy. Part V. The neutral mesons. *Phil. Mag.* 41:701–724, 1950.

[134] A.J. Seriff *et al.* Cloud-chamber observations of the new unstable cosmic-ray particles. *Phys. Rev.* 78:290–291, 1950.

[135] R. Bjorklund *et al.* High energy photons from proton-nucleon collisions. *Phys. Rev.* 77:213–218, 1950.

[136] B.P. Gregory, B. Rossi, and J.H. Tinlot. Production of gamma-rays in nuclear inter-actions of cosmic rays. *Phys. Rev.* 77:299–300, 1950.

[137] L. Biermann. Kometenschweife und solare Korpuskularstrahlung. *Z. Astrophys.* 29:274–286, 1951.

[138] R. Armenteros *et al.* CXIII. The properties of neutral V-particles. *Phil. Mag.* 42:1113–1135, 1951.

[139] R. Armenteros *et al.* LVI. The properties of charged V-particles. *Phil. Mag.* 43:597–611, 1952.

[140] P.F.A. Klinkenberg. Tables of nuclear shell structure. *Rev. Mod. Phys.* 24:63–73, 1952.

[141] B. Rossi. *High-energy particles.* Prentice-Hall, Englewood Cliffs, N.J., 1952.

[142] H.A. Bethe. Molière's theory of multiple scattering. *Phys. Rev.* 89:1256–1266, 1953.

[143] M. Danysz and J. Pniewski. Delayed disintegration of a heavy nuclear fragment: I. *Phil. Mag.* 44:348–350, 1953.

[144] A. Bonetti *et al.* Observation of the decay at rest of a heavy particle. *Nuovo Cim.* 10:345–346, 1953.

[145] A. Bonetti *et al.* On the existence of unstable charged particles of hyperprotonic mass. *Nuovo Cim.* 10:1736–1743, 1953.

[146] P. Bassi, G. Clark and B. Rossi. Distribution of arrival times of air shower particles. *Phys. Rev.* 92:441–451, 1953.

[147] L.F. Curtiss. *The Geiger-Müller Counter.* U.S. Nat. Bureau of Standards Circular 490. 1954.

[148] E. Fermi. Galactic magnetic fields and the origin of cosmic radiation. *Astrophys. J.* 119:1–6, 1954.

[149] J.G. Wilson and K. Greisen. *Progress in cosmic ray physics, Vol. III.* North Holland, 1956.

[150] G. Clark *et al.* An experiment on air showers produced by high-energy cosmic rays. *Nature,* 180:353–356, 1957.

[151] S. Olbert and R. Stora. Theory of high-energy N-component cascades. *Annals of Physics,* 1:247–269, 1957.

[152] K. Kamata and J. Nishimura. The lateral and the angular structure functions of electron showers. *Progr. Theor. Phys. Suppl.* 6:93–155, 1958.

[153] G.V. Kulikov and G.B. Khristiansen. On the size spectrum of extensive air showers. *JETP,* 35:441–444, 1959.

[154] S.L. Miller and H.C. Urey. Organic compound synthesis on the primitive Earth. *Science,* 130:245–251, 1959.

[155] S. Fukui *et al.* A study on the structure of the extensive air shower. *Prog. Theor. Phys. Suppl.* 16:1–53, 1960.

[156] S.L. Glashow. Resonant scattering of antineutrinos. *Phys. Rev.* 118:316–317, 1960.

[157] K. Greisen. Cosmic ray showers. *Ann. Rev. Nucl. Sci.* 10:63–108, 1960.

[158] M.A. Markov. On high energy neutrino physics. In *10th International Conference on High Energy Physics, Calgary, Canada,* 578–581. 1960.

[159] E.N. Parker. The hydrodynamic theory of of solar corpuscular radiation and stellar winds. *Astrophys. J.* 132:821–826, 1960.

[160] H.W. Babcock. The topology of the Sun's magnetic field and the 22-years cycle. *Astrophys. J.* 133:572, 1961.

[161] G. Clark *et al.* Cosmic-ray air showers at sea level. *Phys. Rev.* 122:637–654, 1961.

[162] H.R. Allan *et al.* The distribution of energy in extensive air showers and the shower size spectrum. *Proc. Phys. Soc.* 79:1170–1182, 1962.

[163] L.D. Landau and E.M. Lifshitz. *The classical theory of fields.* Addison-Wesley, 1962.

[164] R.L. Gluckstern. Uncertainties in track momentum and direction, due to multiple scattering and measurement errors. *Nucl. Instrum. Meth. A*, 24:381–389, 1963.

[165] E.N. Parker. Dynamical theory of the solar wind. *Space Sci. Rev.* 4:666–708, 1965.

[166] N.L. Grigorov *et al.* Some problems and perspectives in cosmic-ray studies. *Space Sci. Rev.* 5:167–209, 1966.

[167] K. Greisen. End to the cosmic-ray spectrum? *Phys. Rev. Lett.* 16:748–750, 1966.

[168] F. Kahn and I. Lerche. Radiation from cosmic ray air showers. *Proc. Royal Soc. A*, 289:206–213, 1966.

[169] B. Rossi. *Cosmic rays.* George Allen and Unwin Ltd., London, 1966.

[170] G. Rudstam. Systematics of spallation yields. *Z. Naturforschg. A*, 21.7:1027–1041, 1966.

[171] G.T. Zatsepin and V.A. Kuzmin. Upper limit of the spectrum of cosmic rays. *J. Exp. Theor. Phys. Lett.* 4:78–80, 1966.

[172] A.N. Bunner. *Cosmic ray detection by atmospheric fluorescence.* Cornell Univ. PhD Thesis Feb. 1967. 1967. URL: https : / / inspirehep . net / files / 0e16fc97801a9c92fc51a7422d09b012.

[173] A.D. Sakharov. Violation of CP invariance, C asymmetry, and baryon asymmetry of the universe. *Pisma Zh. Eksp. Teor. Fiz.* 5:32–35, 1967.

[174] R.M. Tennent. The Haverah Park extensive air shower array. *Proc. Phys. Soc.* 92:622–630, 1967.

[175] L.J. Gleeson and W.I. Axford. The Compton-Getting effect. *Astrophys. Space Sci.* 2:432–437, 1968.

[176] N.A. Bicket and G.A. Gary. *Celestial Coordinate Transformations.* NASA Technical memorandum, Report No. 53943. 1969.

[177] T. Hara *et al.* Observation of air showers of energies above 10^{18} eV. *Conf. Proc. C*, 690825:361–367, 1969.

[178] G.R. Blumenthal and R.J. Gould. Bremsstrahlung, synchrotron radiation, and Compton scattering of high-energy electrons traversing dilute gases. *Rev. Mod. Phys.* 42:237–270, 1970.

[179] N.L. Grigorov *et al.* Investigation of energy spectrum of primary cosmic particles with high and super-high energies of space station PROTON. *Yad. Fiz.* 11:1058–1069, 1970.

[180] V.C. Rubin and Jr. W. K. Ford. Rotation of the Andromeda nebula from a spectroscopic survey of emission regions. *Astrophys. J.* 158:379–403, 1970.

[181]   N.L. Grigorov *et al.* Energy spectrum of cosmic ray α-particles in $5 \times 10^{10} - 10^{12}$ eV/nucleon energy range. In *12th International Cosmic Ray Conference, Hobart, Australia, Volume 5*, 1760–1768. 1971.

[182]   N.L. Grigorov *et al.* Energy spectrum of primary cosmic rays in the 10^{11}–10^{15}eV energy range according to the data of Proton-4 measurements. In *12th International Cosmic Ray Conference, Hobart, Australia, Volume 5*, 1746–1751. 1971.

[183]   N.L. Grigorov *et al.* On irregularity in the primary cosmic ray spectrum in the 10^{12} eV energy range. In *12th International Cosmic Ray Conference, Hobart, Australia, Volume 5*, 1752–1759. 1971.

[184]   J.D. Sullivan. Geometrical factor and directional response of single and multi-element particle telescopes. *Nucl. Instrum. Meth.* 95:5–11, 1971.

[185]   G.R. Thomas and D.M. Willis. Analytical derivation of the geometrical factor of a particle detector having circular or rectangular geometry. *J. Phys. E*, 5:260–263, 1971.

[186]   W.L. Kraushaar *et al.* High-energy cosmic gamma-ray observations from the OSO-3 satellite. *Astrophys. J.* 177:341–363, 1972.

[187]   J. Benecke *et al.* Rapidity gap separation and study of single diffraction dissociation in p-p collisions at 12 GeV/c and 24 GeV/c. *Nucl. Phys. B*, 76:29–47, 1974.

[188]   C.F. Hall. Pioneer 10. *Science*, 183:301–302, 1974.

[189]   H.K. Paetzold, G. Pfotzer and E. Schopper. "Erich Regener als Wegbereiter der extraterrestrischen Physik". In *Zur Geschichte der Geophysik*. Ed. by H. Birett *et al.* Berlin, Heidelberg, New York: Springer, 1974, pp. 167–188.

[190]   E. Juliusson. Charge composition and energy spectra of cosmic-ray nuclei at energies above 20 GeV per nucleon. *Astrophys. J.* 191:331–348, 1974.

[191]   E. Longo and I. Sestili. Monte Carlo calculation of photon-initiated electromagnetic showers in lead glass. *Nucl. Instrum. Meth.* 128:283–307, 1975.

[192]   W.M. Morse *et al.* π^+-p, K^+-p and p-p topological cross-sections and inclusive interactions at 100 GeV using a hybrid bubble chamber–spark chamber system and a tagged beam. *Phys. Rev. D*, 15:66, 1977.

[193]   T.K. Gaisser and A.M. Hillas. Reliability of the method of constant intensity cuts for reconstructing the average development of vertical showers. In *15th International Cosmic Ray Conference, Budapest, Hungary.* Vol. 8, 353–357. 1977.

[194]   C.E. Kohlhase and P.A. Penzo. Voyager mission description. *Space Sci. Rev.* 21:77–101, 1977.

[195]   H.R. Carlton. *Electric Conductivity of Moist Air.* US Army Armament Research and Development Command, Report AD-A056 235. 1978.

[196]   R.L. Golden *et al.* A magnetic spectrometer for cosmic ray studies. *Nucl. Instrum. Meth.* 148:179–185, 1978.

[197]   J.A. Lezniak and W.R. Webber. The charge composition and energy spectra of cosmic-ray nuclei from 3000 MeV per nucleon to 50 GeV per nucleon. *Astrophys. J.* 223:676–696, 1978.

[198]   R. Benson and J. Linsley. Satellite observation of cosmic ray air showers. In *17th International Cosmic Ray Conference, Paris, France.* 1981.

[199]   J.-S. Young *et al.* The elemental and isotopic composition of cosmic rays - silicon to nickel. *Astrophys. J.* 246:1014–1030, 1981.

[200] J.R. Jokipii and B. Thomas. Effects of drift on the transport of cosmic rays IV. Modulation by a wavy interplanetary current sheet. *Astrophys. J.* 243:1115–1122, 1981.

[201] G. Minagawa. The abundances and energy spectra of cosmic ray iron and nickel at energies from 1 to 10 GeV per AMU. *Astrophys. J.* 248:847–855, 1981.

[202] R.K. DeKosky. William Crookes and the quest for absolute vacuum in the 1870s. *Ann. Sci.* 40:1–18, 1983.

[203] P.L. Galison. The discovery of the muon and the failed revolution against Quantum Electrodynamics. *Centaurus*, 26:262, 1983.

[204] A. Kondor. Method of convergent weights – An iterative procedure for solving Fredholm's integral equations of the first kind. *Nucl. Instrum. Meth.* 216:177–181, 1983.

[205] M. Milgrom. A modification of the Newtonian dynamics - Implications for galaxies. *Astrophys. J.* 270:371–383, 1983.

[206] M. Milgrom. A modification of the Newtonian dynamics - Implications for galaxy systems. *Astrophys. J.* 270:384–389, 1983.

[207] M. Milgrom. A modification of the Newtonian dynamics as a possible alternative to the hidden mass hypothesis. *Astrophys. J.* 270:365–370, 1983.

[208] V. Blobel. Unfolding methods in high-energy physics experiments. In *1984 CERN School of Computing.* 1984.

[209] A. Breakstone *et al.* Charged multiplicity distribution in p-p interactions at ISR energies. *Phys. Rev. D*, 30:528–535, 1984.

[210] R.L. Golden *et al.* A measurement of the absolute flux of cosmic ray electrons. *Astrophys. J.* 287:622–632, 1984.

[211] E. Witten. Cosmic separation of phases. *Phys. Rev. D*, 30:272–285, 1984.

[212] T.S. van Albada *et al.* Distribution of dark matter in the spiral galaxy NGC 3198. *Astrophys. J.* 295:305–313, 1985.

[213] H. Fesefeldt. *The Simulation of Hadronic Showers – Physics and Applications.* PITHA 85/02. 1985.

[214] R.M. Baltrusaitis *et al.* (Fly's Eye Coll.) The Utah Fly's Eye detector. *Nucl. Instrum. Meth. A,* 240:410–428, 1985.

[215] B. Rossi. "Arcetri, 1928–1932". In *Early History of Cosmic Ray Studies.* Ed. by Y. Sehido and H. Elliot (Edts.) Dordrecht: D. Reidel, 1985, pp. 53–73.

[216] G.M. Webb, M.A. Forman, and W.I. Axford. Cosmic-ray acceleration at stellar wind terminal shocks. *Astrophys. J.* 298:684–709, 1985.

[217] G.J. Alner *et al.* Scaling violations in multiplicity distributions at 200 GeV and 900 GeV. *Phys. Lett. B,* 167:476–480, 1986.

[218] E.N. Alekseev *et al.* Possible detection of a neutrino signal on 23 February 1987 at the Baksan underground scintillation telescope of the Institute of Nuclear Research. *JETP Lett.* 45:589–592, 1987.

[219] K. Hirata *et al.* Observation of a neutrino burst from the supernova SN1987A. *Phys. Rev. Lett.* 58:1490–1493, 1987.

[220] M. Aglietta *et al.* On the Event Observed in the Mont Blanc Underground Neutrino Observatory during the Occurrence of Supernova 1987a. *Europhys. Lett.* 3:1315–1320, 1987.

[221] R.M. Bionta *et al.* Observation of a neutrino burst in coincidence with supernova 1987A in the Large Magellanic Cloud. *Phys. Rev. Lett.* 58:1494–1496, 1987.

[222] W. Kunkel *et al.* Supenova 1987A in the Large Magellanic Cloud. *IAU Circular*, 4316:1, 1987.

[223] P.H. Fowler. The π discovery. In *40 Years of Particle Physics, Proc. Int. Conf. to Celebrate the 40th Anniversary of the Discoveries of the π- and V-Particles, Bristol, UK.* 1987.

[224] G.D. Rochester. The discovery of V-particles. In *40 Years of Particle Physics, Proc. Int. Conf. to Celebrate the 40th Anniversary of the Discoveries of the π- and V-Particles, Bristol, UK.* 1987.

[225] J.F. Ormes *et al.* (Edts.) *Report of the Astromag Definition Team – The Particle Astrophysics Magnet Facility.* NASA Goddard Space Flight Center. 1988.

[226] M.A. Green *et al.* Astromag: A superconducting particle astrophysics magnet facility for the space station. *IEEE Trans. Magn.* 23:1240–1243, 1988.

[227] G. Kanbach *et al.* The project EGRET (energetic gamma-ray experiment telescope) on NASA's Gamma-Ray Observatory GRO. *Space Sci. Rev.* 49:69–84, 1988.

[228] NRC Task Group on Solar and Space Physics. *Solar and Space Science in the Twenty-First Century – Imperatives for the Decades 1995 to 2015.* The National Academies Press, Washington DC, 1988.

[229] T.F. Tascione. *Introduction to the Space Environment.* Krieger Publishing Company, 1988.

[230] I.P. Ivanenko *et al.* Energy spectrum of primary cosmic-ray particles at 1–100 TeV from data from the Sokol package. *JETP Lett.* 49:222–224, 1989.

[231] M. Aglietta *et al.* The EAS-TOP array at $E_0 = 10^{14} - 10^{16}$ eV: Stability and resolutions. *Nucl. Instrum. Meth. A*, 277:23–28, 1989.

[232] W.R. Binns *et al.* Abundances of ultraheavy elements in the cosmic radiation: Results from HEAO 3. *Astrophys. J.* 346:997, 1989.

[233] Y. Kawamura *et al.* "Quasidirect" observations of cosmic-ray primaries in the energy region 10^{12}–10^{14} eV. *Phys. Rev. D*, 40:729–753, 1989.

[234] W.R. Binns. Large Isotope Spectrometer for Astromag. In *AIP Conference: Particle Astrophysics, Greenbelt, Maryland, USA.* Vol. 203, 83–88. 1990.

[235] J.J. Engelmann *et al.* Charge composition and energy spectra of cosmic-ray nuclei for elements from Be to Ni. Results from HEAO-3-C2. *Astron. Astroph.* 233:96–111, 1990.

[236] L.M. Barbier *et al.* Astromag: Current capabilities and status. *Nucl. Phys. B - Proc. Suppl.* 14:3–21, 1990.

[237] R.L. Golden. WiZard – An experiment to measure the cosmic rays including antiprotons, positrons, nuclei and to conduct a search for primordial antimatter. In *AIP Conference: Particle Astrophysics, Greenbelt, Maryland, USA.* Vol. 203, 76–82. 1990.

[238] W. Vernon Jones. Particle astrophysics magnet facility for space station Freedom. *Acta Astronautica*, 21:505–512, 1990.

[239] R.A. Mewald and E.C. Stone. Cosmic ray studies with an interstellar probe. In *AIP Conference: Particle Astrophysics, Greenbelt, Maryland, USA.* Vol. 203, 264–267. 1990.

[240] G.F. Smoot. The Astromag facility. In *AIP Conference: Particle Astrophysics, Greenbelt, Maryland, USA*. Vol. 203, 67–75. 1990.

[241] E.S. Seo *et al.* Measurement of cosmic-ray proton and helium spectra during the 1987 solar minimum. *Astrophys. J.* 378:763–772, 1991.

[242] R.L. Golden *et al.* Performance of a balloon-borne magnet spectrometer for cosmic ray studies. *Nucl. Instrum. Meth. A*, 306:366–377, 1991.

[243] W.R. Webber *et al.* A measurement of the cosmic-ray ^{2}H and ^{3}He spectra and ^{2}H/^{4}He and ^{3}He/^{4}He ratios in 1989. *Astrophys. J.* 380:230–234, 1991.

[244] G.R. Lynch and O.I. Dahl. Approximations to multiple Coulomb scattering. *Nucl. Instrum. Meth. B*, 58:6–10, 1991.

[245] M. De Maria, M.G Ianniello, and A. Russo. The discovery of cosmic rays: Rivalries and controversies between Europe and the United States. *HSPS*, 22:165–192, 1991.

[246] E.S. Seo. *Measurement of Galactic Cosmic-Ray Proton and Helium Spectra during the 1987 Solar Minimum*. Louisiana State University Thesis UMI-92-00090. 1991.

[247] J.-A. Esposito *et al.* The ALICE instrument and the measured cosmic ray elemental abundances. *Astropart. Phys.* 1:33–45, 1992.

[248] J.A. Simpson *et al.* The Ulysses cosmic ray and solar particle investigation. *Astron. Astrophys. Suppl. Series*, 92:365–399, 1992.

[249] M.A. Green *et al.* The Astromag superconducting magnet facility configured for a free-flying satellite. *Cryogenics*, 32:91–97, 1992.

[250] M. Koshiba. Observational neutrino astrophysics. *Physics Reports*, 220:229–381, 1992.

[251] B. Dolgoshein. Transition radiation detectors. *Nucl. Instrum. Meth. A*, 326:434–469, 1993.

[252] A.Y. Varkovitskaya *et al.* Energy spectra of primary protons and other nuclei at energies of 10 TeV/particle to 100 TeV/particle. *JETP Lett.* 57:469–472, 1993.

[253] C.H. Tsao *et al.* Scaling algorithm to calculate heavy-ion spallation cross sections. *Phys. Rev. C*, 47:1257–1262, 1993.

[254] D.J. Thompson *et al.* Calibration of the Energetic Gamma-Ray Experiment Telescope (EGRET) for the Compton Gamma-Ray Observatory. *Astrophys. J. Suppl.* 86:629–656, 1993.

[255] I.P. Ivanenko *et al.* Energy spectra of cosmic rays above 2 TeV as measured by the "Sokol" apparatus. *23rd International Cosmic Ray Conference, Calgary, Canada*, 2:17–20, 1993.

[256] J.W. Mitchell *et al.* (IMAX) Isotope Matter-Antimatter Experiment. In *23rd International Cosmic Ray Conference, Calgary, Canada*. Vol. 1, 519–522. 1993.

[257] M. Ichimura *et al.* Observation of heavy cosmic-ray primaries over the wide energy range from ∼100 GeV/particle to ∼100 TeV/particle: Is the celebrated "knee" actually so prominent? *Phys. Rev. D*, 48:1949–1975, 1993.

[258] O. Adriani *et al.* (L3 Coll.) Results from the L3 experiment at LEP. *Phys. Rept.* 236:1–146, 1993.

[259] V.I. Zatsepin. Energy spectra and composition of primary cosmic rays above 10^{12} eV. In *23rd International Cosmic Ray Conference, Calgary, Canada*, 439–446. 1993.

[260] J. Cortázar. *Cuentos Completos*. Alfaguara, Madrid, 1994.

[261] N. Hayashida *et al.* Observation of a very energetic cosmic ray well beyond the predicted 2.7K cutoff in the primary energy spectrum. *Phys. Rev. Lett.* 73:3491–3494, 1994.

[262] T.A. Gabriel *et al.* Energy dependence of hadronic activity. *Nucl. Instrum. Meth. A*, 338:336–347, 1994.

[263] W.R. Leo. *Techniques for Nuclear and Particle Physics Experiments*. Springer, 1994.

[264] P. Todd. Cosmic radiation and evolution of life on Earth: Roles of environment, adaptation and selection. *Adv. Space Res.* 14:313–321, 1994.

[265] A. Burkert. The structure of dark matter halos in dwarf galaxies. *Astrophys. J. Lett.* 447:L25, 1995.

[266] G. D'Agostini. A multidimensional unfolding method based on Bayes' theorem. *Nucl. Instrum. Meth. A*, 362:487–498, 1995.

[267] D.J. Bird *et al.* (Fly's Eye Coll.) Detection of a cosmic ray with measured energy well beyond the expected spectral cutoff due to cosmic microwave radiation. *Astrophys. J.* 1995.

[268] M. G. Kivelson and C. T. Russell. *Introduction to Space Physics*. Cambridge University Press, 1995.

[269] G.D. Lafferty and T.R. Wyatt. Where to stick your data points: The treatment of measurements within wide bins. *Nucl. Instrum. Meth. A*, 355:541–547, 1995.

[270] O. Adriani *et al.* (WiZARD Coll.) The magnetic spectrometer PAMELA for the study of cosmic antimatter in space. In *24th International Cosmic Rays Conference, Rome, Italy*. 1995.

[271] H.V. Cane, I.G. Richardson, and I.T. von Rosenvinge. Cosmic ray decreases: 1964–1994. *J. Geophys. Res*, 101:21561–21572, 1996.

[272] B.W. Carroll and D.A. Ostlie. *An Introduction to Modern Astrophysics*. Addison-Wesley, 1996.

[273] D. O'Sullivan *et al.* Investigation of $Z \geq 70$ cosmic ray nuclei on the LDEF mission. *Rad. Meas.* 26:889–892, 1996.

[274] R. Gandhi *et al.* Ultrahigh-energy neutrino interactions. *Astropart. Phys.* 5:81–110, 1996.

[275] O. Piccioni. "The Discovery of the muon". In *History of Original Ideas and Basic Discoveries in Particle Physics*. Ed. by H.B. Newman and Th. Ypsilantis. New York: Plenum, 1996, pp. 143–159.

[276] H.V. Cane, I.G. Richardson, and G. Wibberenz. Helios 1 and 2 observations of particle decreases, ejecta, and magnetic clouds. *J. Geophys. Res.* 102.A7:7075–7086, 1997.

[277] J.W. Cronin, T.K. Gaisser, and S.P. Swordy. Cosmic rays at the energy frontier. *Scientific American*, 276:44–49, 1997.

[278] G. Baur *et al.* Observation of antihydrogen production in flight at CERN. *Hyperfine Interact.* 109:191–203, 1997.

[279] J.F. Navarro *et al.* A universal density profile from hierarchical clustering. *Astrophys. J.* 490:493–508, 1997.

[280] P.L. Galison. Marietta Blau: Between nazis and nuclei. *Physics Today*, 50:42–48, 1997.

[281]   J.-P. Meyer, L. O'C. Drury, and D.C. Ellison. Galactic cosmic rays from supernova remnants. I. A cosmic-ray composition controlled by volatility and mass-to-charge ratio. *Astrophysi. J.* 487:182–196, 1997.

[282]   J. Renn, T. Sauer, and J. Stachel. The origin of gravitational lensing: A postscript to Einstein's 1936 *Science* paper. *Science*, 275:184–186, 1997.

[283]   T. Stanev. Ultra-high-energy cosmic rays and the large-scale structure of the galactic magnetic field. *Astroph.-J.* 479:290–295, 1997.

[284]   A.G. Riess *et al.* (Supernova Search Team Coll.) Observational evidence from supernovae for an accelerating universe and a cosmological constant. *Astron. J.* 116:1009–1038, 1998.

[285]   O. Reimer *et al.* The cosmic-ray ^{3}He/^{4}He ratio from 200 MeV per nucleon^{-1} to 3.7 GeV per nucleon^{-1}. *Astrophys. J.* 496:490–502, 1998.

[286]   T.I. Gombosi. *Physics of the Space Environment.* Cambridge University Press, 1998.

[287]   K. Asakimori *et al.* (JACEE Coll.) Cosmic-ray proton and helium spectra: Results from the JACEE experiment. *Astrophys. J.* 502:278–283, 1998.

[288]   I.V. Moskalenko and A.W. Strong. Production and propagation of cosmic ray positrons and electrons. *Astrophys. J.* 493:694–707, 1998.

[289]   A.W. Strong and I.V. Moskalenko. Propagation of cosmic-ray nucleons in the galaxy. *Astrophys. J.* 509:212–228, 1998.

[290]   W.R. Webber. A new estimate of the local interstellar energy density and ionization rate of galactic cosmic rays. *Astrophys. J.* 506:329–334, 1998.

[291]   T.A. Armstrong *et al.* (E864 Coll.) Search for neutral strange quark matter in high-energy heavy ion collisions. *Phys. Rev. C*, 59:1829–1833, 1999.

[292]   R.C. Hartman *et al.* The third EGRET catalog of high-energy gamma-ray sources. *Astrophys. J. Suppl.* 123:79–202, 1999.

[293]   O. Ganel, E.S. Seo, and J.Z. Wang. On the use of low Z targets in space based hadron calorimetry. In *26th International Cosmic Ray Conference, Salt Lake City, United States.* Vol. 5, 33–36. 1999.

[294]   T. Hams *et al.* (ISOMAX Coll.) The ISOMAX magnetic rigidity spectrometer. In *26th International Cosmic Ray Conference, Salt Lake City, United States.* Vol. 3, 121–124. 1999.

[295]   J.D. Jackson. *Classical Electrodynamics (3rd Edition).* J. Wiley & Sons, Inc., 1999.

[296]   D.V. Reames. Particle acceleration at the Sun and in the heliosphere. *Space Sci. Rev.* 90:413–491, 1999.

[297]   S. Perlmutter *et al.* (Supernova Cosmology Project Coll.) Measurements of Ω and Λ from 42 high-redshift supernovae. *Astrophys. J.* 565:517–586, 1999.

[298]   Y. Ajima *et al.* (BESS Coll.) A superconducting solenoidal spectrometer for a balloon-borne experiment. *Nucl. Instrum. Meth. A*, 443:71–100, 2000.

[299]   H.V. Cane. Coronal mass ejections and Forbush decreases. *Space Sci. Rev.* 93:55–77, 2000.

[300]   C. Song *et al.* Energy estimation of UHE cosmic rays using the atmospheric fluorescence technique. *Astropart. Phys.* 14:7–13, 2000.

[301]   H.L. Lai *et al.* Global QCD analysis of parton structure of the nucleon: CTEQ5 parton distributions. *Eur. Phys. J. C*, 12:375–392, 2000.

[302]   J. Adams *et al.* Particle energy determination device for the International Space Station using a new approach to cosmic ray spectral measurements (TUS-M Mission). *AIP Conf. Proc.* 504:175–180, 2000.

[303]   M. Hof *et al.* ISOMAX: A balloon-borne instrument to measure cosmic ray isotopes. *Nucl. Instrum. Meth. A*, 454:180–185, 2000.

[304]   W. Menn *et al.* The absolute flux of protons and helium at the top of the atmosphere Using IMAX. *Astrophys. J.* 533:281–297, 2000.

[305]   J.F. Bottolier *et al.* Assessing exposure to cosmic radiation during long-haul flights. *Radiat. Res.* 153:526–532, 2000.

[306]   J. Joyce. *A Portrait of the Artist as a Young Man.* Oxford Univ. Press, 2000.

[307]   J. Lyon. Solar wind-magnetosphere-ionosphere system. *Science*, 288:1987–1991, 2000.

[308]   M. Nagano and A.A. Watson. Observations and implications of the ultrahigh-energy cosmic rays. *Rev. Mod. Phys.* 72:689–732, 2000.

[309]   N.D. Marsh and H. Svensmark. Low cloud properties influenced by cosmic rays. *Phys. Rev. Lett.* 85:5004–5007, 2000.

[310]   T. Ondoh and K. Marubashi, eds. *Wave Summit Course: Science of Space Environment.* Ohmsha, Ltd., 2000.

[311]   W. Friedberg *et al.* Radiation exposure during air travel: Guidance provided by the FAA for air carrier crews. *Health Phys.* 79:591–595, 2000.

[312]   H. Yoshii *et al.* (BASJE Coll.) Present experiment of BASJE group at Mt. Chacaltaya. *Nuovo Cim.* 24:507–512, 2001.

[313]   R. Beck. Galactic and extragalactic magnetic fields. *Space Sci. Rev.* 99:243–260, 2001.

[314]   T.A. Armstrong *et al.* (E864 Coll.) Search for strange quark matter produced in relativistic heavy ion collisions. *Phys. Rev. C*, 63:054903, 2001.

[315]   W.R. Binns *et al.* Galactic cosmic ray neon isotopic abundances measured by the cosmic ray isotope spectrometer CRIS on ACE. *Adv. Space Res.* 27:767–772, 2001.

[316]   K.M. Ferrière. The interstellar environment of our galaxy. *Rev. Mod. Phys.* 73:1031–1066, 2001.

[317]   G. Clark, F. Nagase and J. Linsley. Minoru Oda. *Physics Today*, 54:74–75, 2001.

[318]   P.K.F. Grieder. *Cosmic Rays at Earth: Researcher's Reference Manual and Data Book.* Elsevier, 2001.

[319]   D.E. Groom, N.V. Mokhov, and S.I. Striganov. Muon stopping power and range tables 10 MeV to 100 TeV. *Atom. Data Nucl. Data Tabl.* 78:183–356, 2001.

[320]   J.W. den Herder *et al.* The reflection grating spectrometer on board XMM-Newton. *Astron. Astroph.* 365:L7–L17, 2001.

[321]   S.P. Swordy. The energy spectra and anisotropies of cosmic rays. *Space Sci. Rev.* 99:85–94, 2001.

[322]   M.C. Todd and D.R. Kniveton. Changes in cloud cover associated with Forbush decreases of galactic cosmic rays. *J. Geophys. Res.* 106:32031–32041, 2001.

[323]   M. Circella *et al.* (WiZard Coll.) Measurements of primary cosmic-ray hydrogen and helium by the WiZard collaboration. *Adv. Space Res.* 27:755–760, 2001.

[324]   M. Aguilar *et al.* (AMS Coll.) The Alpha Magnetic Spectrometer (AMS) on the International Space Station: Part I – results from the test flight on the space shuttle. *Physics Reports*, 366:331–405, 2002.

[325]   A. Yamamoto *et al.* (BESS Coll.) BESS and its future prospect for polar long duration flights. *Advances in Space Research*, 30:1253–1262, 2002.

[326]   B. Blau *et al.* The superconducting magnet system of AMS-02 – a particle physics detector to be operated on the International Space Station. *IEEE Transactions on Applied Superconductivity*, 12:349–352, 2002.

[327]   M. Boezio *et al.* A high granularity imaging calorimeter for cosmic ray physics. *Nucl. Instrum. Meth. A*, 487:407–422, 2002.

[328]   J.L. Han. Magnetic fields in our Galaxy: How much do we know? (II) Halo fields and the global field structure. *API Conf. Proc.* 609:96–101, 2002.

[329]   J.S. Smith. Overview of the Ultra Long Duration Balloon project. *Adv. Space Res.* 30:1205–1213, 2002.

[330]   L.M. Widrow. Origin of galactic and extragalactic magnetic fields. *Rev. Mod. Phys.* 74:775–823, 2002.

[331]   A. Heger *et al.* How massive stars end their life. *Astrophys. J.* 591:288–300, 2003.

[332]   B. Baret *et al.* Secondary proton flux induced by cosmic ray interactions with the atmosphere. *Phys. Rev. D*, 68:053009, 2003.

[333]   D.J. McComas *et al.* The three-dimensional solar wind around solar maximum. *Geophys. Res. Lett.* 30:2003GL017136, 2003.

[334]   E.J. Smith *et al.* The Sun and heliosphere at solar maximum. *Science*, 302:1165:1169, 2003.

[335]   M. Boezio *et al.* The cosmic-ray proton and helium spectra measured with the CAPRICE98 balloon experiment. *Astropart. Phys.* 19:583–604, 2003.

[336]   M. Motoki *et al.* Precise measurements of atmospheric muon fluxes with the BESS spectrometer. *Astropart. Phys.* 19:113–126, 2003.

[337]   T Antoni *et al.* The cosmic-ray experiment KASCADE. *Nucl. Instrum. Meth. A*, 513:490–510, 2003.

[338]   K. Bernlöhr *et al.* The optical system of the H.E.S.S. imaging atmospheric Cherenkov telescopes, Part I: layout and components of the system. *Astropart. Phys.* 20:111–1128, 2003.

[339]   R. Cornlis *et al.* The optical system of the H.E.S.S. imaging atmospheric Cherenkov telescopes, Part II: mirror alignment and point spread function. *Astropart. Phys.* 20:129–143, 2003.

[340]   W.R. Webber et al. Updated formula for calculating partial cross sections for nuclear reactions of nuclei with $Z \leq 28$ and $E > 150$ MeV nucleon^{-1} in hydrogen targets. *Astrophys. J. Suppl.* 144:153–167, 2003.

[341]   A.R. Bell. Turbulent amplification of magnetic field and diffusive shock acceleration of cosmic rays. *Mon. Not. Roy. Astron. Soc.* 353:550–558, 2004.

[342]   J. Diemand, B. Moore, and J. Stadel. Convergence and scatter of cluster density profiles. *Mon. Not. Roy. Astron. Soc.* 353:624, 2004.

[343]   H.M. Barbosa *et al.* Determination of the calorimetric energy in extensive air showers. *Astropart. Phys.* 22:159–166, 2004.

[344]   L. Anchordoqui *et al.* High energy physics in the atmosphere: Phenomenology of cosmic ray air showers. *Annals Phys.* 314:145–207, 2004.

[345]   M Ambriola *et al.* The transition radiation detector of the PAMELA space mission. *Nucl. Instrum. Meth. A*, 522:77–80, 2004.

[346] M. S. Clowdsley *et al.* Radiation protection quantities for near Earth environments. In *Proceedings of Space 2004*, 1–6. 2004.

[347] P. Papini *et al.* High-energy deuteron measurement with the CAPRICE98 experiment. *Astrophys. J.* 615:259–274, 2004.

[348] S. Haino *et al.* Measurements of primary and atmospheric cosmic-ray spectra with the BESS-TeV spectrometer. *Phys. Lett. B*, 594:35–46, 2004.

[349] T. Yoshida *et al.* BESS-Polar experiment. *Advances in Space Research*, 33:1755–1762, 2004.

[350] F. Kennrich *et al.* (VERITAS Coll.) VERITAS: the Very Energetic Radiation Imaging Telescope Array System. *New Astronomy Reviews*, 48:345–349, 2004.

[351] N.L. Grigorov and E.D. Tolstaya. The spectrum of cosmic-ray particles and their prigin. *JETP*, 98:643–650, 2004.

[352] T. Hams *et al.* (ISOMAX Coll.) Measurement of the abundance of radioactive ^{10}Be and other light isotopes in cosmic radiation up to 2 GeV nucleon^{-1} with the balloon-borne instrument ISOMAX. *Astrophys. J.* 611:892–905, 2004.

[353] M. M. Nozaki. BESS-Polar. *Nucl. Instrum. Meth. B*, 214:110–115, 2004.

[354] M. Markevitch. Chandra observation of the most interesting cluster in the Universe 2004. arXiv: 0511345v1 [astro-ph].

[355] I. G. Richardson. Energetic particles and corotating interaction regions in the solar wind. *Space Sci. Rev.* 111:267–376, 2004.

[356] G.B. Rybicki and A.P. Lightman. *Radiative Processes in Astrophysics*. Wiley-VCH, 2004.

[357] T. Asaka, S. Blanchet, and M. Shaposhnikov. The νMSM, dark matter and neutrino masses. *Phys. Lett B*, 631:151–156, 2005.

[358] T. Asaka and M. Shaposhnikov. The νMSM, dark matter and baryon asymmetry of the universe. *Phys. Lett B*, 620:17–26, 2005.

[359] R. Beck and M. Krause. Revised equipartition and minimum energy formula for magnetic field strength estimates from radio synchrotron observations. *Astron. Nachr.* 326:414–427, 2005.

[360] D. Ferenc (MAGIC Coll.) The MAGIC gamma-ray observatory. *Nucl. Instrum. Meth. A*, 553:274–281, 2005.

[361] F. Dyson. The death of a star. *Nature*, 438:1086, 2005.

[362] C.R. Canizares *et al.* The Chandra high-energy transmission grating: Design, fabrication, ground calibration, and 5 years in flight. *Pub. Astron. Soc. Pac.* 117:1144–1171, 2005.

[363] D.N. Burrows *et al.* The Swift X-ray telescope. *Space Sci. Rev.* 120:165–195, 2005.

[364] O. Ganel *et al.* Beam tests of the balloon-borne ATIC experiment. *Nucl. Instrum. Meth. A*, 552:409–419, 2005.

[365] V. A. Derbina *et al.* Cosmic-ray spectra and composition in the energy range of 10–1000 TeV per particle obtained by the RUNJOB experiment. *Astrophys. J.* 628:L41–L44, 2005.

[366] J. Madsen. Strangelet propagation and cosmic ray flux. *Phys. Rev. D*, 71:014026, 2005.

[367] J. Matthews. A Heitler model of extensive air showers. *Astropart. Phys.* 22:387–397, 2005.

[368] D.F. Smart and M.A. Shea. A review of geomagnetic cutoff rigidities for earth-orbiting spacecraft. *Adv. Space Res.* 36:2012–2020, 2005.

[369] L.W. Townsend. Critical analysis of active shielding methods for space radiation protection. In *2005 IEEE Aerospace Conference, Big Sky, MT, USA*, 724–730. 2005.

[370] G.A. de Nolfo *et al.* (Ace Coll.) Observations of the Li, Be and B isotopes and constraints on the cosmic-ray propagation. *Adv. Space Res.* 38:1558–1564, 2006.

[371] E. Blackman. Giants of physics found white-dwarf mass limits. *Nature*, 440:148, 2006.

[372] D. Boteler. The super storms of August/September 1859 and their effects on the telegraph system. *Adv. Space Res.* 38:159–172, 2006.

[373] P.R. Chowdhury and D.N. Basu. Nuclear matter properties with the re-evaluated coefficients of liquid drop model. *Acta Phys. Polon. B*, 37:1833–1846, 2006.

[374] F.A. Cucinotta and M. Durante. Cancer risk from exposure to galactic cosmic rays: implications for space exploration by human beings. *The Lancet Oncology*, 7:431–435, 2006.

[375] A.W. Graham *et al.* Empirical models for dark matter halos. III. The Kormendy relation and the $\log(\rho_e)$-$\log(R_e)$ relation. *Astron. J.* 132:2711–2716, 2006.

[376] D. Crowe *et al.* A direct empirical proof of the existence of dark matter. *Astrophys. J.* 648:L109–L113, 2006.

[377] J.L. Han *et al.* Pulsar rotation measures and the large-scale structure of the galactic magnetic field. *Astroph. J.* 642:868–881, 2006.

[378] P. Erdogdu *et al.* The dipole anisotropy of the 2 Micron All-Sky Redshift Survey. *Mon. Not. R. Astron. Soc.* 368:1515–1526, 2006.

[379] V. Aynutdinov *et al.* Search for a diffuse flux of high-energy extraterrestrial neutrinos with the NT200 neutrino telescope. *Astropart. Phys.* 25:140–150, 2006.

[380] B.K. Lubsandorzhiev. On the history of photomultiplier tube invention. *Nucl. Instrum. Meth. A*, 567:236–238, 2006.

[381] A. De Rújula. A cannonball model of cosmic rays. *Nucl. Phys. B*, 151:23–32, 2006.

[382] Y. Shikaze *et al.* (BESS Coll.) Measurements of 0.2 to 20 GeV/n cosmic-ray proton and helium spectra from 1997 through 2002 with the BESS spectrometer. *Astropart. Phys.* 28:154–167, 2007.

[383] A. Achterberg *et al.* Multi-year search for a diffuse flux of muon neutrinos with AMANDA-II. *Phys. Rev. D*, 76. [Erratum: Phys.Rev.D 77, 089904 (2008)]:042008, 2007.

[384] H.S. Ahn *et al.* The Cosmic Ray Energetics And Mass (CREAM) instrument. *Nucl. Instrum. Meth. A*, 579:1034–1053, 2007.

[385] F. Aharonian *et al.* (H.E.S.S. Coll.) First ground-based measurement of atmospheric Cherenkov light from cosmic rays. *Phys. Rev. D*, 75:042004, 2007.

[386] J. Kirkby. Cosmic rays and climate. *Surv. Geophys.* 28:333–375, 2007.

[387] Hamamatsu Photonics K.K. *Photomultiplier Tubes: Basics and Applications, 3rd edition*. 2007. URL: https://www.hamamatsu.com/resources/pdf/etd/PMT_handbook_v3aE.pdf.

[388]   J. Albert *et al.* (MAGIC Coll.) Unfolding of differential energy spectra in the MAGIC experiment. *Nucl. Instrum. Meth. A*, 583:494–506, 2007.

[389]   ICRP publication No. 103. *The 2007 Recommendations of the International Commission on Radiological Protection.* Ann. ICRP 37 (2–4). 2007.

[390]   D. Overbye. *Long-Awaited Cosmic-Ray Detector May Be Shelved.* The New York Times, April 3. 2007.

[391]   L. Bonechi *et al.* (PAMELA Coll.) Status of the PAMELA silicon tracker. *Nucl. Instrum. Meth. A*, 570:281–285, 2007.

[392]   T. Pulkkinen. Space weather: Terrestrial perspective. *Living Rev. Solar Phys.* 4:1, 2007.

[393]   A.W. Strong, I.V. Moskalenko, and V.S. Ptuskin. Cosmic-ray propagation and interactions in the galaxy. *Ann. Rev. Nucl. Part. Sci.* 57:285–327, 2007.

[394]   Columbia Accident Investigation Board (CAIB). *Columbia Crew Survival Investigation Report.* NASA/SP-2008-565. 2008. URL: https://www.nasa.gov/pdf/298870main_SP-2008-565.pdf.

[395]   E.P. Cherenkova. The discovery of the Cherenkov radiation. *Nucl. Instrum. Meth. A*, 595. Ed. by A. Bressan *et al.*:8–11, 2008.

[396]   F.A. Cucinotta. Physical and biological dosimetry analysis from International Space Station astronauts. *Radiat. Res.* 170:127–138, 2008.

[397]   A. Dar and A. De Rújula. A theory of cosmic rays. *Phys. Pep.* 466:179–241, 2008.

[398]   C. Evoli *et al.* Cosmic ray nuclei, antiprotons and gamma rays in the galaxy: a new diffusion model. *J. Cosm. Astropart. Phys.* 2008:018, 2008.

[399]   K. Abe *et al.* Measurement of cosmic-ray low-energy antiproton spectrum with the first BESS-Polar Antarctic flight. *Phys. Lett. B*, 670:103–108, 2008.

[400]   M. Ave *et al.* Composition of primary cosmic-ray nuclei at high energies. *Astrophys. J.* 678:262–273, 2008.

[401]   M. Unger *et al.* Reconstruction of longitudinal profiles of ultra-high energy cosmic ray showers from fluorescence and Cherenkov light measurements. *Nucl. Instrum. Meth. A*, 588:433–441, 2008.

[402]   F. Aharonian *et al.* (H.E.S.S. Coll.) The energy spectrum of cosmic-ray electrons at TeV energies. *Phys. Rev. Lett.* 101:261104, 2008.

[403]   J.J. Hester. The Crab nebula: An astrophysical chimera. *Ann. Rev. Astron. Astrophys.* 46:127–155, 2008.

[404]   M. Kachelriess. Lecture notes on high energy cosmic rays. In *17th Jyvaskyla Summer School.* 2008.

[405]   Y.N. Kharzeev. Use of silica aerogels in Cherenkov counters. *Phys. Part. Nucl.* 39:107–135, 2008.

[406]   D. Overbye. *Fighting to Launch Cosmic-Ray Detector.* The New York Times, February 29. 2008.

[407]   R. Panek. *The Father of Dark Matter Still Gets No Respect.* Discover Magazine, Dec. 31. 2008. URL: https://www.discovermagazine.com/the-sciences/the-father-of-dark-matter-still-gets-no-respect.

[408]   J. Abraham *et al.* (Pierre Auger Coll.) Observation of the suppression of the flux of cosmic rays above 4×10^{19} eV. *Phys. Rev. Lett*, 101:061101, 2008.

[409] R.U. Abbasi *et al.* (HiRes Coll.) First observation of the Greisen-Zatsepin-Kuzmin suppression. *Phys. Rev. Lett.* 100:101101, 2008.

[410] M. Shaposhnikov. Sterile neutrinos in cosmology and how to find them in the lab. *J. Phys.: Conf. Series*, 136:022045, 2008.

[411] H. Kawai *et al.* (TA Coll.) Telescope Array experiment. *Nucl. Phys. B Proc. Suppl.* 175–176:221–226, 2008.

[412] S. Weinberg. *Cosmology.* Oxford Univ. Press, 2008.

[413] J.S. George *et al.* (ACE Coll.) Elemental composition and energy spectra of galactic cosmic rays during solar cycle 23. *Astroph. J.* 698:1666–1681, 2009.

[414] A.D. Panov *et al.* Energy spectra of abundant nuclei of primary cosmic rays from the data of ATIC-2 experiment: Final results. *Bull. Russ. Acad. Sci. Phys.* 73:554–567, 2009.

[415] B.F. Rauch *et al.* Cosmic ray origin in OB associations and preferential acceleration of refractory elements: evidence from abundances of elements $_{26}$Fe through $_{34}$Se. *Astrophys. J.* 697:2083–2088, 2009.

[416] C. Labanti *et al.* Design and construction of the mini-calorimeter of the AGILE satellite. *Nucl. Instrum. Meth.* A, 598:470–479, 2009.

[417] D.J. McComas *et al.* Global observations of the interstellar interaction from the Interstellar Boundary Explorer (IBEX). *Science*, 326:959–962, 2009.

[418] J.N. Arkani-Hamed *et al.* A theory of dark matter. *Phys. Rev.* D 79:015014, 2009.

[419] K. Han *et al.* Search for stable strange quark matter in lunar soil. *Phys. Rev. Lett.* 103:092302, 2009.

[420] M. Tavani *et al.* The AGILE mission. *Astron. Astroph.* 502:995–1013, 2009.

[421] O. Adriani *et al.* A new measurement of the antiproton-to-proton flux ratio up to 100 GeV in the cosmic radiation. *Phys. Rev. Lett.* 102:051101, 2009.

[422] W. B. Atwood *et al.* The Large Area Telescope on the Fermi gamma-ray space telescope mission. *Astrophys. J.* 697:1071–1102, 2009.

[423] R. Abbasi *et al.* The IceCube data acquisition system: Signal capture, digitization, and timestamping. *Nucl. Instrum. Meth.* A, 601:294–316, 2009.

[424] F. Aharonian *et al.* (H.E.S.S. Coll.) Probing the ATIC peak in the cosmic-ray electron spectrum with H.E.S.S. *Astron. Astrophys.* 508:561, 2009.

[425] J.C. McPhee and J.B. Charles (Edts.) *Human Health and Performance Risks of Space Exploration Missions.* NASA SP-2009-3405, 2009.

[426] V.T. Sarychev. Electromagnetic field of a rotating magnetic dipole and electric-charge motion in this field. *Radiophys. and Quantum Electr.* 52:900–907, 2009.

[427] D.F. Smart and M.A. Shea. Fifty years of progress in geomagnetic cutoff rigidity determinations. *Adv. Space Res.* 44:1107–1123, 2009.

[428] P. Spillantini. CR from space based observatories: History, results and perspectives of the PAMELA mission. In *9th Baikal Summer School on Physics of Elementary Particles and Astrophysics, Bol'shie Koty, Russia*, 213–234. 2009.

[429] H. S. Ahn *et al.* (CREAM Coll.) Discrepant hardening observed in cosmic-ray elemental spectra. *Astrophys. J.* 714:L89–L93, 2010.

[430] C.D. Dermer and S. Razzaque. Acceleration of ultra-high-energy cosmic rays in the colliding shells of blazars and gamma-ray bursts: Constraints from the FERMI gamma-ray space telescope. *Astrophys. J.* 724:1366–1372, 2010.

[431] A. Bulgarelli *et al.* The AGILE silicon tracker: Pre-launch and in-flight configuration. *Nucl. Instrum. Meth. A*, 614:213–226, 2010.

[432] G. Dobler *et al.* The Fermi haze: a gamma-ray counterpart to the microwave haze. *Astrophys. J.* 717:825–842, 2010.

[433] J. Navarro *et al.* The diversity and similarity of simulated cold dark matter halos. *Mon. Not. R. Astron. Soc.* 402:21–34, 2010.

[434] T. Delahaye *et al.* Galactic electrons and positrons at the Earth. *Astron. and Astrophys.* 524:A51, 2010.

[435] V. Aynutdinov *et al.* Baikal neutrino project: History and prospects. *Russ. Phys. J.* 53:601–610, 2010.

[436] W.D Apel *et al.* The KASCADE-Grande experiment. *Nucl. Instrum. Meth. A*, 620:202–216, 2010.

[437] M. Ackermann *et al.* (Fermi LAT Coll.) Searches for cosmic-ray electron anisotropies with the Fermi Large Area Telescope. *Phys. Rev. D*, 82:092003, 2010.

[438] G. Gaudio, M. Livan, and R. Wigmans. The art of calorimetry. In *Radiation and Particle Detectors*. Ed. by S. Bertolucci, U. Bottigli, and P. Oliva, 31–77. IOS Press, 2010.

[439] N. Agafonova *et al.* (OPERA Coll.) Observation of a first ν_τ candidate in the OPERA experiment in the CNGS beam. *Phys. Lett. B*, 691:138–145, 2010.

[440] D. Overbye. *A Costly Quest for the Dark Heart of the Cosmos*. The New York Times, November 16. 2010.

[441] J. Abraham *et al.* (Pierre Auger Coll.) The fluorescence detector of the Pierre Auger Observatory. *Nucl. Instrum. Meth. A*, 620:227–251, 2010.

[442] I.G. Richardson and H.V. Cane. Near-earth interplanetary coronal mass ejections during solar cycle 23 (1996 – 2009): Catalog and summary of properties. *Solar Phys.* 264:189–237, 2010.

[443] M. Shaposhnikov and C. Wetterich. Asymptotic safety of gravity and the Higgs boson mass. *Phys. Lett. B*, 683:196–200, 2010.

[444] H. Soltner and P. Blümler. Dipolar Halbach magnet stacks made from identically shaped permanent magnets for magnetic resonance. *Concepts in Magnetic Resonance Part A*, 36A:211–222, 2010.

[445] M. Su, T.R. Slatyer, and D.P. Finkbeiner. Giant gamma-ray bubbles from Fermi-LAT: AGN activity or bipolar galactic wind? *Astrophys. J.* 724:1044–1082, 2010.

[446] K. Blum. Cosmic ray propagation time scales: lessons from radioactive nuclei and positron data. *J. Cosm. Astropart. Phys.* 2011:037–037, 2011.

[447] T.H.-J. Mathes (Pierre Auger Coll.) The HEAT telescopes of the Pierre Auger Observatory: Status and first data. In *32nd International Cosmic Ray Conference, Beijing, China*. Vol. 3, 153–156. 2011.

[448] B.R. Dawson. *Comment on Japanese Detection of Air Fluorescence Light from a Cosmic Ray Shower in 1969*. arXiv:1112.5686v1 [physics.hist-ph]. 2011.

[449] R. Durrer. What do we really know about dark energy? *Phil. Trans. Roy. Soc. Lond. A*:5102–5114, 2011.

[450] A. A. Abdo *et al.* FERMI Large Area Telecope observations of Markarian 421: The missing piece of its spectral energy distribution. *Astrophys. J.* 736:131, 2011.

[451] A. Bhattacharya *et al.* The Glashow resonance at IceCube: Signatures, event rates and *pp* vs. *pγ* interactions. *JCAP*, 10:017, 2011.

[452] A. Obermeier *et al.* Energy spectra of primary and secondary cosmic-ray nuclei measured with TRACER. *Astrophys. J.* 742:14, 2011.

[453] E.G. Adelberger *et al.* Solar fusion cross sections II: The pp chain and the CNO cycles. *Rev. Mod. Phys.* 83:195–245, 2011.

[454] K. Lübelsmeyer *et al.* Upgrade of the Alpha Magnetic Spectrometer (AMS-02) for long term operation on the International Space Station (ISS). *Nucl. Instrum. Meth. A*, 654:639–648, 2011.

[455] M. Cirelli *et al.* PPPC 4 DM ID: A poor particle physicist cookbook for dark matter indirect detection. *JCAP*, 03. [Erratum: JCAP 10, E01 (2012)]:051, 2011.

[456] O. Adriani *et al.* Cosmic-ray electron flux measured by the PAMELA experiment between 1 and 625 GeV. *Phys. Rev. Lett.* 106:201101, 2011.

[457] O. Adriani *et al.* PAMELA measurements of cosmic-ray proton and helium spectra. *Science*, 332:69–72, 2011.

[458] R. Trotta *et al.* Constraints on cosmic-ray propagation models from a global Bayesian analysis. *Astrophys. J.* 729:106, 2011.

[459] C. Farnier, R. Walter, and J.-C. Leyder. Eta Carinae: a very large hadron collider. *Astron. Astrophys.* 526:A57, 2011.

[460] R.A. Fricke. *Günther & Tegetmeyer 1901–1958: Instrumente für die Wissenschaft aus Braunschweig*. Selbstverlag, 2011.

[461] J. Knuesel. The photographic emulsion technology of the OPERA experiment on its way to find the $\nu_\mu \to \nu_\tau$ oscillation. *Nucl. Phys. B (Proc. Suppl.)* 215:66–68, 2011.

[462] M.S. Longair. *High Energy Astrophysics (3rd Edition)*. Cambridge Univ. Press, 2011.

[463] S. Ostapchenko. Monte Carlo treatment of hadronic interactions in enhanced pomeron scheme: I. QGSJET-II model. *Phys. Rev. D*, 83:014018, 2011.

[464] I. G. Richardson and H. V. Cane. Galactic cosmic ray intensity response to interplanetary coronal mass ejections/magnetic clouds in 1995–2009. *Solar Phys.* 270:609–627, 2011.

[465] V. Trimble. *Kenneth I. Greisen: A Biographical Memoir*. National Academy of Sciences. 2011. URL: http://www.nasonline.org/publications/biographical-memoirs/memoir-pdfs/greisen-kenneth.pdf.

[466] N. Akchurin and R. Wigmans. Hadron calorimetry. *Nucl. Instrum. Meth. A*, 666:80–97, 2012.

[467] R. Catinaud. Which physics for a new institute? Albert Gockel, Joseph Kowalski and the early years of the Fribourg Institute of Physics. *SPS Communications*, 36:24–27, 2012.

[468] B. Coste et al. Constraining galactic cosmic-ray parameters with $Z \le 2$ nuclei. *Astron. Astrophys.* 539:A88, 2012.

[469] Ch. Spiering (Edt.) Cosmic rays, gamma rays and neutrinos: a survey of 100 years of research. *Eur. Phys. J. H*, 37:320–565, 2012.

[470] A.E. Vladimirov *et al.* Testing the origin of high-energy cosmic rays. *Astrophys. J.* 752:68, 2012.

[471]   I. Kravchenko *et al.* Updated results from the RICE experiment and future prospects for ultra-high energy neutrino detection at the South Pole. *Phys. Rev. D*, 85:062004, 2012.

[472]   K. Abe *et al.* Measurement of the cosmic-ray antiproton spectrum at solar minimum with a long-duration balloon flight over Antarctica. *Phys. Rev. Lett.* 108:051102, 2012.

[473]   K. Abe *et al.* Search for antihelium with the BESS-Polar spectrometer. *Phys. Rev. Lett.* 108:131301, 2012.

[474]   K. Mitsuda *et al.* The high-resolution X-ray microcalorimeter spectrometer, SXS, on Astro-H. *J. Low Temp. Phys.* 167:795–802, 2012.

[475]   M. Dan *et al.* How the merger of two white dwarfs depends on their mass ratio: Orbital stability and detonations at contact. *Mon. Not. R. Astron. Soc.* 422:2417–2428, 2012.

[476]   M. Ackermann *et al.* (Fermi LAT Coll.) Measurement of separate cosmic-ray electron and positron spectra with the Fermi Large Area Telescope. *Phys. Rev. Lett.* 108:011103, 2012.

[477]   R.G.A. Fricke and K. Schlegel. 100th anniversary of the discovery of cosmic radiation: the role of Günther and Tegetmeyer in the development of the necessary instrumentation. *Hist. Geo Space Sci.* 3:151–158, 2012.

[478]   T.K. Gaisser. Spectrum of cosmic-ray nucleons, kaon production, and the atmospheric muon charge ratio. *Astropart. Phys.* 35:801–806, 2012.

[479]   A. Garg. *Classical Electromagnetism in a Nutshell.* Princeton Univ. Press, 2012.

[480]   D. Hanna. Early muon-physics measurements with cosmic rays. *Physics in Canada*, 68:7, 2012.

[481]   K.-H. Kampert and A.A. Watson. Extensive air showers and ultra high-energy cosmic rays: a historical review. *Eur. Phys. J. H*, 37:359–412, 2012.

[482]   J. Lacki. Albert Gockel: from atmospheric electricity to cosmic radiation. *SPS Communications*, 38:25–29, 2012.

[483]   B.S. Meyer. Webnucleo.org. *PoS (NIC XII)*:096, 2012.

[484]   D. Müller. Direct observation of galactic cosmic rays. *Eur. Phys. J. H*, 37:413–458, 2012.

[485]   G. Steigman, B. Dasgupta, and J.F. Beacom. Precise relic WIMP abundance and its impact on searches for dark matter annihilation. *Phys. Rev. D*, 86:023506, 2012.

[486]   M. Su and D.P. Finkbeiner. Evidence for gamma-ray jets in the Milky Way. *Astrophys. J.* 753:61, 2012.

[487]   H. Tokuno *et al.* (TA Coll.) New air fluorescence detectors employed in the Telescope Array experiment. *Nucl. Instrum. Meth. A*, 676:54–65, 2012.

[488]   T. Abu-Zayyad *et al.* (TA Coll.) The surface detector array of the Telescope Array experiment. *Nucl. Instrum. Meth. A*, 689:87–97, 2012.

[489]   N. Tomassetti. Entering the cosmic ray precision era. *Astrophys. J. Lett.* 752:L13, 2012.

[490]   N. Tomassetti. Propagation of H and He cosmic ray isotopes in the galaxy: Astrophysical and nuclear uncertainties. *Astrophys. Space Sci.* 342:131–136, 2012.

[491]   R. Wigmans. "Calorimeters". In *Handbook of Particle Detection and Imaging.* Ed. by C. Grupen and I. Buvat. Springer, 2012.

[492] I. Cholis and D. Hooper. Dark matter and pulsar origins of the rising cosmic ray positron fraction in light of new data from AMS. *Phys. Rev. D*, 88:023013, 2013.

[493] C. Adloff *et al.* The AMS-02 lead-scintillating fibres electromagnetic calorimeter. *Nucl. Instrum. Meth. A*, 714:147–154, 2013.

[494] K.-A. Lave *et al.* Galactic cosmic-ray energy spectra and composition during the 2009–2010 solar minimum period. *Astrophys. J.* 770:117, 2013.

[495] L. Bergstrom *et al.* New limits on dark matter annihilation from AMS cosmic ray positron data. *Phys. Rev. Lett.* 111:171101, 2013.

[496] R. Abbasi *et al.* An improved method for measuring muon energy using the truncated mean of dE/dx. *Nucl. Instrum. Meth. A*, 703:190–198, 2013.

[497] R. Abbasi *et al.* IceTop: The surface component of IceCube. *Nucl, Instrum. Meth. A*, 700:188–220, 2013.

[498] J. van Es *et al. AMS02 Tracker Thermal Control Cooling System commissioning and operational results.* 2013. URL: https://core.ac.uk/download/pdf/53034866.pdf.

[499] J.R. Formaggio and G.P. Zeller. From eV to EeV: Neutrino cross sections across energy scales. *Rev. Mod. Phys.* 84:1307, 2013.

[500] M.G. Aartsen *et al.* (IceCube Coll.) Evidence for high-energy extraterrestrial neutrinos at the IceCube detector. *Science*, 342:1242856, 2013.

[501] M.G. Aartsen *et al.* (IceCube Coll.) Measurement of the cosmic ray energy spectrum with IceTop-73. *Phys. Rev. D*, 88:042004, 2013.

[502] M.G. Aartsen *et al.* (IceCube Coll.) Observation of cosmic ray anisotropy with the IceTop air shower array. *Astrophys. J.* 765:55, 2013.

[503] J.F. Krizmanic, J.W. Mitchell, and R.E. Streitmatter (OWL Coll.) Optimization of the Orbiting Wide-angle Light collectors (OWL) mission for charged-particle and neutrino astronomy. In *33rd International Cosmic Ray Conference, Rio de Janeiro, Brazil*, 1085. 2013.

[504] M. Walter. Early cosmic ray research with balloons. *Nucl. Phy. B (Proc. Suppl.)* 239–240:11–18, 2013.

[505] P. Massey and M.M. Hanson. "Astronomical spectroscopy". In *Planets, Stars and Stellar Systems*. Ed. by T.D. Oswalt and H.E. Bond. Springer, 2013, pp. 35–98.

[506] S. Ostapchenko. QGSJET-II: Physics, recent improvements, and results for air showers. *EPJ Web Conf.* 52:02001, 2013.

[507] M.J. Owens and R.J. Forsyth. The heliospheric magnetic field. *Living Rev. Solar Phys.* 10:5, 2013.

[508] O. Adriani *et al.* (PAMELA Coll.) Cosmic-ray positron energy spectrum measured by PAMELA. *Phys. Rev. Lett.* 111:081102, 2013.

[509] O. Adriani *et al.* (PAMELA Coll.) Measurement of the flux of primary cosmic ray antiprotons with energies of 60 MeV to 350 GeV in the PAMELA experiment. *JETP Lett.* 96:621–627, 2013.

[510] O. Adriani *et al.* (PAMELA Coll.) Time dependence of the proton flux measured by PAMELA during the July 2006 – December 2009 solar minimum. *Astrophys. J.* 765:91–98, 2013.

[511] M. S. Potgieter. Solar modulation of cosmic rays. *Living Rev. Solar Phys.* 10:3, 2013.

[512] R. Lewin Sime. Marietta Blau: Pioneer of photographic nuclear emulsions and particle physics. *Phys. Persp.* 5:3–32, 2013.

[513]   T. Abu-Zayyad *et al.* (TA Coll.) The cosmic-ray energy spectrum observed with the surface detector of the Telescope Array experiment. *Astrophys. J. Lett.* 786:L1, 2013.

[514]   C. Winteler. *Light Element Production in the Big Bang and the Synthesis of Heavy Elements in 3D MHD Jets from Core-Collapse Supernovae.* Univ. Basel PhD Thesis. 2013. URL: https://edoc.unibas.ch/29895/1/thesis_winteler.pdf.

[515]   M. Aguilar *et al.* (AMS Coll.) Electron and positron fluxes in primary cosmic rays measured with the Alpha Magnetic Spectrometer on the International Space Station. *Phys. Rev. Lett.* 113:121102, 2014.

[516]   P. Billoir (Pierre Auger Coll.) The Cherenkov surface detector of the Pierre Auger Observatory. *Nucl. Instrum. Meth. A,* 766:78–82, 2014.

[517]   J. Conrad. Indirect detection of WIMP dark matter: A compact review. In *Interplay between Particle and Astroparticle Physics (IPA 2014), London, UK.* 2014. URL: https://arxiv.org/pdf/1411.1925.pdf.

[518]   D. Atri and A.L. Melott. Cosmic rays and terrestrial life: A brief review. *Astropart. Phys.* 53:186–190, 2014.

[519]   K.A. Olive *et al.* Review of particle physics. *Chin. Phys. C,* 38:090001, 2014.

[520]   L. Feng *et al.* AMS-02 positron excess: new bounds on dark matter models and hint for primary electron spectrum hardening. *Phys. Lett. B,* 728:250–255, 2014.

[521]   M.G. Aartsen *et al.* Improvement in fast particle track reconstruction with robust statistics. *Nucl. Instrum. Meth. A,* 736:143–149, 2014.

[522]   S. Adrián-Martinez *et al.* Searches for point-like and extended neutrino sources close to the galactic centre using the ANTARES neutrino telescope. *Astrophys. J. Lett.* 786:L5, 2014.

[523]   S. N. Zhang *et al.* The high energy cosmic-radiation detection (HERD) facility onboard China's space station. In *Space Telescopes and Instrumentation 2014: Ultraviolet to Gamma Ray.* Ed. by T.Takahashi, J.-W.A. den Herder, and M. Bautz. Vol. 9144. International Society for Optics and Photonics, 293 –301. SPIE, 2014.

[524]   A. Ibarra, A.S. Lamperstorfer, and J. Silk. Dark matter annihilations and decays after the AMS-02 positron measurements. *Phys. Rev. D,* 89:063539, 2014.

[525]   J. Bolmont *et al.* The camera of the fifth H.E.S.S. telescope. Part I: System description. *Nucl. Instrum. Meth. A,* 761:46–57, 2014.

[526]   R. Kappl and M.W. Winkler. The cosmic ray antiproton background for AMS-02. *J. Cosm. Astropart. Phys.* 2014:051, 2014.

[527]   R. Kissmann. PICARD: A novel code for the galactic cosmic ray propagation problem. *Astropart. Phys.* 55:37–50, 2014.

[528]   P. Lipari. Cosmic rays and hadronic interactions. *Comptes Rendus Physique,* 15:357–366, 2014.

[529]   M. Ackermann *et al.* (Fermi-LAT Coll.) The spectrum and morphology of the Fermi bubbles. *Astrophys. J.* 793:64, 2014.

[530]   M. Longair. C.T.R. Wilson and the cloud chamber. *Astropart. Phys.* 53:55–60, 2014.

[531]   M. Walter, Ch. Spiering and J. Knapp (Edts.) 100 years of cosmic rays: The anniversary of their discovery by V.F. Hess. *Astropart. Phys.* 53:1–190, 2014.

[532]   D. Maurin, F. Melot, and R. Taillet. A database of charged cosmic rays. *Astron. Astrophys.* 569:A32, 2014.

[533] J.M.C. Montanus. An extended Heitler-Matthews model for the full hadronic cascade in cosmic air showers. *Astropart. Phys.* 59:4–11, 2014.

[534] G. Neuneck. Physiker im Ersten Weltkrieg. *Wissenschaft & Frieden*, 3:41–45, 2014.

[535] O. Adriani *et al.* (PAMELA Coll.) Measurement of boron and carbon fluxes in cosmic rays with the PAMELA experiment. *Astrophys. J.* 791:93, 2014.

[536] P.A.R. Ade *et al.* (Planck Coll.) Planck 2013 results. XVI. Cosmological parameters. *Astron. and Astrophys.* 571:A16, 2014.

[537] M. Pohl. Particle detection technology for space-borne astroparticle experiments. In *Technology and Instrumentation in Particle Physics*. Proceedings of Science, 2014. URL: https://pos.sissa.it/213/013/pdf.

[538] J. Rosado, F. Blanco, and F. Arqueros. On the absolute value of the air-fluorescence yield. *Astropart. Phys.* 55:51–62, 2014.

[539] V.V. Prosin *et al.* (Tunka-133 Coll.) Tunka-133: Results from 3 year operation. *Nucl. Instrum. Meth. A*, 756:94–101, 2014.

[540] A. Aab *et al.* (Auger Coll.) The Pierre Auger cosmic ray observatory. *Nucl. Instrum. Meth. A*, 798:173–213, 2015.

[541] R. Aloisio, P. Blasi, and P.D. Serpico. Nonlinear cosmic ray galactic transport in the light of AMS-02 and Voyager data. *Astron. Astrophys.* 583:A95, 2015.

[542] M. Aguilar *et al.* (AMS Coll.) Precision Measurement of the helium flux in primary cosmic rays of rigidities 1.9 GV to 3 TV with the Alpha Magnetic Spectrometer on the International Space Station. *Phys. Rev. Lett.* 115:211101, 2015.

[543] A. Aab *et al.* (Auger Coll.) Large scale distribution of ultra high energy cosmic rays detected at the Pierre Auger Observatory with zenith angles up to 80°. *Astrophys. J.* 802:111, 2015.

[544] E. Aver, K.A. Olive, and E.D. Skillman. The effects of He I λ10830 on helium abundance determinations. *J. Cosm. Astropart. Phys.* 2015:011–011, 2015.

[545] A.S. Brown. Baade and Zwicky: "Supernovae," neutron stars, and cosmic rays. *Proc. Natl. Acad. Sci. USA*, 112:1241–1242, 2015.

[546] D. Caprioli. Cosmic-ray acceleration and propagation. *PoS*, ICRC2015:008, 2015.

[547] J.C. David. Spallation reactions: A successful interplay between modelling and applications. *Eur. Phys. J. A* 51:68, 2015.

[548] A. Coc *et al.* New reaction rates for improved primordial D/H calculation and the cosmic evolution of deuterium. *Phys. Rev. D*, 92:123526, 2015.

[549] D. Besson *et al.* First cosmogenic neutrino limits from the ARA testbed station at South Pole. *PoS*, ICRC2015:1105, 2015.

[550] E. Atkin *et al.* The NUCLEON space experiment for direct high energy cosmic rays investigation in TeV–PeV energy range. *Nucl. Instrum. Meth. A*, 770:189–196, 2015.

[551] J. Buch *et al.* PPPC 4 DM secondary: A poor particle physicist cookbook for secondary radiation from dark matter. *JCAP*, 09:037, 2015.

[552] K.P. Arunbabu *et al.* How are Forbush decreases related to interplanetary magnetic field enhancements? *Astron. Astrophys.* 580:A41, 2015.

[553] T. Pierog *et al.* EPOS LHC: Test of collective hadronization with data measured at the CERN Large Hadron Collider. *Phys. Rev. C*, 92:034906, 2015.

[554] D.H. Hathaway. The solar cycle. *Living Rev. Solar Phys.* 12:4, 2015.

[555]   K. Rawlins *et al.* (IceCube Coll.) Latest results on cosmic ray spectrum and composition from three years of IceTop and IceCube. *PoS*, ICRC2015:334, 2015.

[556]   J.H. Adams *et al.* (JEM-EUSO Coll.) The JEM-EUSO instrument. *Exp. Astron.* 40:19–44, 2015.

[557]   J.H. Adams *et al.* (JEM-EUSO Coll.) The JEM-EUSO mission: An introduction. *Exp. Astron.* 40:3–17, 2015.

[558]   S. Schoo *et al.* (Kaskade Grande Coll.) The energy spectrum of cosmic rays in the range from 10^{14} to 10^{18}eV. *PoS*, ICRC2015:263, 2015.

[559]   S.-J. Lin, Q. Yuan, and X.-J. Bi. Quantitative study of the AMS-02 electron/positron spectra: Implications for pulsars and dark matter properties. *Phys. Rev. D*, 91:063508, 2015.

[560]   M. Aguilar *et al.* (AMS Coll.) Precision measurement of the proton flux in primary cosmic rays from rigidity 1 GV to 1.8 TV with the Alpha Magnetic Spectrometer on the International Space Station. *Phys. Rev. Lett.* 114:171103, 2015.

[561]   N. Agafonova *et al.* (OPERA Coll.) Discovery of τ. *Phys. Rev. Lett.* 115:121802, 2015.

[562]   O. Adriani *et al.* (PAMELA Coll.) New upper limit on strange quark matter abundance in cosmic rays with the PAMELA space experiment. *Phys. Rev. Lett.* 115:111101, 2015.

[563]   I. Valiño *et al.* (Pierre Auger Coll.) The flux of ultra-high energy cosmic rays after ten years of operation of the Pierre Auger Observatory. *PoS*, ICRC2015:271, 2015.

[564]   P.D. Serpico. Possible physics scenarios behind cosmic-ray "anomalies". *PoS*, ICRC2015:009, 2015.

[565]   M. Spurio. *Particles and Astrophysics: A Multi-Messenger Approach.* Springer, 2015.

[566]   D. Ivanov *et al.* (TA Coll.) TA spectrum summary. *PoS*, ICRC2015:349, 2015.

[567]   N. Tomassetti. Cosmic-ray protons, nuclei, electrons, and antiparticles under a two-halo scenario of diffusive propagation. *Phys. Rev. D*, 92:081301, 2015.

[568]   M. Unger, G.R. Farrar, and L.A. Anchordoqui. Origin of the ankle in the ultrahigh energy cosmic ray spectrum, and of the extragalactic protons below it. *Phys. Rev. D*, 92:123001, 2015.

[569]   M. Aguilar *et al.* (AMS Coll.) Antiproton flux, antiproton-to-proton flux ratio, and properties of elementary particle fluxes in primary cosmic rays measured with the Alpha Magnetic Spectrometer on the International Space Station. *Phys. Rev. Lett.* 117:091103, 2016.

[570]   Y.-H. Chen, K. Cheung, and P.Y. Tseng. Dark matter with multiannihilation channels and the AMS-02 positron excess and antiproton data. *Phys. Rev. D*, 93:015015, 2016.

[571]   Z.Y. Zhang *et al.* (DAMPE Coll.) The calibration and electron energy reconstruction of the BGO ECAL of the DAMPE detector. *Nucl. Instrum. Meth. A*, 836:98–104, 2016.

[572]   M. D'Angelo, P. Blasi, and E. Amato. Grammage of cosmic rays around galactic supernova remnants. *Phys. Rev. D*, 94:083003, 2016.

[573]   E.M. Dunne *et al.* Global atmospheric particle formation from CERN CLOUD measurements. *Science*, 354:119–124, 2016.

[574]   A. Aab *et al.* Prototype muon detectors for the AMIGA component of the Pierre Auger Observatory. *J. Inst.* 11:P02012–P02012, 2016.

[575]   A.C. Cummings *et al.* Galactic cosmic rays in the local interstellar medium: Voyager-1 observations and model results. *Astrophys. J.* 831:18, 2016.

[576]   C. Corti *et al.* Solar modulation of the local interstellar spectrum with Voyager 1, AMS-02, PAMELA and BESS. *Astrophys. J.* 829:8, 2016.

[577]   G. Jóhannesson *et al.* Bayesian analysis of cosmic-ray propagation: evidence against homogeneous diffusion. *Astrophys. J.* 824:16, 2016.

[578]   J. Aleksić *et al.* The major upgrade of the MAGIC telescopes, Part I: The hardware improvements and the commissioning of the system. *Astropart. Phys.* 72:61–75, 2016.

[579]   J. Aleksić *et al.* The major upgrade of the MAGIC telescopes, Part II: A performance study using observations of the Crab Nebula. *Astropart. Phys.* 72:76–94, 2016.

[580]   K. Abe *et al.* Measurements of cosmic-ray proton and helium spectra from the BESS-Polar long-duration balloon flights over Antarctica. *Astrophys. J.* 822:65, 2016.

[581]   M. Di Mauro *et al.* Dark matter vs. astrophysics in the interpretation of AMS-02 electron and positron data. *JCAP*, 05:031, 2016.

[582]   P. Kuhl *et al.* Annual cosmic ray spectra from 250 MeV up to 1.6 GeV from 1995–2014 measured with the electron proton helium instrument onboard SOHO. *Solar Physics*, 291:965–974, 2016.

[583]   P. Van Dokkum *et al.* A high stellar velocity dispersion and ~ 100 globular clusters for the ultra-diffuse galaxy Dragonfly 44. *Astrophys. J. Lett.* 828:L6, 2016.

[584]   R. Alves Batista *et al.* CRPropa 3—a public astrophysical simulation framework for propagating extraterrestrial ultra-high energy particles. *J. Cosm. Astropart. Phys.* 2016:038–038, 2016.

[585]   R.P. Murphy *et al.* Galactic cosmic ray origin and OB associations: Evidence from SuperTIGER observation of elements $_{26}$Fe through $_{40}$Zr. *Astrophys. J.* 831:148, 2016.

[586]   S. Adrián-Martinez *et al.* Letter of intent for KM3NeT 2.0. *J. Phys. G*, 43:084001, 2016.

[587]   S. Thoudam *et al.* Cosmic-ray energy spectrum and composition up to the ankle: The case for a second galactic component. *Astron. and Astrophys.* 595:A33, 2016.

[588]   F. Aharonian *et al.* The quiescent intracluster medium in the core of the Perseus cluster. *Nature*, 535:117–121, 2016.

[589]   M.G. Aartsen *et al.* (IceCube Coll.) Anisotropy in cosmic-ray arrival directions in the southern hemisphere based on six years of data from the IceCube detector. *Astrophys. J.* 826:220, 2016.

[590]   J. Feng and H.-H. Zhang. Pulsar interpretation of lepton spectra measured by AMS-02. *Eur. Phys. J. C* 76:229, 2016.

[591]   J. Nielsen, A. Guffanti and S. Sarkar. Marginal evidence for cosmic acceleration from type Ia supernovae. *Sci. Rep.* 6:35596, 2016.

[592]   M.I. Panasyuk *et al.* (JEM-EUSO Coll.) Ultra high energy cosmic ray detector KLYPVE on board the Russian segment of the ISS. *PoS*, ICRC2015:669, 2016.

[593]   O. Adriani *et al.* (PAMELA Coll.) Time dependence of the electron and positron components of the cosmic radiation measured by the PAMELA experiment between July 2006 and December 2015. *Phys. Rev. Lett.* 116:241105, 2016.

[594] S. Ansoldi *et al.* (MAGIC Coll.) Teraelectronvolt pulsed emission from the Crab pulsar detected by MAGIC. *Astron. Astrophys.* 585:A133, 2016.

[595] J. Huang *et al.* (ASγ Coll.) Measurement of high energy cosmic rays by the new Tibet hybrid experiment. In *35th International Cosmic Ray Conference, Busan, Korea*, 484. 2017.

[596] J. Berdugo et al. Determination of the rigidity scale of the Alpha Magnetic Spectrometer. *Nucl. Instrum. Meth. A*, 869:10–14, 2017.

[597] C. Bustard, E.G. Zweibel, and C. Cotter. Cosmic ray acceleration by a versatile family of galactic wind termination shocks. *Astrophys. J.* 835:72, 2017.

[598] Y. Asaoka *et al.* (CALET Coll.) Energy calibration of CALET onboard the International Space Station. *Astroparticle Physics*, 91:1–10, 2017.

[599] B.W. Carroll and D.A. Ostie. *An Introduction to Modern Astrophysics (2nd edition)*. Cambridge Univ. Press, 2017.

[600] A. Cov and E. Vangioni. Primordial nucleosynthesis. *Int. J. Mod. Phys.* 26:1741002, 2017.

[601] G. Ambrosi *et al.* (DAMPE Coll.) Direct detection of a break in the teraelectronvolt cosmic-ray spectrum of electrons and positrons. *Nature*, 552:63–66, 2017.

[602] J. Chang *et al.* (DAMPE Coll.) The DArk Matter Particle Explorer mission. *Astrop. Phys.* 95:6–24, 2017.

[603] E. Pian *et al.* Origin of the heavy elements in binary neutron-star mergers from a gravitational-wave event. *Nature*, 551:80–84, 2017.

[604] E. Pian *et al.* Spectroscopic identification of *r*-process nucleosynthesis in a double neutron-star merger. *Nature*, 551:67–70, 2017.

[605] E. Troja *et al.* The X-ray counterpart to the gravitational-wave event GW170817. *Nature*, 551:71–74, 2017.

[606] A. Neronov *et al.* Sensitivity of a proposed space-based Cherenkov astrophysical-neutrino telescope. *Phys. Rev. D*, 95:023004, 2017.

[607] D. Hooper *et al.* HAWC observations strongly favor pulsar interpretations of the cosmic-ray positron excess. *Phys. Rev. D*, 96:103013, 2017.

[608] G. Cinelli *et al.* European annual cosmic-ray dose rate: estimation of population exposure. *Journal of maps*, 13:812–821, 2017.

[609] J.H. Adams *et al.* White paper on EUSO-SPB2 2017. arXiv: 1703.04513 [hep-ex].

[610] K. Abe *et al.* The results from BESS-Polar experiment. *Adv. Space Res.* 60:806–814, 2017.

[611] M. Matsuura *et al.* ALMA spectral survey of supernova 1987A – molecular inventory, chemistry, dynamics and explosive nucleosynthesis. *Mon. Not. Roy. Astron. Soc.* 469:3347–3362, 2017.

[612] M.J. Boschini *et al.* Energy calibration of CALET onboard the International Space Station. *Astrophys. J.* 840:115, 2017.

[613] P.A. Klimov *et al.* The TUS detector of extreme energy cosmic rays on board the Lomonosov satellite. *Space Sci. Rev.* 212:1687–1703, 2017.

[614] P.L. Biermann *et al.* The nature and origin of ultra-high energy cosmic ray particles. *Frascati Phys. Ser.* 64:103–121, 2017.

[615] R. Engel *et al.* The hadronic interaction model Sibyll – past, present and future. *EPJ Web Conf.* 145:08001, 2017.

[616] V. Bindi *et al.* Overview of galactic cosmic ray solar modulation in the AMS-02 era. *Adv. Space Res.* 60:865–878, 2017.

[617] Y. Génolini *et al.* Indications for a high-rigidity break in the cosmic-ray diffusion coefficient. In *35th International Cosmic Ray Conference, Busan, Korea*, 268. 2017.

[618] F. Aharonian *et al.* Hitomi constraints on the 3.5 keV line in the Perseus galaxy cluster. *Astrophys. J. Lett*, 837:L15, 2017.

[619] S. Abdollahi *et al.* (Fermi-LAT Coll.) Cosmic-ray electron-positron spectrum from 7 GeV to 2 TeV with the Fermi Large Area Telescope. *Phys. Rev. D*, 95:082007, 2017.

[620] R.G.A. Fricke and K. Schlegel. Julius Elster and Hans Geitel – Dioscuri of physics and pioneer investigators in atmospheric electricity. *Hist. Geo. Space. Sci.* 8:1–7, 2017.

[621] I. Arcavi *et al.* Optical emission from a kilonova following a gravitational-wave-detected neutron-star merger. *Nature*, 551:64–66, 2017.

[622] D. Kerszberg. *Etude du fond diffus galactique des électrons et positrons et étude des performances de la seconde phase de l'expérience H.E.S.S.* PhD Thesis, Université Pierre et Marie Curie – Paris VI. 2017.

[623] A. Kounine et al. Precision measurement of 0.5 GeV – 3 TeV electrons and positrons using the AMS electromagnetic calorimeter. *Nucl. Instrum. Meth. A*, 869:110–117, 2017.

[624] B.P. Abbott *et al.* (LIGO, Virgo, and other coll.) Multi-messenger observations of a binary neutron star merger. *Astrophys. J.* 848:L12, 2017.

[625] B.P. Abbott *et al.* (LIGO Scientific and Virgo Coll.) GW170817: Observation of gravitational waves from a binary neutron star inspiral. *Phys. Rev. Lett.* 119:161101, 2017.

[626] J. Lippuner and L.F. Roberts. SkyNet: A modular nuclear reaction network library. *Astrophys. J. Suppl.* 233:18, 2017.

[627] M. Aguilar *et al.* (AMS Coll.) Observation of identical rigidity dependence of He, C and O cosmic rays at high rigidities by the Alpha Magnetic Spectrometer on the International Space Station. *Phys. Rev. Lett.* 119:251101, 2017.

[628] K.D. Makwana, R. Keppens, and G. Lapenta. Two-way coupling of magnetohydrodynamic simulations with embedded particle-in-cell simulations. *Computer Phys. Comm.* 221:81–94, 2017.

[629] Z. Osmanov and F.M. Rieger. Pulsed VHE emission from the Crab pulsar in the context of magnetocentrifugal particle acceleration. *Mon. Not. R. Astron. Soc.* 464:1347–1352, 2017.

[630] A.M. Galper *et al.* (PAMELA coll.) The Pamela experiment: A decade of cosmic ray physics in space. *J. Phys. Conf. Series*, 798:012033, 2017.

[631] O. Adriani *et al.* (PAMELA Coll.) Ten years of PAMELA in space. *Riv. Nuovo Cim.* 40:473–522, 2017.

[632] A. Aab *et al.* (Pierre Auger Coll.) Combined fit of spectrum and composition data as measured by the Pierre Auger Observatory. *JCAP*, 04. [Erratum: JCAP 03, E02 (2018)]:038, 2017.

[633]   A. Aab *et al.* (Pierre Auger Coll.) Observation of a large-scale anisotropy in the arrival directions of cosmic rays above 8×10^{18} eV. *Science*, 357:1266–1270, 2017.

[634]   J. Pétri. Radiation from an off-centred rotating dipole in vacuum. *Mon. Not. R. Astron. Soc.* 463:1240–1268, 2017.

[635]   S.J. Smartt *et al.* A kilonova as the electromagnetic counterpart to a gravitational-wave source. *Nature*, 551:75–79, 2017.

[636]   M.W. Winkler. Cosmic ray antiprotons at high energies. *J. Cosm. Astropart. Phys.* 2017:048, 2017.

[637]   Y.S. Yoon *et al.* (CREAM-III Coll.) Proton and helium spectra from the CREAM-III flight. *Astrophys. J.* 839:5, 2017.

[638]   E. Amato and P. Blasi. Cosmic ray transport in the galaxy: A review. *Advances in Space Research*, 62. Origins of Cosmic Rays:2731–2749, 2018.

[639]   M. Aguilar *et al.* (AMS Coll.) Observation of fine time structures in the cosmic proton and helium fluxes with the Alpha Magnetic Spectrometer on the International Space Station. *Phys. Rev. Lett.* 121:051101, 2018.

[640]   M. Aguilar et al. (AMS Coll.) Observation of complex time structures in the cosmic-ray electron and positron fluxes with the Alpha Magnetic Spectrometer on the International Space Station. *Phys. Rev. Lett.* 121:051102, 2018.

[641]   Y. Bai, J. Berger, and S. Lu. Supersymmetric resonant dark matter: A thermal model for the AMS-02 positron excess. *Phys. Rev. D*, 97:115012, 2018.

[642]   G. Bertone and D. Hooper. History of dark matter. *Rev. Mod. Phys.* 90:045002, 2018.

[643]   R. Bütikofer. "Ground-based measurements of energetic particles by neutron monitors". In. *Solar Particle Radiation Storms Forecasting and Analysis: The HESPERIA HORIZON 2020 Project and Beyond.* Ed. by Olga E. Malandraki and Norma B. Crosby. Springer International Publishing, 2018, pp. 95–111.

[644]   O. Adriani *et al.* (CALET Coll.) Extended measurement of the cosmic-ray electron and positron spectrum from 11 GeV to 4.8 TeV with the Calorimetric Electron Telescope on the International Space Station. *Phys. Rev. Lett.* 120:261102, 26 2018.

[645]   B. Cerutti. Particle acceleration and radiation in pulsars: New insights from kinetic simulations. *Nucl. Part. Phys. Proc.* 297–299:85–90, 2018.

[646]   I. Cholis, T. Karwal, and M. Kamionkowski. Features in the spectrum of cosmic-ray positrons from pulsars. *Phys. Rev. D*, 97:123011, 2018.

[647]   CTA Consortium. *Science with the Cherenkov Telescope Array.* World Scientific, 2018.

[648]   J.J. Zang *et al.* (DAMPE Coll.) Measurement of absolute energy scale of ECAL of DAMPE with geomagnetic rigidity cutoff. *PoS*, ICRC2017:197, 2018.

[649]   A. Bruno *et al.* Solar energetic particle events observed by the PAMELA mission. *Astrophys. J.* 862:97, 2018.

[650]   B. Adams *et al.* Letter of Intent: A new QCD facility at the M2 beam line of the CERN SPS (COMPASS++/AMBER) 2018. arXiv: 1808.00848 [hep-ex].

[651]   B. Coté *et al.* The origin of *r*-process elements in the Milky Way. *Astrophys. J.* 855:99, 2018.

[652]   D.J. Knipp *et al.* On the little-known consequences of the 4 August 1972 ultra-fast coronal mass ejecta: Facts, commentary and call to action. *Space Weather*, 16:1635–1643. 2018.

[653]   M. Aartsen *et al.* Multimessenger observations of a flaring blazar coincident with high-energy neutrino IceCube-170922A. *Science*, 361:eaat1378, 2018.

[654]   M. Matteucci *et al.* Proton fluxes measured by the PAMELA experiment from the minimum to the maximum solar activity for solar cycle 24. *Astrophys J. Lett.* 854:L2, 2018.

[655]   M.J. Boschini *et al.* Deciphering the local interstellar spectra of primary cosmic ray species with HelMod. *Astrophys. J.* 858:61, 2018.

[656]   S. Acharya *et al.* Production of deuterons, tritons, ^{3}He nuclei and their antinuclei in pp collisions at $\sqrt{s} = 0.9$, 2.76 and 7 TeV. *Phys. Rev.* D 97:024615, 2018.

[657]   Y. Genolini *et al.* Current status and desired precision of the isotopic production cross sections relevant to astrophysics of cosmic rays: Li, Be, B, C, and N. *Phys. Rev. C*, 98:034611, 2018.

[658]   M. Felcini. Searches for dark matter particles at the LHC. In *53rd Rencontres de Moriond on Cosmology, La Thuile, Italy.* 2018. eprint: `arXiv1809.06341`.

[659]   J. Feng and H.-H. Zhang. Dark matter search in space: Combined analysis of cosmic ray antiproton-to-proton flux ratio and positron flux measured by AMS-02. *Astrophys. J.* 858:116, 2018.

[660]   T.K. Gaisser, R. Engel, and E. Resconi. *Cosmic Rays and Particle Physics, 2nd Edition.* Cambridge Univ. Press, 2018.

[661]   N.Y. Ganushkina, M.W. Liemohn, and S. Dubyagin. Current systems in the Earth's magnetosphere. *Rev. Geophys.* 56:309–332, 2018.

[662]   S. Grimm, R. Engel, and D. Veberic. Heitler-Matthews model with leading-particle effect. *PoS*, ICRC2017:299, 2018.

[663]   E. Jaupart, E. Parizot, and D. Allard. Contribution of the galactic centre to the local cosmic-ray flux. *Astron. Astrophys.* 619:A12, 2018.

[664]   M. Krauss. *The Cosmic-Ray Electron Anisotropy Measured with H.E.S.S. and Characterization of a Readout System for the SST Cameras of CTA.* PhD Thesis, Universität Erlangen. 2018.

[665]   M. Aguilar *et al.* (AMS Coll.) Observation of new properties of secondary cosmic rays lithium, beryllium and boron by the Alpha Magnetic Spectrometer on the International Space Station. *Phys. Rev. Lett.* 120:021101, 2018.

[666]   N. Agafonova *et al.* (OPERA Coll.) Final results of the OPERA experiment on ν_τ appearance in the CNGS neutrino beam. *Phys. Rev. Lett.* 120:211801, 2018. Erratum: Phys. Rev. Lett. 121 (2018) 139901.

[667]   J.F. Ormes. Cosmic rays and climate. *Adv. in Space Res.* 62:2880–2891, 2018.

[668]   R. Munini *et al.* (PAMELA Coll.) Evidence of energy and charge sign dependence of the recovery time for the 2006 December Forbush event measured by the PAMELA experiment. *Astrophys. J.* 853:76, 2018.

[669]   A. Aab *et al.* (Pierre Auger Coll.) An indication of anisotropy in arrival directions of ultra-high-energy cosmic rays through comparison to the flux pattern of extragalactic gamma-ray sources. *Astrophys. J.* 853:L29, 2018.

[670]   A. Aab *et al.* (Pierre Auger Coll.) Large-scale cosmic-ray anisotropies above 4 EeV measured by the Pierre Auger Observatory. *Astrophys. J.* 868:4, 2018.

[671]   D. Wittkowski *et al.* (Pierre Auger Coll.) Reconstructed properties of the sources of UHECR and their dependence on the extragalactic magnetic field. *PoS*, ICRC2017:563, 2018.

[672] A. Reinert and M. W. Winkler. A precision search for WIMPs with charged cosmic rays. *J. Cosm. Astropart. Phys.* 2018:055, 2018.

[673] J. Seo, H. Kang, and D. Ryu. The contribution of stellar winds to cosmic ray production. *J. Korean Astron. Soc.* 51:37–48, 2018.

[674] P.D. Serpico. Entering the cosmic ray precision era. *J. Astrophys. Astr.* 39:41, 2018.

[675] R.U. Abbasi *et al.* (TA Coll.) Testing a reported correlation between arrival directions of ultra-high-energy cosmic rays and a flux pattern from nearby starburst galaxies using Telescope Array data. *Astrophys. J.* 867:L27, 2018.

[676] R.U. Abbasi *et al.* (TA Coll.) The cosmic-ray energy spectrum between 2 PeV and 2 EeV observed with the TALE detector in monocular mode. *Astrophys. J.* 865:74, 2018.

[677] V. Tatischeff and S. Gabici. Particle acceleration by supernova shocks and spallogenic nucleosynthesis of light elements. *Ann. Rev. Nucl. Part. Sci.* 68:377–404, 2018.

[678] A. Archer *et al.* (VERITAS Coll.) Measurement of the iron spectrum in cosmic rays by VERITAS. *Phys. Rev. D*, 98:022009, 2018.

[679] V.V. Zerkin and B. Pritychenko. The experimental nuclear reaction data (EXFOR): Extended computer database and web retrieval system. *Nucl. Instrum. Meth. A*, 888:31–43, 2018.

[680] C.-R. Zhu, Q. Yuan, and D.-M. Wei. Studies on cosmic ray nuclei with Voyager, ACE and AMS-02: I. Local interstellar spectra and solar modulation. *Astrophys. J.* 863:119, 2018.

[681] V.N. Zirakashvili and V.S. Ptuskin. Cosmic ray acceleration in magnetic circumstellar bubbles. *Astropart. Phys.* 98:21–27, 2018.

[682] A. Watson. The highest-energy cosmic-rays – the past, the present and the future. In *UHECR 2018, Paris, France.* Vol. 210, 00001. 2019. URL: https://doi.org/10.1051/epjconf/201921000001.

[683] M. Ahlers. The dipole anisotropy of galactic cosmic rays. *J. Phys. Conf Ser.* 1181:012004, 2019.

[684] T. Sako *et al.* (ALPACA Coll.) ALPACA air shower array to explore 100 TeV gamma-ray sky in Bolivia. In *36th International Cosmic Ray Conference, Madison, WI, U.S.A.* Vol. 358, 779. 2019.

[685] M. Aguilar et al. (AMS Coll.) Properties of cosmic helium isotopes measured by the Alpha Magnetic Spectrometer. *Phys. Rev. Lett.* 123:181102, 2019.

[686] M. Amenomori *et al.* (ASγ Coll.) First detection of photons with energy beyond 100 TeV from an astrophysical source. *Phys. Rev. Lett.* 123:051101, 2019.

[687] F. Capel and D.J. Mortlock. Impact of using the ultrahigh-energy cosmic ray arrival energies to constrain source associations. *Mon. Not. Roy. Astron. Soc.* 484:2324–2340, 2019.

[688] L. Cazon. Probing high-energy hadronic interactions with extensive air showers. *PoS*, ICRC2019:005, 2019.

[689] W. Hanlon (TA Coll.) Telescope Array 10 year composition. *PoS*, ICRC2019:280, 2019.

[690] D. Ivanov. TA spectrum. *EPJ Web Conf.* 210:01001, 2019.

[691]   D. Ivanov *et al.* (TA Coll.) Energy spectrum measured by the Telescope Array experiment. In *36th International Cosmic Ray Conference, Madison, WI, U.S.A.* 298. Proceedings of Science, 2019.

[692]   D. Watson *et al.* Identification of strontium in the merger of two neutron stars. *Nature,* 574:497–500, 2019.

[693]   Q. An *et al.* (DAMPE Coll.) Measurement of the cosmic-ray proton spectrum from 40 GeV to 100 TeV with the DAMPE satellite. *Sci. Adv.* 5:eaax3793, 2019.

[694]   O. Deligny. Measurements and implications of cosmic ray anisotropies from TeV to trans-EeV energies. *Astropart. Phys.* 104:13–41, 2019.

[695]   F. Donato, M. Korsmeier, and M. DiMauro. Production cross sections of cosmic antiprotons in the light of new data from the NA61 and LHCb experiments. *PoS*, ICRC2019:061, 2019.

[696]   A. Bruno *et al.* Spectral analysis of the September 2017 solar energetic particle events. *Space Weather*, 17:419–437, 2019.

[697]   A. Cuoco *et al.* Scrutinizing the evidence for dark matter in cosmic-ray antiprotons. *Phys. Rev. D*, 99:103014, 2019.

[698]   C. Corti *et al.* Numerical modelling of galactic cosmic-ray proton and helium observed by AMS-02 during the solar maximum of solar cycle 24. *Astrophys. J.* 871:253, 2019.

[699]   D. Koll *et al.* Interstellar ^{60}Fe in Antarctica. *Phys. Rev. Lett.* 123:072701, 2019.

[700]   L. Vlahos *et al.* Sources of solar energetic particles. *Phil. Trans. R. Soc. A*, 377:20180095. 2019.

[701]   M. Orcinha *et al.* Observation of a time lag in solar modulation of cosmic rays in the heliosphere. *J. Phys. Conf. Ser.* 1181:012013, 2019.

[702]   N. Tomassetti *et al.* Numerical modelling of cosmic-ray transport in the heliosphere and interpretation of the proton-to-helium ratio in Solar Cycle 24. *Adv. Space Res.* 64:2477–2489, 2019.

[703]   O. Adriani *et al.* The CALOCUBE project for a space based cosmic ray experiment: design, construction, and first performance of a high granularity calorimeter prototype. *J. Inst.* 14:P11004, 2019.

[704]   R. Alves Batista *et al.* Open questions in cosmic-ray research at ultrahigh energies. *Frontiers in Astronomy and Space Sciences*, 6 2019.

[705]   S. Gabici *et al.* The origin of galactic cosmic rays: Challenges to the standard paradigm. *Int. J. Mod. Phys. D*, 28:1930022, 2019.

[706]   S. Schael *et al.* AMS-100: The next generation magnetic spectrometer in space – An international science platform for physics and astrophysics at Lagrange point 2. *Nucl. Instrum. Meth. A*, 944:162561, 2019.

[707]   S.M. Krimigis *et al.* Energetic charged particle measurements from Voyager 2 at the heliopause and beyond. *Nature Astron.* 3:997–1006, 2019.

[708]   T. Pierog *et al.* EPOS 3 and air Showers. *EPJ Web Conf.* 210:02008, 2019.

[709]   V. Grebenyuk *et al.* Energy spectra of abundant cosmic-ray nuclei in the NUCLEON experiment. *Adv. Space Res.* 64:2546–2558, 2019.

[710]   X. Luo *et al.* A numerical study of cosmic proton modulation using AMS-02 observations. *Astrophys. J.* 878:6, 2019.

[711] Y. Génolini *et al.* Cosmic-ray transport from AMS-02 boron to carbon ratio data: Benchmark models and interpretation. *Phys. Rev. D*, 99:123028, 2019.

[712] ESA. *Luca to lead most challenging spacewalks since Hubble repairs.* 2019. URL: `http://www.esa.int/Science_Exploration/Human_and_Robotic_Exploration/`.

[713] C. Evoli, R. Aloisio, and P. Blasi. Galactic cosmic rays after the AMS-02 observations. *Phys. Rev. D*, 99:103023, 2019.

[714] H. Abdalla *et al.* (H.E.S.S. Coll.) H.E.S.S. and Suzaku observations of the Vela X pulsar wind nebula. *Astron. Astrophys.* 627:A100, 2019.

[715] J.R. Hörandel. "Radio detection of air showers with LOFAR and AERA". In. *Proceedings of International Symposium for Ultra-High Energy Cosmic Rays (UHECR2014).* 2019, pp. 108–115.

[716] M.G. Aartsen *et al.* (IceCube Coll.) Cosmic ray spectrum and composition from PeV to EeV using 3 years of data from IceTop and IceCube. *Phys. Rev. D*, 100:082002, 2019.

[717] A. Ishihara. The IceCube upgrade – Design and science goals. *PoS*, ICRC2019:1031, 2019.

[718] W.D. Apel *et al.* (KASCADE-Grande Coll.) Search for large-scale anisotropy in the arrival direction of cosmic rays with KASCADE-Grande. *Astrophys. J.* 870:91, 2019.

[719] M. Aguilar *et al.* (AMS Coll.) Towards understanding the origin of cosmic-ray electrons. *Phys. Rev. Lett.* 122:101101, 2019.

[720] M. Aguilar *et al.* (AMS Coll.) Towards understanding the origin of cosmic-ray positrons. *Phys. Rev. Lett.* 122:0411102, 2019.

[721] N. Konovalova. Emulsion detector for the future experiment SHiP at CERN. *Persp. Sci.* 12:100401, 2019.

[722] O. Adriani *et al.* (CALET Coll.) Direct measurement of the cosmic-ray proton spectrum from 50 GeV to 10 TeV with the Calorimetric Electron Telescope on the International Space Station. *Phys. Rev. Lett.* 122:181102, 2019.

[723] A. De Rújula. The cosmic-ray spectra: News on their knees. *Phys. Lett. B*, 790:444–452, 2019.

[724] M. Schumann. Direct detection of WIMP dark matter: Concepts and status. *J. Phys. G*, 46:103003, 2019.

[725] S. Torii and P.S. Marrocchesi (CALET Coll.) The CALorimetric Electron Telescope (CALET) on the International Space Station. *Adv. Space Res.* 64:2531–2537, 2019.

[726] M. Unger. New results from the cosmic-ray program of the NA61/SHINE facility at the CERN SPS. *PoS*, ICRC2019:446, 2019.

[727] J. Wu and H. Chen. Revisit cosmic ray propagation by using 1H, 2H, 3He and 4He. *Phys. Lett. B*, 789:292–299, 2019.

[728] Y. Asaoka *et al.* (CALET Coll.) The CALorimetric Electron Telescope (CALET) on the International Space Station: Results from the first two years on orbit. *J. Phys.: Conf. Ser.* 1181:012003, 2019.

[729] R.-Z. Yang and F. Aharonian. Interpretation of the excess of antiparticles within a modified paradigm of galactic cosmic rays. *Phys. Rev. D*, 100:063020, 2019.

[730] S. Acharya *et al.* (ALICE Coll.) Measurement of the low-energy antideuteron inelastic cross section. *Phys. Rev. Lett.* 125:162001, 2020.

[731] M. Aguilar *et al.* (AMS Coll.) Properties of neon, magnesium, and silicon primary cosmic rays: Results from the Alpha Magnetic Spectrometer. *Phys. Rev. Lett.* 124:211102, 2020.

[732] M. Charlton, S. Eriksson, and G.M. Shore. *Antihydrogen and Fundamental Physics.* Springer, 2020.

[733] G. Cataldi (Pierre Auger Coll.) The AugerPrime Upgrade of the Pierre Auger Observatory. *PoS*, ICHEP2020:727, 2020.

[734] L.W. Piotrowski (JEM-EUSO Coll.) Results and status of the EUSO-TA detector. *PoS*, ICRC2019:388, 2020.

[735] F. Dyson. *The Power of Morphological Thinking.* New York Review of Books, January 16. 2020.

[736] C. Evoli *et al.* AMS-02 beryllium data and its implication for cosmic ray transport. *Phys. Rev. D*, 101:023013, 2020.

[737] A. Takenaka *et al.* Search for proton decay via $p \to e^+\pi^0$ and $p \to \mu^+\pi^0$ with an enlarged fiducial volume in Super-Kamiokande I–IV. *Phys. Rev. D*, 102:112011, 2020.

[738] A.J. Deason *et al.* The edge of the galaxy. *Mon. Not. Roy. Astron. Soc.* 496:3929–3942, 2020.

[739] D. Maurin *et al.* Cosmic-ray database update: Ultra-high energy, ultra-heavy, and antinuclei cosmic-ray data (CRDB v4.0). *Universe*, 6:102, 2020.

[740] F. Riehn *et al.* Hadronic interaction model Sibyll 2.3d and extensive air showers. *Phys. Rev. D*, 102:063002, 2020.

[741] K. Ebinger *et al.* PUSHing core-collapse supernovae to explosions in spherical symmetry. IV. Explodability, remnant properties, and nucleosynthesis yields of low-metallicity stars. *Astrophys. J.* 888:91, 2020.

[742] M.J. Boschini *et al.* Deciphering the local interstellar spectra of secondary nuclei with the Galprop/HelMod framework and a hint for primary lithium in cosmic rays. *Astrophys. J.* 889:167, 2020.

[743] M.J. Boschini *et al.* Inference of the local interstellar spectra of cosmic-ray nuclei $Z \le 28$ with the Galprop-HelMod framework. *Astrophys. J. Suppl.* 250:27, 2020.

[744] N. Marcelli *et al.* Time dependence of the flux of helium nuclei in cosmic rays measured by the PAMELA experiment between 2006 July and 2009 December. *Astrophys. J. Lett.* 893:145, 2020.

[745] N. Weinrich *et al.* Combined analysis of AMS-02 (Li,Be,B)/C, N/O, 3He, and 4He data. *Astron. Astrophys.* 639:A131, 2020.

[746] Q.Yuan *et al.* Nearby source interpretation of differences among light and medium composition spectra in cosmic rays. *Frontiers of Physics*, 16:24501, 2020.

[747] R. Beck *et al.* Synthesizing observations and theory to understand galactic magnetic fields: Progress and challenges. *Galaxies*, 8:4, 2020.

[748] T. Ashley *et al.* Mapping outflowing gas in the Fermi bubbles: A UV absorption survey of the galactic nuclear wind. *Astrophys. J.* 898:128, 2020.

[749] Z.-Q. Huang *et al.* Examining the secondary product origin of cosmic-ray positrons with the latest AMS-02 data. *Astrophys. J.* 895:53, 2020.

[750] J.J. Ethier and E.R. Nocera. Parton distributions in nucleons and nuclei. *Ann. Rev. Nucl. Part. Sci.* 70:43–76, 2020.

[751] S. Abdollahi *et al.* (Fermi LAT Coll.) Fermi Large Area Telescope fourth source catalog. *Astrophys. J. Suppl.* 247:33, 2020.

[752] M. Hesse and P.A. Cassak. Magnetic reconnection in the space sciences: Past, present and future. *J. Geophys. Res.* 125:e2018JA025935, 2020.

[753] M.G. Aartsen *et al.* (IceCube Coll.) Cosmic ray spectrum from 250 TeV to 10 PeV using IceTop. *Phys. Rev. D*, 102:122001, 2020.

[754] M. Kachelriess, S. Ostapchenko, and J. Tjemsland. Revisiting cosmic ray antinuclei fluxes with a new coalescence model. *JCAP*, 08:048, 2020.

[755] D. Kramer. Questions surround NASA's shutdown of an international cosmic-ray instrument. *Physics Today*, 72:30–32, 2020.

[756] P. Lipari and S. Vernetto. The shape of the cosmic ray proton spectrum. *Astropart. Phys.* 120:102441, 2020.

[757] K. Lodders. "Solar elemental abundances". In *The Oxford Research Encyclopedia of Planetary Science*. Ed. by P. Read. Oxford Univ. Press, 2020.

[758] M. Agostini *et al.* (Borexino Coll.) Experimental evidence of neutrinos produced in the CNO fusion cycle in the Sun. *Nature*, 587:577–582, 2020.

[759] M. Marshall. *The Genesis Quest: The Geniuses and Eccentrics on a Journey to Uncover the Origin of Life on Earth*. Weidenfeld & Nicolson, 2020.

[760] J.H. Matthews, A.R. Bell, and K.M. Blundell. Particle acceleration in astrophysical jets. *New Astron. Rev.* 89:101543, 2020.

[761] P. Mertsch. Test particle simulations of cosmic rays. *Astrophys. Space Sci.* 365:135, 2020.

[762] N. Globus and R.D. Blandford. The chiral puzzle of life. *Astrophys. J. Lett.* 895:L11:1–14, 2020.

[763] O. Adriani *et al.* (CALET Coll.) Direct measurement of the cosmic-ray carbon and oxygen spectra from 10 GeV/n to 2.2 TeV/n with the Calorimetric Electron Telescope on the International Space Station. *Phys. Rev. Lett.* 125:251102, 2020.

[764] M. Agostini *et al. (P-ONE Coll.)* The Pacific Ocean Neutrino Experiment. *Nature Astron.* 4:913–915, 2020.

[765] P.A. Zyla *et al.* (Particle Data Group). Review of particle physics. *Prog. Theor. Exp. Phys.* 2020:083C01, 2020.

[766] A. Aab *et al.* (Pierre Auger Coll.) Cosmic-ray anisotropies in right ascension measured by the Pierre Auger Observatory. *Astrophys. J.* 891:142, 2020.

[767] A. Aab *et al.* (Pierre Auger Coll.) Measurement of the cosmic-ray energy spectrum above 2.5×10^{18} eV using the Pierre Auger Observatory. *Phys. Rev. D*, 102:062005, 2020.

[768] A. Aab *et al.* (Pierre Auger Coll.) Reconstruction of events recorded with the surface detector of the Pierre Auger Observatory. *J. Inst.* 15:P10021, 2020.

[769] K. Abe *et al.* (T2K Coll.) Constraint on the matter-antimatter symmetry-violating phase in neutrino oscillations. *Nature*, 530:339–344, 2020.

[770] R.U. Abbasi *et al.* (TA Coll.) Search for large-scale anisotropy on arrival directions of ultra-high-energy cosmic rays observed with the Telescope Array experiment. *Astrophys. J.* 898:L28, 2020.

[771] Q. Yan et al. Measurements of nuclear interaction cross sections with the Alpha Magnetic Spectrometer on the International Space Station. *Nucl. Phys. A*, 996:121712, 2020.

[772] M. Aguilar *et al.* (AMS Coll.) Properties of a new group of cosmic nuclei: Results from the Alpha Magnetic Spectrometer on sodium, aluminum, and nitrogen. *Phys. Rev. Lett.* 127:021101, 2021.

[773] M. Aguilar *et al.* (AMS Coll.) Properties of heavy secondary fluorine cosmic rays: Results from the Alpha Magnetic Spectrometer. *Phys. Rev. Lett.* 126:081102, 2021.

[774] M. Aguilar *et al.* (AMS Coll.) Properties of Iron primary cosmic rays: Results from the Alpha Magnetic Spectrometer. *Phys. Rev. Lett.* 126:0411104, 2021.

[775] M. Aguilar *et al.* (AMS Coll.) The Alpha Magnetic Spectrometer (AMS) on the International Space Station: Part II – Results from the first seven years. *Physics Reports*, 894:1–116, 2021.

[776] M. Aguilar et al. (AMS Coll.) Periodicities in the daily proton fluxes: Results from the Alpha Magnetic Spectrometer. *Phys. Rev. Lett.* 127:271102, 2021.

[777] L.A. Fusco *et al.* (ANTARES Coll.) Search for a diffuse flux of cosmic neutrinos with the ANTARES neutrino telescope. *J. Inst.* 16:C10001, 2021.

[778] P. Brogi and K. Kobayashi *et al.* (CALET Coll.) Measurement of the energy spectrum of cosmic-ray helium with CALET on the International Space Station. *PoS*, ICRC2021:101, 2021.

[779] O. Adriani *et al.* (CALET Coll.) Measurement of the iron spectrum in cosmic rays from $10\,\mathrm{GeV}/n$ to $2.0\,\mathrm{TeV}/n$ with the Calorimetric Electron Telescope on the International Space Station. *Phys. Rev. Lett.* 126:241101, 2021.

[780] L. Chen and B. Zhang. Analytical solution of magnetically dominated astrophysical jets and winds: Jet launching, acceleration, and collimation. *Astrophys. J.* 906:105, 2021.

[781] B.A. Clark. The IceCube-Gen2 neutrino observatory. *J. Inst.* 16:C10007, 2021.

[782] M.E. Bertaina (JEM-EUSO Coll.) An overview of the JEM-EUSO program and results. *PoS*, ICRC2021:406, 2021.

[783] R. Zanin *et al.* (CTA Coll.) CTA – the world's largest ground-based gamma-ray observatory. *PoS*, ICRC2021:005, 2021.

[784] F. Alemanno *et al.* (DAMPE Coll.) Measurement of the cosmic ray helium spectrum from 70 GeV to 80 TeV with the DAMPE space mission. *Phys. Rev. Lett.* 126:201102, 2021.

[785] S. Dupourqué, L. Tibaldo, and P. von Ballmoos. Constraints on the antistar fraction in the solar system neighborhood from the 10-year Fermi Large Area Telescope gamma-ray source catalog. *Phys. Rev. D*, 103:083016, 2021.

[786] A.P. Ravi *et al.* Spectral evolution of the X-ray remnant of SN 1987A: A high-resolution Chandra HETG study. *Astrophys. J.* 922:140, 2021.

[787] C. Hill *et al.* Performance of the D-Egg optical sensor for the IceCube-Upgrade. *PoS*, ICRC2021:1042, 2021.

[788] E. Greco *et al.* Indication of a pulsar wind nebula in the hard X-ray emission from SN 1987A. *Astrophys. J. Lett.* 908:L45, 2021.

[789] E. Kuulkers *et al.* INTEGRAL reloaded: Spacecraft, instruments and ground system. *New Astron. Rev.* 93:101629, 2021.

[790] F. Kahlhoefer *et al.* Constraining dark matter annihilation with cosmic ray antiprotons using neural networks. *JCAP*, 12:037, 2021.

[791] H.G. Zhang *et al.* Performance of the ISS-CREAM calorimeter in a calibration beam test. *Astropart. Phys.* 130:102583, 2021.

[792] I. Belolaptikov *et al.* Neutrino telescope in lake Baikal: Present and nearest future. *PoS*, ICRC2021:002, 2021.

[793] J. Borowka *et al.* The acoustic module for the IceCube upgrade. *PoS*, ICRC2021:1059, 2021.

[794] J.C. Algaba *et al.* Broadband multi-wavelength properties of M87 during the 2017 Event Horizon Telescope campaign. *Astrophys. J. Lett.* 911:L11, 2021.

[795] M. Ajello *et al.* Fermi Large Area Telescope performance after 10 years of operation. *Astrophys. J. Suppl.* 256:12, 2021.

[796] M. Gerontidou *et al.* Space weather: Terrestrial perspective. *Adv. Space Res.* 67:2231–2240, 2021.

[797] M. Vecchi *et al.* Combined analysis of AMS-02 secondary-to-primary ratios: Universality of cosmic-ray propagation and consistency of nuclear cross-sections. *PoS*, ICRC2021:174, 2021.

[798] M. Werhahn *et al.* Cosmic rays and non-thermal emission in simulated galaxies. I. Electron and proton spectra compared to Voyager-1 data. *Mon. Not. Roy. Astron. Soc.* 505:3273, 2021.

[799] M. Werhahn *et al.* Cosmic rays and non-thermal emission in simulated galaxies. II. γ-ray maps, spectra and the far-infrared-γ-ray relation. *Mon. Not. Roy. Astron. Soc.* 505:3295, 2021.

[800] M. Werhahn *et al.* Cosmic rays and non-thermal emission in simulated galaxies: III. Probing cosmic ray calorimetry with radio spectra and the FIR-radio correlation. *Mon. Not. Roy. Astron. Soc.* 508:4072, 2021.

[801] M.G. Aartsen *et al.* Detection of a particle shower at the Glashow resonance with IceCube. *Nature*, 591:220–224, 2021.

[802] M.G. Aartsen *et al.* IceCube-Gen2: The window to the extreme universe. *J. Phys. G*, 48:060501, 2021.

[803] M.J. Boschini *et al.* The discovery of a low-energy excess in cosmic-ray iron: Evidence of the past supernova activity in the local bubble. *Astrophys. J.* 913:5, 2021.

[804] N. Park *et al.* Cosmic-ray isotope measurements with HELIX. *PoS*, ICRC2021:091, 2021.

[805] R. Battiston *et al.* High precision particle astrophysics as a new window on the universe with an Antimatter Large Acceptance Detector In Orbit (ALADInO). *Exper. Astron.* 51. [Erratum: Exper.Astron. 51, 1331–1332 (2021)]:1299–1330, 2021.

[806] R. Diehl *et al.* Steady-state nucleosynthesis throughout the galaxy. *New Astronomy Reviews*, 92:101608, 2021.

[807] S. Bacholle *et al.* Mini-EUSO mission to study earth UV emissions on board the ISS. *Astrophys. J. Suppl.* 253:36, 2021.

[808] S. Ghosh *et al.* Leptophilic-portal dark matter in the light of AMS-02 positron excess. *Phys. Rev. D*, 104:075016, 2021.

[809] T. Bringmann *et al.* Precise dark matter relic abundance in decoupled sectors. *Phys. Lett. B*, 817:136341, 2021.

[810] V. Tatischeff *et al.* The origin of galactic cosmic rays as revealed by their composition. *Mon. Not. Roy. Astron. Soc.* 508:1321–1345, 2021.

[811] Z. Cao *et al.* Ultrahigh-energy photons up to 1.4 petaelectronvolts from 12 γ-ray galactic sources. *Nature*, 594:33–36, 2021.

[812] D.A. Godzieba, D. Radice, and S. Bernuzzi. On the maximum mass of neutron stars and GW190814. *Astrophys. J.* 908:122, 2021.

[813] M. Hanasz, A.W. Strong, and P. Girichidis. Simulations of cosmic ray propagation. *Living Rev. Comp. Astrophys.* 7:2, 2021.

[814] F. McNally *et al.* (IceCube Coll.) Observation of cosmic ray anisotropy with nine years of IceCube data. *PoS*, ICRC2021:320, 2021.

[815] D.H. Kang *et al.* (KASCADE-Grande Coll.) Results from the KASCADE-Grande data analysis. *PoS*, ICRC2021:313, 2021.

[816] K. Kobayasi and P.S. Marrochesi *et al.* (CALET Coll.) Extended measurement of the proton spectrum with CALET on the International Space Station. *PoS*, ICRC2021:098, 2021.

[817] S. Komissarov and O. Porth. Numerical simulation of jets. *New Astron. Rev.* 92:101610, 2021.

[818] V. Kozhevnikov. Meissner effect: History of development and novel aspects. *J. Supercond. Nov. Magn.* 34:1979–2009, 2021.

[819] A.A. Lagutin and N.V. Volkov. Features of the energy spectra of primary and secondary nuclei of cosmic rays: A consistent astrophysical interpretation. *Bull. Russ. Acad. Sci.* 85:375–378, 2021.

[820] Z. Liu. "Measurement of cosmic-ray magnesium and aluminum fluxes with the Alpha Magnetic Spectrometer on the International Space Station". Thesis No. 5571, University of Geneva. PhD thesis. 2021.

[821] M. A. Malkov and I.V. Moskalenko. The TeV cosmic-ray bump: A message from the Epsilon Indi or Epsilon Eridani Star? *Astrophys. J.* 911:151, 2021.

[822] M. Di Mauro and M.W. Winkler. Multimessenger constraints on the dark matter interpretation of the Fermi-LAT Galactic center excess. *Phys. Rev. D*, 103:123005, 2021.

[823] A.V. Olinto and J.F. Krizmanic (POEMMA Coll.) The roadmap to the POEMMA mission. *PoS*, ICRC2021:976, 2021.

[824] Y. Tsunsesada *et al.* (Pierre Auger and TA Coll.) Joint analysis of the energy spectrum of ultra-high-energy cosmic rays measured at the Pierre Auger Observatory and the Telescope Array. *PoS*, ICRC2021:337, 2021.

[825] A.M. Botti *et al.* (Pierre Auger Coll.) Status and performance of the underground muon detector of the Pierre Auger Observatory. *PoS*, ICRC2021:233, 2021.

[826] E. Guido *et al.* (Pierre Auger Coll.) Combined fit of the energy spectrum and mass composition across the ankle with the data measured at the Pierre Auger Observatory. *PoS*, ICRC2021:311, 2021.

[827] F. Schlüter *et al. (Pierre Auger Coll.)* Expected performance of the AugerPrime Radio Detector. *PoS*, ICRC2021:262, 2021.

[828] T. Fodran *et al. (Pierre Auger Coll.)* First Results from the AugerPrime Radio Detector. *PoS*, ICRC2021:270, 2021.

[829] M. Potts and C. Jui (TA Coll.) Monocular energy spectrum using the TAx4 fluorescence detector. *PoS*, ICRC2021:343, 2021.

[830] T.P. Ray and J. Ferreira. Jets from young stars. *New Astron. Rev.* 93:101615, 2021.

[831] R. Abbasi *et al.* (TA Coll.) Current status and prospects of surface detector of the TAx4 experiment. *PoS*, ICRC2021:203, 2021.

[832] T. Fujii *et al.* (TA Coll.) Telescope Array anisotropy summary. *PoS*, ICRC2021:392, 2021.

[833] F.-K. Thielemann. Origin of the elements; Part I: From H to Fe, Ni and Zn. *SPG Mitteilungen*, 63:24–31, 2021.

[834] F.-K. Thielemann. Origin of the elements; Part II: From Fe to Pb and the Actinides. *SPG Mitteilungen*, 64:29–39, 2021.

[835] B.F. Rauch *et al.* (TIGRISS Coll.) Determination of expected TIGERISS observations. *PoS*, ICRC2021:088, 2021.

[836] J. Wei. "Measurement of the isotopic composition of cosmic-ray beryllium with the Alpha Magnetic Spectrometer on the International Space Station". Thesis No. 5582, University of Geneva. PhD thesis. 2021.

[837] M. Aguilar *et al.* (AMS Coll.) Properties of daily helium fluxes. *Phys. Rev. Lett.* 128:231102, 23 2022.

[838] O.M. Bitter and D. Hooper. Constraining the Milky Way's pulsar population with the cosmic-ray positron fraction 2022. arXiv: 2205.05200 [astro-ph.HE].

[839] O. Adriani *et al.* (CALET Coll.) Direct measurement of the nickel spectrum in cosmic rays in the energy range from $8.8\,\text{GeV}/n$ to $240\,\text{GeV}/n$ with CALET on the International Space Station. *Phys. Rev. Lett.* 128:131103, 2022.

[840] O. Adriani *et al.* (CALET Coll.) Observation of spectral structures in the flux of cosmic-ray protons from 50 GeV to 60 TeV with the Calorimetric Electron Telescope on the International Space Station. *Phys. Rev. Lett.* 129:101102, 2022.

[841] D. Kyratzis (HERD Coll.) Overview of the HERD space mission. *Physica Scripta*, 97:054010, 2022.

[842] J. Stasielak (Pierre Auger Coll.) AugerPrime - The upgrade of the Pierre Auger Observatory. *Int. J. Mod. Phys. A*, 37:2240012, 2022.

[843] J.H. Adams (EUSO Coll.) A review of the EUSO-balloon pathfinder for the JEM-EUSO program. *Space Sci. Rev.* 218:3, 2022.

[844] C. Chung *et al.* The development of SiPM-based fast time-of-flight detector for the AMS-100 experiment in space. *Instruments*, 6:14, 2022.

[845] E. Bueno *et al.* Transport parameters from AMS-02 F/Si data and fluorine source abundance 2022. arXiv: 2208.01337v1 [astro-ph.HE].

[846] F. Calore *et al.* AMS-02 antiprotons and dark matter: Trimmed hints and robust bounds. *SciPost Phys.* 12:163, 2022.

[847] M. Boschini *et al.* Spectra of cosmic-ray sodium and aluminum and unexpected aluminum excess. *Astrophys. J.* 933:147, 2022.

[848] M. Vecchi *et al.* The rigidity dependence of galactic cosmic-ray fluxes and its connection with the diffusion coefficient 2022. arXiv: 2203.06479 [astro-ph.HE].

[849] M.J. Boschini *et al.* A hint of a low-energy excess in cosmic-ray fluorine. *Astrophys. J.* 925:108, 2022.

[850] N. Marcelli *et al.* Helium fluxes measured by the PAMELA experiment from the minimum to the maximum solar activity for solar cycle 24. *Astrophys. J. Lett.* 925:L24, 2022.

[851] O. Adriani *et al.* Design of an Antimatter Large Acceptance Detector In Orbit (ALADInO). *Instruments*, 6:19, 2022.

[852] O. Adriani *et al.* Light yield non-proportionality of inorganic crystals and its effect on cosmic-ray measurements. *J. Inst.* 17:P08014, 2022.

[853] P. Klimov *et al.* Status of the K-EUSO orbital detector of ultra-high energy cosmic rays. *Universe*, 8:88, 2022.

[854] R. Abbasi *et al.* Improved characterization of the astrophysical muon-neutrino flux with 9.5 years of IceCube data. *Astrophys. J.*50, 2022.

[855] Y. Harikane *et al.* A search for H-dropout Lyman break galaxies at $z \sim 12$–16. *Astrophys. J.* 929:1, 2022.

[856] P. Kempski and E. Quataert. Reconciling cosmic ray transport theory with phenomenological models motivated by Milky-Way data. *Mon. Not. Roy. Astron. Soc.* 514:657–674, 2022.

[857] M.A. Malkov and I.V. Moskalenko. On the origin of observed cosmic-ray spectrum below 100 TV. *Astrophys. J.* 933:78, 2022.

[858] J.-S. Niu. Hybrid origins of the cosmic-ray nucleus spectral hardening at a few hundred GV. *Astrophys. J.* 932:37, 2022.

[859] S. Poluianov and O. Batalla. Cosmic-ray atmospheric cutoff energies of polar neutron monitors. *Adv. Space Res.* in press 2022.

[860] A.A. Watson. Further evidence for an increase of the mean mass of the highest-energy cosmic-rays with energy. *J. High Energy Astrophys.* 33:14–19, 2022.

Index

Printed in the United States
by Baker & Taylor Publisher Services

Printed in the United States
by Baker & Taylor Publisher Services